CAMBRIDGE ILLUSTRATED DICTIONARY OF ASTRONOMY

This lavishly illustrated new dictionary, written by an experienced writer and consultant on astronomy, provides an essential guide to the universe for amateur astronomers of all ages. It can be used both as a comprehensive reference work, and as a fascinating compendium of facts to dip into.

Around 1300 carefully selected and cross-referenced entries are complemented by hundreds of beautiful color illustrations, taken from space missions, the Hubble Space Telescope, and other major observatories on Earth and in space. Distinguished stellar illustrator Wil Tirion has drawn 20 new star maps especially for inclusion here. A myriad of named astronomical objects, constellations, observatories, and space missions are described in detail, as well as biographical sketches for 70 of the most luminous individuals in the history of astronomy and space science. Acronyms and specialist terms are clearly explained, making for the most thorough and carefully assembled reference resource that teachers and enthusiasts of astronomy will ever need.

JACQUELINE MITTON trained as an astronomer at both Oxford and Cambridge Universities. She is the author or co-author of over 20 astronomy books for both children and adults, and has also been a consultant or contributor to many other reference books. She has been editor of the *Journal of the British Astronomical Association*, and the annual *Handbook of the British Astronomical Association*. As Press Officer of the Royal Astronomical Society, she made regular contributions to TV and radio about astronomical developments. She continues to keep up-to-date with recent astronomical advances.

CAMBRIDGE ILLUSTRATED DICTIONARY

of **Astronomy**

JACQUELINE MITTON

CAMBRIDGE
UNIVERSITY PRESS

CAMBRIDGE UNIVERSITY PRESS
Cambridge, New York, Melbourne, Madrid, Cape Town, Singapore, São Paulo, Delhi

Cambridge University Press
The Edinburgh Building, Cambridge CB2 8RU, UK

Published in the United States of America by Cambridge University Press, New York

www.cambridge.org
Information on this title: www.cambridge.org/9780521823647

© J. Mitton 2007

First published 2007

Printed in Malaysia

A catalogue record for this publication is available from the British Library

ISBN 978-0-521-82364-7 hardback

Preface

There is always something new in astronomy. Exciting discoveries follow one after another at a dizzying pace, thanks to the batteries of giant telescopes perched on mountain tops and equipped with the latest technological innovations, observatories orbiting high above the troublesome atmosphere, and spacecraft exploring the worlds of the solar system from close quarters. Keeping abreast of it all can be a challenge!

For this illustrated A-to-Z, I have made an up-to-date selection of 1 300 entries covering hundreds of named astronomical objects as well as the terms and abbreviations most commonly encountered in astronomy. I have also included biographical entries on 70 people who have made significant contributions to the development of astronomy. Three hundred entries are illustrated, nearly all in color.

The idea for an illustrated dictionary grew from the dictionary I originally compiled in 1988–90, the most recent edition of which was published by Cambridge University Press in 2001. But this is a new book with a fresh style, which I hope will appeal to a wide range of readers young and old – not just as a reference source in which to look things up, but also as a book full of fascinating facts and beautiful pictures to dip into anytime.

Using the book

The alphabetical order takes no account of word breaks or hyphens. Entries beginning with a Greek letter are treated as if the letter were spelled out.

Words printed in italics and preceded by the symbol ➤ have their own entries, but not all possible cross-references are indicated in this way. The symbol ➤ preceding a word or words in italics means "see also."

Acknowledgements

I am deeply grateful to the numerous individuals who have provided me with advice and information and indebted to the countless reference sources I have consulted since I began to compile my dictionary database in 1988. It is impossible to list them all but I would particularly like to thank my husband, Simon, for his support and for his assistance in compiling the biographical entries. Any errors or omissions, however, I accept as being my own responsibility. I would like to thank the many organizations that have freely made their superb pictures available and those who have given me individual permission to use their copyright images. A full list of credits can be found at the end of the book. Finally, I would like to thank Cambridge University Press for their continuing support during the preparation of this book.

Jacqueline Mitton
November 2006

A

aberration An effect that makes the observed position of a star slightly different from its true position. It results from a combination of the finite speed of the starlight and the motion through space of the observer on Earth. Most aberration is due to Earth's yearly motion in orbit around the Sun and is called annual aberration. A much smaller contribution from Earth's daily rotation is called diurnal aberration.

absolute magnitude A number that gives the true, relative brightness of an astronomical body, ignoring the dimming effect of distance. The absolute magnitude of a star is the ➤ *magnitude* it would appear to be if it were 10 ➤ *parsecs* away. The absolute magnitude of a planet, asteroid or comet is the ➤ *apparent magnitude* it would have if it were at a distance of 1 AU from both the Sun and Earth, with its disk fully illuminated.

absolute zero The lowest possible temperature. It is the zero point of the Kelvin temperature scale used in science. Its equivalent on the Celsius scale is $-273.16\,°C$.

absorption line A sharp dip in a continuous ➤ *spectrum*. Absorption lines look like narrow gaps in a spectrum. They are seen in the spectra of the majority of stars. In the case of the Sun, they are known as ➤ *Fraunhofer lines*.

　　Atoms create these dark lines by absorbing radiation. Each chemical element creates a unique pattern of lines. By measuring the strengths of absorption lines it is possible to deduce the abundance of the various elements, though the lines are also affected by temperature, density and other factors. ➤➤ *emission line*.

absorption nebula A dark interstellar cloud that blocks the light from stars and galaxies lying behind it. Absorption nebulae range in size from small ➤ *globules* to large clouds visible to the naked eye. Absorption nebulae contain both dust and gas, and the temperatures in them are low enough for simple molecules to form. Much of what is known about these nebulae comes from observing their infrared and radio radiation, which, unlike visible light, can pass through them. ➤➤ *molecular cloud*.

accelerating universe The concept that the expansion of the universe is speeding up. Evidence that the universe is now expanding at an ever faster rate first came from measurements made in the late 1990s of the distances to very remote galaxies in which there were ➤ *supernova* explosions. Astronomers estimate that the expansion of the universe began to accelerate about 5 billion

An absorption nebula. Lanes of absorbing cold dust obscure the light of stars in the Milky Way in this infrared image.

years ago when the power of ➤ *dark energy* to propel the universe apart became greater than the power of gravity to hold back the expansion.

accretion disk A disk that forms around a spinning object, such as a star or ➤ *black hole*, when its gravity draws in material from a companion star or from the ➤ *interstellar medium*.

Achernar (Alpha Eridani) The brightest star in the constellation Eridanus, representing the River Eridanus. Its name comes from Arabic and means "the end of the river." It marks the extreme southern point of the constellation. Achernar is a ➤ *B star* of magnitude 0.5 and is 144 light years away.

588 Achilles The first of the ➤ *Trojan asteroids* to be identified. Discovered by Max Wolf in 1906, its diameter is about 116 km (72 miles).

achondrite A type of stony ➤ *meteorite* that crystallized from molten rock. Unlike ➤ *chondrites*, achondrites do not contain small mineral spheres known as ➤ *chondrules*.

An artist's impression of the accretion disk that forms around a black hole as it draws material from a companion star.

Acrux (Alpha Crucis) The brightest star in the constellation Crux. To the naked eye it looks like a single white star of magnitude 0.9, but a telescope shows two ➤ B stars, of magnitudes 1.4 and 1.9, separated by 4.4 arc seconds. The spectrum of the brighter one shows it has a very close companion so there are at least three stars in this system, which is 320 light years away.

active galactic nucleus (AGN) A small central region in a galaxy where exceptionally large amounts of energy are being generated. The only way such a concentrated source of power can be explained is by matter falling into a supermassive ➤ black hole. Active galaxies are categorized by their appearance and the nature of the radiation they emit. ➤ Quasars, ➤ Seyfert galaxies, ➤ radio galaxies, ➤ N galaxies and ➤ blazars are all examples.

AGNs have high-speed jets of material shooting out from them. The black hole is surrounded by a ring of dust and gas at right angles to the jets. The differences between the various categories of AGN can be accounted for by the level of their power output and the angle from which they are viewed. In radio galaxies, the ring is edge-on, hiding the light from the disk of hot material swirling into the black hole. In quasars and Seyfert galaxies, the ring is oriented so we can see the light emitted by the hot, glowing disk. Blazars are thought to have jets pointing directly at Earth.

active galaxy A galaxy with an ➤ active galactic nucleus at its center.

active optics A method of maintaining the precise shape of the main mirror in a reflecting telescope. A computer continually monitors the quality of the image and feeds the information back to a motorized support system under the mirror. Using active optics means that mirrors can be thinner and more light-weight. The mirror's tendency to change its shape under its own weight as the telescope moves can be corrected in just a few minutes.

active region A region in the outer layers of the Sun where there is ➤ solar activity. Active regions develop where strong magnetic fields break through from below. ➤ Sunspots, ➤ plages and ➤ flares are all evidence of an active region. The radiation given off is normally enhanced across the whole of the electromagnetic spectrum, from X-rays to radio waves, except in sunspots

The active galaxy Centaurus A. This X-ray image shows a jet being fired from the center.

Active optics. The computer-controlled supports under the 3.5-meter primary mirror of the WIYN Telescope at the Kitt Peak Observatory.

themselves, where the temperature is reduced and less light is emitted. There is a large variation in the size and duration of active regions: they may last from several hours up to a few months. Electrically charged particles and the enhanced ultraviolet and X-radiation from active regions affect the ➤ interplanetary medium and Earth's upper atmosphere.

Adams, John Couch (1819–92) John Couch Adams is chiefly remembered for predicting the existence and position of the planet Neptune in 1845 by analyzing the way Uranus had departed from its expected orbit since it was discovered in 1781. A brilliant mathematician, he worked on the motion of the Moon and planets. He became a professor at the University of Cambridge in 1858 and director of the University Observatory in 1860.

Unfortunately for Adams, in 1845 neither George Biddell ➤ *Airy*, the Astronomy Royal, nor the Director of the Cambridge Observatory, Professor James Challis, treated his prediction of a new planet with any seriousness or real urgency. Unknown to them, Urbain J. J. ➤ *Leverrier* in France had independently made the same calculation. Acting on Leverrier's prediction, Johann Galle in Berlin discovered Neptune in 1846. Only later, and after a good deal of controversy, did Adams received credit for his work when Challis and John ➤ *Herschel* pointed out that his prediction was made before Leverrier's.

adaptive optics A technique for improving the image an astronomical telescope makes by compensating for changes in the quality of ➤ *seeing*. A small, very thin, flexible mirror placed a short distance in front of the focus of the telescope corrects for the distortion. An image sensor detects the amount of distortion and feeds the information to a microprocessor, which controls actuators to bend the corrector mirror. The system has to respond to changes in less than a hundredth of a second.

Adhara (Epsilon Canis Majoris) The second-brightest star in the constellation Canis Major. Lying 425 light years away, it is a giant ➤ *B star* of magnitude 1.5 with an eighth-magnitude companion. Derived from Arabic, Adhara means "the virgins," which may come from a name given to a triangle of stars of which Adhara is one.

2101 Adonis A small asteroid, discovered in 1936 by Eugene Delporte, which came within 2 million km (1.4 million miles) of Earth in 1937. After that it was lost until 1977 when it was recovered following a recomputation of its orbit. It is a member of the ➤ *Apollo* asteroid group and is about 2 km across.

Adrastea A small inner moon of Jupiter discovered by David Jewitt in 1979. It measures $26 \times 20 \times 16$ km ($16 \times 12 \times 10$ miles).

Advanced Electro-Optical System Telescope (AEOS Telescope) A US Air Force 3.67-m telescope at the Air Force Maui Optical Station in Hawaii. It is the world's largest telescope capable of tracking rapidly moving objects, such as satellites. Though built principally for military purposes, it is also used for astronomical research.

aerobot A scientific experiment package deployed in a planetary atmosphere using a balloon.

aerobraking The controlled use of atmospheric drag to reduce the speed of a satellite and modify its orbit.

aeronomy The study of physical and chemical processes in the upper atmosphere of Earth, or of any planet.

Agena An alternative name for the star ➤ *Hadar*.

Ahnighito meteorite ➤ *Cape York meteorite*.

airglow Faint light given out by Earth's own atmosphere. From space, it appears as a ring of greenish light around the Earth. It is caused by the Sun's radiation. The brightest airglow comes from a layer approximately 10–20 km (about 10 miles) thick at a height of around 100 km (60 miles). Glowing oxygen and sodium atoms are major contributors. Airglow does not include ➤ *thermal radiation*, ➤ *auroras*, lightning and ➤ *meteor* trains.

air shower A cascade of high-energy, electrically charged particles in the atmosphere, triggered by the collision of a ➤ *cosmic ray* particle with the nucleus of a gas atom. The nucleus that has been hit emits a number of fast-moving particles. They in turn strike other nuclei, which eject yet more particles.

Airy, Sir George Biddell (1801–92) Airy was one of the most prominent figures in British astronomy during the nineteenth century. He began his career at Cambridge University, where he became a professor of mathematics in 1826, then professor of astronomy and Director of the Observatory two years later. He was appointed ➤ *Astronomer Royal* in 1835 and held that position for 46 years until he retired at the age of 80. He was both a very able mathematician and a practical scientist who tackled many varied problems relating to astronomy and geophysics. He also had exceptional organizing ability. As Astronomer Royal, he re-equipped the Royal Observatory at Greenwich and improved the range and efficiency of its work.

Airy published hundreds of papers but he is now remembered most for his work in optics, his detailed analysis of the orbits of Earth and Venus, and estimating the mean density of Earth in an experiment involving pendulums at the top and bottom of a mineshaft. He was also responsible for establishing Greenwich Mean Time as the official time throughout Britain and the ➤ *transit circle* he had built at Greenwich in 1850 was later chosen as the zero point of longitude for the world.

Aitne A small outer moon of Jupiter discovered in 2001. Its diameter is about 3 km (2 miles).

albedo The proportion of the light falling on a body or surface that is reflected. Albedo is given either as a decimal between 0 (perfectly absorbing) and 1 (perfectly reflecting) or as an equivalent percentage. For a planet or asteroid, the ratio between the total amount of reflected light and the incident light is called the Bond albedo. The reflecting qualities of planetary bodies are also measured by their geometric albedo. Geometric albedo is formally defined as the ratio between the brightness of the body as viewed from the direction of the Sun and the brightness of a hypothetical white sphere of the same size and at the same distance that is diffusely reflecting sunlight.

719 Albert A small asteroid discovered from the Vienna Observatory in 1911 by Johann Palisa, who named it after a benefactor of the Observatory, Baron Albert Freiherr von Rothschild. It was subsequently lost for almost 80 years, but recovered by chance in May 2000. It is only about 2–4 km (2 miles) across but makes relatively close approaches to Earth of between 30 and 46 million km (19 and 29 million miles) every 30 years. 1911, the discovery year, was one such occasion.

Albiorix A small outer moon of Saturn in a very elliptical orbit. It was discovered in 2000 and its diameter is estimated to be 26 km (16 miles).

Albireo (Beta Cygni) The second brightest star in the constellation Cygnus. Visual observers regard it as one of the most beautiful double stars. The brightest of the pair is a giant, yellow–orange ➤ *K star* of magnitude 3.2 and is itself a double too close to be split in a telescope. Its companion is a bluish ➤ *B star* of magnitude 5.4. The two stars are separated by 34 arc seconds and are 380 light years away.

Alcaid Alternative form of ➤ *Alkaid*.

Alcor (80 Ursae Majoris) A fourth-magnitude ➤ *A star* very close to ➤ *Mizar*, one of the bright stars forming the "tail" of the Great Bear. The two stars are separated by 11.5 arc minutes on the sky and both can easily be seen by the naked eye. Though they look like a double star, their distances are not known accurately enough to say whether they form a real binary system or not. Alcor is about 81 light years away.

Alcyone (Eta Tauri) The brightest member of the ➤ *Pleiades* star cluster in the constellation Taurus. Alcyone is a ➤ *B star* of magnitude 2.9.

Aldebaran (Alpha Tauri) The brightest star in the constellation Taurus. Its Arabic name means "the follower." Aldebaran is a giant ➤ *K star* of magnitude 0.9. Although it appears in the sky to be part of the Hyades star cluster, it is not in fact a cluster member, lying only half as far away at a distance of 65 light years.

Alderamin (Alpha Cephei) The brightest star in the constellation Cepheus. It is an ➤ *A star* of magnitude 2.7 lying 49 light years away. The name, which is of Arabic origin, means "the right arm."

Algenib (Gamma Pegasi) One of the four stars marking the corners of the ➤ *Square of Pegasus*. It is a ➤ *B star* of magnitude 2.8 and is 335 light years away. The name comes from Arabic and means "the side." The star Alpha Persei, more usually known as ➤ *Mirfak*, is also sometime called Algenib.

Algieba (Gamma Leonis) A second-magnitude star, which is the third brightest in the constellation Leo. Viewed through a telescope, Algieba is a ➤ *visual binary*, consisting of two yellowish giant stars separated by 4 arc seconds. Their individual magnitudes are 2.6 and 3.8, and they take more than 500 years to complete one orbit around each other. Algieba is 126 light years away. Its Arabic name means "the forehead."

Algol (The Demon Star; Beta Persei) An ➤ *eclipsing binary* system in the constellation Perseus, which is one of the best-known of all variable stars. Algol varies between magnitudes 2.2 and 3.5 over a period of 2.87 days because the two stars regularly cross in front of each other as viewed from Earth.

The brighter member of the system is a ➤ *B star* and the fainter one a much larger but far cooler ➤ *G star*. As the G star cuts off light from its more brilliant companion, their combined brightness declines over 4 hours, reaching a minimum that lasts only 20 minutes. The eclipse of the dimmer star by its partner causes a dip in brightness of only 0.06 magnitude, which is not detectable by eye. Regular variations in the spectrum of Algol over a period of 1.862 years reveal the presence of a third, more distant star in the system.

The spectrum also reveals evidence for ➤ *mass transfer* between the two close companions, which are separated by less than one tenth the distance between the Sun and Earth. Observations that Algol is a radio star erratically flaring up to 20 times its normal radio brightness support the idea that mass transfer is taking place.

Alioth (Epsilon Ursae Majoris). The brightest star in the constellation Ursa Major, the Greek letters in this case being allotted in order of position rather than of brightness. Alioth is an ➤ *A star* of magnitude 1.8 and is 81 light years away.

Alkaid (Eta Ursae Majoris) A star in Ursa Major, at the end of the bear's "tail." It is a ➤ *B star* of magnitude 1.9. The Arabic name means "chief of the mourners," for the Arabs saw the constellation as a bier rather than a bear. Its distance is 100 light years.

Allan Hills A region in Antarctica from where large numbers of meteorites have been recovered. The meteorites become concentrated in the area by natural movements in the ice sheet, and are relatively easy to identify against the ice.

Allen Telescope Array A project of the ➤ *SETI Institute* and the University of California, Berkeley, to construct a radio telescope in the form of an array of dish antennas each 6.1 meters (200 feet) across. The main purpose is to seek signals from possible extraterrestrial civilizations, but the array will be available for conventional radio astronomy as well. Construction has begun at Hat Creek Observatory. The target is a total of 350 dishes sometime between 2015 and 2020.

Allende meteorite A meteorite of the ➤ *carbonaceous chondrite* type, which fell in Mexico in 1969. More than 2 tonnes of material was scattered over an area 48 km by 7 km (30 by 4 miles).

ALMA ➤ *Atacama Large Millimeter Array.*

Almagest A large astronomical work written by the Greek astronomer ➤ *Ptolemy* (Claudius Ptolemaeus), who worked in Alexandria between about AD 127 and 151. *Almagest* is an Arabic corruption of Greek, meaning "The Greatest," though Ptolemy's original title was *The Mathematical Collection.* It is one of the most important works on astronomy ever written. Ptolemy included a star

The Allen Telescope Array. An artist's impression of the array when complete.

catalog and dealt with the motion of the Moon and planets. The rules set out for calculating the future positions of the planets on the basis of an Earth-centered universe were used for centuries.

almanac A book of tables giving the future positions of the Moon, planets and other celestial objects. An almanac normally covers one calendar year.

Alnath An alternative spelling of the star name ➤ Elnath.

Alnilam (Epsilon Orionis) One of the three bright stars forming Orion's belt. It is a ➤ supergiant ➤ B star of magnitude 1.7, estimated to be 1340 light years away. "Alnilam" comes from the Arabic for "string of pearls."

Alnitak (Zeta Orionis) One of the three bright stars forming Orion's belt. Its Arabic name means "the girdle." Alnitak is a ➤ supergiant ➤ O star of magnitude 1.8 and is about 800 light years away.

Alpha Centauri The brightest star in the constellation Centaurus and the nearest bright star to the Sun, at a distance of 4.36 light years. It is a ➤ visual binary star with an orbital period of 80 years. It consists of a ➤ G star and a ➤ K star, which have a combined magnitude of 0.27. The eleventh-magnitude star Proxima Centauri, though two degrees away on the sky, is thought to be associated with this star system because it has a similar motion in space. Proxima, a dim ➤ M star, is the nearest star to the Sun at a distance of 4.24 light years. Alpha Centauri is also called by the Arabic name Rigil Kentaurus (sometimes Rigel, or shortened to Rigil Kent), which means "the foot of the Centaur." Another alternative name is Toliman.

Alphard (Alpha Hydrae) The brightest star in the constellation Hydra. Its Arabic name means "the solitary one of the serpent." It is a ➤ *K star* of magnitude 2.0 lying 175 light years away.

Alphekka (Gemma; Alpha Coronae Borealis) The brightest star in the constellation Corona Borealis. It is an ➤ *A star* of magnitude 2.2. The Arabic name, also spelt Alphecca, means "bright one." This star is sometimes called by the Latin name Gemma, the "jewel" in the crown. Its distance is 75 light years.

Alpheratz (Sirrah; Alpha Andromedae) The brightest star in the constellation Andromeda, also marking one corner of the ➤ *Square of Pegasus*. It was formerly considered to belong to Pegasus and was designated Delta Pegasi. Alpheratz is an ➤ *A star* of magnitude 2.1 and is 97 light years away.

Alphonsus A lunar crater, 118 km (73 miles) in diameter. A prominent ridge runs across the center, almost along a north–south line, through a central peak about 1 km high. Temporary reddish clouds were observed there in 1958 and 1959, possibly due to the release of gas from the rocks.

Alpine Valley (Vallis Alpes) A flat-bottomed valley on the Moon, running for 150 km (95 miles). It crosses the lunar Alps and connects the Mare Frigoris with the Mare Imbrium.

ALSEP Abbreviation for Apollo Lunar Science Experiment Package, an experimental set-up deployed on the Moon by astronauts during the manned ➤ *Apollo program* (1969–72). One was left by every mission except the first. All the experiments were turned off in 1978.

The Apollo 16 ALSEP with astronaut John Young.

Al Sufi, Abd Al-Rahman (903–986) Al Sufi worked in Persia and Baghdad. He was the first person to revise Ptolemy's ➤ *Almagest* and to relate Greek constellation and star names to traditional Arabic ones. His *Book on the Constellations of Fixed Stars* was published in 964 and included drawings of each constellation. He was the first person to record seeing the ➤ *Andromeda Galaxy*, describing it as a "small cloud." He also wrote on the use of ➤ *astrolabes*. In the West, he has sometimes been known as Azophi, which is the name given to the crater on the Moon named in his honor.

Altair (Alpha Aquilae) The brightest star in the constellation Aquila. It is an ➤ *A star* of magnitude 0.8 and is one of the closest bright stars at a distance of only 17 light years. Derived from Arabic, the word Altair means "the flying eagle."

altazimuth mount A form of telescope mount that allows the telescope to rotate about two axes – one horizontal and one vertical. It is the simplest type of mount to construct but the telescope must be turned about both axes simultaneously in order to track the motion of celestial objects across the sky. However, computers that can control the motion of a large telescope mean that altazimuth mounts are now used for all new professional instruments.

altitude The height of a celestial object, measured upwards as an angle from the horizon.

aluminizing The process that deposits a thin reflecting layer of aluminum on the glass surface of a telescope mirror.

Amalthea A small inner moon of Jupiter discovered by Edward E. Barnard in 1892. Images obtained by the ➤ *Voyager 1* mission showed it as a red-colored, potato-shaped object. The surface is heavily cratered, the largest depression, Pan, being 90 km (56 miles) across. The red color is thought to be due to sulfur compounds blown off the moon ➤ *Io*. Data from the ➤ *Galileo* spacecraft show that it is like an icy rubble pile with a density less than that of water. Amalthea measures 262 × 146 × 134 km (163 × 91 × 83 miles).

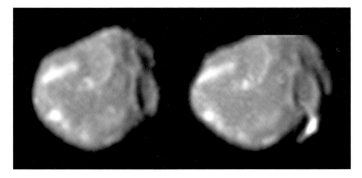

A stereo pair of images of Amalthea taken by the Galileo spacecraft in August and November 1999.

AMANDA A detector for cosmic neutrinos, built into the Antarctic ice at the South Pole. AMANDA stands for "Antarctic Muon And Neutrino Detector Array." The detector consists of photomultiplier tubes arranged in concentric rings between 1300 and 2400 meters (0.8–1.5 miles) below the surface. The photomultipliers detect light flashes created when neutrinos occasionally interact with atomic nuclei in the ice, causing then to emit muons. AMANDA began operation in 1999 and in 2005 was incorporated into a successor project called ➤ *IceCube.* ➤ *neutrino astronomy.*

1221 Amor A small asteroid, about 1–2 km (1 mile) across, which is the prototype of a group of Earth-approaching asteroids. The Amor group travel on orbits that bring them closer than the main ➤ *asteroid belt* with ➤ *perihelion* between the orbits of Mars and Earth. They can cross Mars's orbit but not Earth's. Amor was discovered by Eugène J. Delporte in 1932.

29 Amphitrite A large asteroid discovered from London in 1854 by Albert Marth. It is estimated to be 200 km (125 miles) across.

analemma The figure-of-eight obtained if the Sun's position in the sky is recorded at the same time of day throughout the year. The position of the Sun varies from day to day because Earth's axis is tilted to its orbit around the Sun and because Earth's orbit is elliptical rather than circular.

Ananke A small moon of Jupiter discovered from Mount Wilson Observatory in 1951 by Seth B. Nicholson. It is about 28 km (17 miles) in diameter. It belongs to a family of jovian moons on highly tilted, ➤ *retrograde* orbits around their parent planet.

Andromeda A large but not very conspicuous northern constellation. In classical mythology, Andromeda was the daughter of King Cepheus and Queen Cassiopeia. She was condemned to be sacrificed to a sea monster but rescued by Perseus. In old star atlases, Andromeda was usually shown as a chained woman. The three brightest stars, Alpha (Alpheratz or Sirrah), Beta (Mirach) and Gamma (Alamak) represent her head, hip and foot. Andromeda is most famous for the ➤ *Andromeda Galaxy.*

Andromeda Galaxy (M31; NGC 224) A large ➤ *spiral galaxy*, visible to the unaided eye as a misty patch in the constellation Andromeda. It belongs to the ➤ *Local Group* of galaxies and is similar to our own ➤ *Galaxy* but its disk appears to be at least about half as large again as the Milky Way's and more luminous. Its mass is estimated to be 300–400 billion Suns. Spiral features are not easy to see since we view its disk tilted by only 13° away from edge-on. In a small telescope, only the small central nucleus is visible though the fainter outer parts extend over 3 degrees of sky – more than six times the apparent diameter of the Moon. Several dwarf galaxies belonging to the Local Group are in orbit around the Andromeda Galaxy, notably M32 and NGC 205.

The Andromeda Galaxy was the first object to be recognized as lying beyond the Milky Way when ➤ *Edwin Hubble* estimated its distance in 1923.

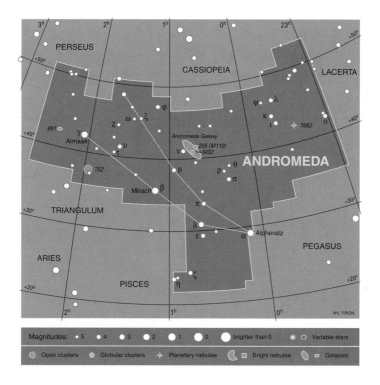

A map of the constellation Andromeda.

Lying 2.4–2.9 million light years away, it is the most distant object visible to the naked eye.

Andromedids A ➤ *meteor shower* associated with Comet ➤ *Biela*. The first recorded appearance of the shower was in 1741. Spectacular meteor storms were observed radiating from near the star Gamma Andromedae on November 27 in both 1872 and 1885, following the break-up of Comet Biela. After a moderate shower in 1904, the Andromedids were not recorded again until a few were identified in 1940. In recent years, about three Andromedids an hour have been detected, around November 14. The shower is also known as the Bielids.

Anglo-Australian Observatory (AAO) An observatory at the Siding Spring site of the ➤ *Mount Stromlo and Siding Spring Observatories* in New South Wales, Australia. It is funded jointly by the UK and Australia to operate the 3.9-m (150-inch) ➤ *Anglo-Australian Telescope* and the 1.2-m (48-inch) Schmidt telescope.

Anglo-Australian Telescope (AAT) A 3.9-m (150-inch) reflecting telescope, owned and funded jointly by the governments of Australia and the UK. It is situated at the ➤ *Siding Spring Observatory* site in New South Wales, Australia. The telescope was constructed in the early 1970s and started scheduled observing in 1975. It was the first telescope to be fully computer controlled.

The Andromeda Galaxy's central region.

angular diameter The apparent diameter of an object measured as an angle in degrees, arc minutes or arc seconds. An object's angular diameter depends on both its true diameter and its distance.

5535 Annefrank A small asteroid in the ➤ *asteroid belt*, which was imaged by the ➤ *Stardust* spacecraft in November 2002. It was discovered in 1942 by Karl Reinmuth in Heidelberg. Stardust passed by Annefrank at a distance of 3300 km (2050 miles) and found that it is about 6 km (4 miles) across.

annular eclipse A solar ➤ *eclipse* when a ring of the Sun's bright disk remains visible. Since the orbits of Earth around the Sun and of the Moon around Earth are elliptical, the distances of the Sun and Moon vary slightly so their apparent sizes vary too. A solar eclipse that would otherwise have been total is annular if the Moon appears smaller than the Sun at the time.

Antares (Alpha Scorpii) The brightest star in the constellation Scorpius. This red ➤ *supergiant* ➤ *M star* is a ➤ *semiregular variable*, fluctuating between magnitudes 0.9 and 1.1 over a five-year timescale. It is rapidly blowing off gas, which has formed a small nebula around the star. The name Antares is derived from Greek and means "rival of Mars." It has a sixth-magnitude blue companion 3 arc seconds away and lies at a distance of 600 light years.

antenna (aerial) A device for collecting or transmitting radio signals. The design of an antenna depends on the wavelength at which it is intended to operate and the strength of the signal. The simplest antenna is a straight rod, or dipole; the commonest type used in radio astronomy is a paraboloid dish.

Antennae Galaxies The popular name of a pair of colliding galaxies, NGC 4038 and 4039. Two long, curved streamers of stars were pulled out of the galaxies by the collision. The galaxies are 48 million light years away and the streamers about a 100 000 light years long.

anthropic principle The idea that the universe must have certain properties to favor the emergence of life. In theory, a large range of universes with different physical properties could exist. The anthropic principle states that only a

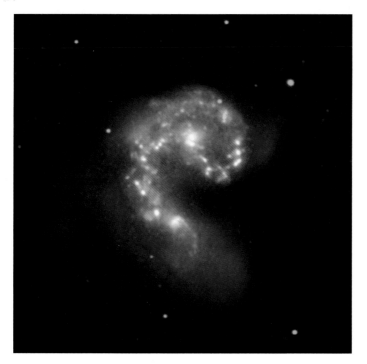

The Antennae Galaxies. This image is a composite of infrared data (shown as red) and visible-light data (shown in blue and green).

proportion of them can have intelligent observers. Since we exist, the universe we inhabit must have characteristics that have permitted us to evolve. This basic form of the anthropic principle is not generally regarded as controversial, and is sometimes called the weak anthropic principle.

The so-called strong anthropic principle is more speculative. This asserts that, because there are so many apparently unconnected coincidences in nature, which together have made it possible for life to develop, the universe *must* give rise to intelligent observers at some stage in its existence.

antitail Part of a comet's ➤ *dust tail* that appears to protrude forwards towards the Sun from the comet's head, sometimes like a spike. The effect is an illusion caused by the way the comet's tail curves and the direction from which we are seeing it.

Antlia (The Air Pump) A small, faint, southern constellation introduced in the mid-eighteenth century by Nicolas L. de Lacaille. It was originally called Antlia Pneumatica.

Antoniadi, Eugène Michel (1870–1944) Antoniadi was a Turkish-born Greek–French astronomer who became famous particularly for his observations of Mars. He worked in Paris from 1893 and became a French citizen in 1928. His books *The Planet Mars*, published in 1930, and *The Planet Mercury*, which followed in 1934, included the best maps of the planets made before the space age. He concluded that the "canals" Percival ➤ *Lowell* had claimed to see on Mars were an optical illusion.

Antoniadi scale A scale of five points, devised by the Greek–French astronomer Eugène ➤ *Antoniadi* that is used by amateur astronomers for describing the quality of ➤ *seeing*. The points on the scale are: I, perfect; II, slight undulations with periods of virtually perfect seeing lasting for several seconds; III, moderately good seeing, though with noticeable air movements; IV, poor seeing making observations difficult; V, very bad seeing that permits no useful observation.

Aoede A small outer moon of Jupiter discovered in 2003. Its diameter is about 4 km (2.5 miles).

Apache Point Observatory An observatory in New Mexico, USA, owned and operated by a consortium of American universities. The principal instrument is a 3.5-m (138-inch) telescope for both optical and infrared observations. A 2.5-m (98-inch) telescope for the ➤ *Sloan Digital Sky Survey*, together with a 0.6-m support telescope for the project, opened in 1997.

apastron The positions of the two members of a ➤ *binary star* system when they are furthest apart in their orbit.

Apennines (Montes Apenninus) A range of mountains on the Moon, forming part of the eastern boundary of the Mare Imbrium.

aperture (symbol *D*) The diameter of the main collecting mirror or lens in a ➤ *telescope*. For a radio telescope, it is the size of the ➤ *antenna*. The aperture is one of the most important characteristics of a telescope since the ability to detect faint objects and resolving power both increase with larger apertures.

aperture synthesis A technique developed by radio astronomers to make maps or images with the ➤ *resolving power* of a very large aperture by combining the observations from a number of smaller ➤ *antennas*. More recently it has also been applied to optical and infrared observations.

In the simplest version, two antennas form a ➤ *radio interferometer*. As the Earth rotates in the course of a day, one antenna is automatically carried right around the other. The effect is like sweeping out a ring. On successive days the separation between the two antennas is changed, so that a different ring is swept out and a large elliptical area is gradually covered. The records are combined in a computer to produce a radio map of the section of sky under observation with the detail resolved as if the telescope aperture were the size of the total area swept out.

In practice, more than two antennas are normally used to speed up the process and give greater flexibility. It is also possible to combine observations made at different sites, separated by thousands of kilometers, to obtain even better resolution.

Beginning in the 1990s, the same principle has been used to obtain high-resolution optical and infrared images. Pioneering instruments include the Cambridge Optical Aperture Synthesis Telescope (COAST) in the UK and the Navy Prototype Optical Interferometer (NPOI) at the US Naval Observatory's site near Flagstaff, Arizona. ➤➤ *interferometer*.

aphelion (pl. aphelia) The point farthest from the Sun in the orbit of a body, such as a planet or comet, that is traveling around the Sun.

apogee The point furthest from Earth in the orbit of the Moon or of an artificial Earth satellite.

1862 Apollo A small asteroid, about 1.4 km (0.9 mile) across, that was discovered in 1932 by Karl W. Reinmuth. It is the prototype of the Apollo group of asteroids with orbits that cross Earth's. Apollo's orbit also crosses that of Venus.

Apollo program An American space program, which in 1969 successfully achieved a manned landing on the Moon. The program consisted of 17 missions in all. Numbers 1 to 6 were unmanned test flights and Apollo 13 was aborted following an explosion on board, though the astronauts were returned safely to Earth. Six Moon landings took place between July 20, 1969 and December 11, 1972. The astronauts collected samples of lunar rocks and soils weighing a total of nearly 400 kilograms (nearly 900 pounds), and took many photographs both on the surface and from lunar orbit. A variety of scientific experiments were carried out on the surface of the Moon, including ones to detect ➤ *cosmic rays* and the ➤ *solar wind*.

The crew of the Apollo program's first mission to land astronauts on the Moon, Apollo 11. Left to right are Neil A. Armstrong, commander; Michael Collins, command module pilot; and Edwin E. Aldrin Jr., lunar module pilot.

The Apollo craft consisted of three modules: the Command Module (CM), the Service Module (SM) and the Lunar Module (LM). The Command and Service Modules (CSM) remained in lunar orbit with one astronaut on board while the other two astronauts made the descent to the Moon's surface in the Lunar Module. The descent stage was left on the Moon when the astronauts returned to lunar orbit by means of the ascent stage, and rejoined the Command and Service Modules. The Service Module was jettisoned shortly before re-entry into the Earth's atmosphere.

Apollo manned Moon landings

Apollo	Astronauts	Landing date	Landing site
11	Armstrong, Aldrin, Collins	July 20, 1969	Mare Tranquillitatis
12	Conrad, Bean, Gordon	November 19, 1969	Oceanus Procellarum
14	Shepard, Mitchell, Roosa	February 5, 1971	Fra Mauro
15	Scott, Irwin, Worden	July 30, 1971	Hadley Rille
16	Young, Duke, Mattingly	April 21, 1972	Cayley–Descartes highland region
17	Cernan, Schmitt, Evans	December 11, 1972	Taurus–Littrow region

Apollo–Soyuz project A joint US–Soviet space project in July 1975 in which an ➤ *Apollo program* Command and Service Module docked with a Soviet Soyuz space station in Earth orbit at an altitude of 225 km (140 miles). The two teams of astronauts were able to visit each others' craft and they performed a number of experiments jointly.

apparent magnitude The relative brightness of a star (or other celestial object) as perceived by an observer. Apparent magnitude depends on both the actual amount of light the body is emitting or reflecting and the distance to the object. The smallest magnitudes correspond to the greatest brightness. ➤ *magnitude*, *absolute magnitude*.

appulse A very close approach of one celestial object to another on the sky, so they seem just to touch without an ➤ *occultation* taking place.

Apus (The Bird of Paradise) A faint constellation near the south celestial pole, probably introduced by sixteenth century navigators. It was included by Johann Bayer in his 1603 star atlas ➤ *Uranometria*.

Aquarius (The Water Carrier) One of the twelve traditional constellations of the ➤ *zodiac*. It is one of the larger constellations but contains no very bright stars.

Aquila (The Eagle) A small but prominent northern constellation. It is said to represent the eagle of classical mythology sent by Jupiter to carry Ganymede to Olympus. It contains one of the brightest stars, ➤ *Altair*.

Ara (The Altar) A small, southerly constellation. Its stars are all faint but it is among the 48 ancient constellations listed by ➤ *Ptolemy*.

arachnoid An informal term for a type of volcanic feature on ➤ *Venus* resembling spiders connected by a web of fractures.

Arche A small outer moon of Jupiter discovered in 2002. Its diameter is 3 km (2 miles).

archeoastronomy The study of how astronomy was done in civilizations and societies of prehistory. Archeoastronomy is particularly concerned with archeological evidence for astronomical knowledge rather than written records. Sites that are studied include the stone-age remains in western Europe, ancient meso-America and the classical Mediterranean civilizations.

Arcturus (Alpha Boötis) The brightest star in the constellation Boötes and the fourth-brightest star in the sky. It is an orange giant ➤ *K star* of magnitude −0.04 lying 37 light years away. The name Arcturus is Greek in origin, and means "bear-watcher." It refers to the fact tht Arcturus seems to follow the Great Bear around the north celestial pole.

Arecibo Observatory A radio astronomy observatory in Puerto Rico, where a dish 305 m (1000 feet) across has been built into a natural bowl shape in hills south of the city of Arecibo. Completed in 1963, the telescope is operated by the National Ionospheric and Astronomy Center of Cornell University in the USA. The reflecting surface cannot be moved, but radio sources can be tracked by

The Arecibo Observatory viewed from the air.

moving the receiver at the focus along a specially designed support. A major refurbishment was completed in 1997. The telescope is larger in area than all the other radio telescopes in the world combined. It is used for radar studies of planets, observing ➤ *pulsars* and the study of hydrogen in distant galaxies. Because of its large collecting area, it can pick up fainter signals than any other dish.

95 Arethusa One of the darkest known asteroids, with an albedo of only a few percent. Its diameter is about 230 km (140 miles). It was discovered in 1867 by Robert Luther working in Düsseldorf.

Argo Navis A very large constellation of the southern sky listed by ➤ *Ptolemy* but no longer recognized officially. It represented the ship of Jason and the Argonauts from Greek mythology. It was so large that astronomers in the nineteenth century started to refer to different parts of the ship. Since 1930,

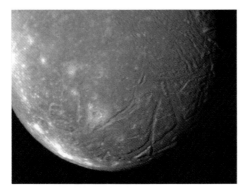

Ariel. A Voyager 2 image made in 1986.

stars that formerly made up Argo have been officially assigned to three separate constellations: ➤ *Carina* (The Keel), ➤ *Puppis* (The Stern or Poop) and ➤ *Vela* (The Sails).

Ariel One of the larger moons of Uranus, with a diameter of 1158 km (720 miles). It was discovered by William Lassell in 1851. Images obtained by the ➤ *Voyager 2* mission in 1986 showed the surface to be heavily cratered and crossed by fault scarps and valleys. Its appearance suggests that there has been considerable geological activity in the past.

Aries (The Ram) A small constellation in the traditional ➤ *zodiac*. It is said to represent the ram with the golden fleece sought by Jason and the Argonauts in classical mythology. Its brightest star is the second-magnitude ➤ *Hamal*. ➤➤ *First Point of Aries*.

Aristarchus A very bright lunar crater surrounded by a pattern of rays. It is 45 km (28 miles) across and has terraces on its inner walls. Temporary reddish glows in Aristarchus have occasionally been reported, perhaps caused by gas being released from the surface rocks. ➤➤ *lunar transient phenomenon*.

Aristarchus of Samos (*c.* **310–230** BC) The Greek astronomer Aristarchus was the first person known to have put forward the idea that Earth and the other planets orbit around the Sun but his theory was rejected by his contemporaries and not resurrected again until the sixteenth century, when it was again proposed by Nicolaus ➤ *Copernicus*. Aristarchus also attempted to measure the size and distance of the Sun and Moon. Though his results were inaccurate because he was unable to make precise enough observations, his methods were correct. He also concluded that the stars were infinitely far away because he could never detect any changes in their positions.

Aristotle (**384–322** BC) Aristotle was one of the greatest of the Greek philosophers. He developed a new style of philosophy which he applied to all aspects of the material world, including biology. His philosophy was so influential and respected, it dominated the way people thought about the universe and the movements of celestial bodies until the sixteenth and seventeenth centuries.

Aristotle was born in Stagira, a Greek colony in Macedonia. At the age of 18, he went to Athens to study under the philosopher ➤ *Plato*. When in his thirties, he returned to his native city and became tutor and advisor to the young prince who would become Alexander the Great. Then in 335 BC he returned to Athens and set up his own school of philosophy.

According to Aristotle, Earth was the center of the universe and the stars and the Sun, Moon and planets went around it in circular orbits. He believed that earthly matter was made of four "elements" – earth, water, fire and air. Earth and water naturally fell down while air and fire rose up. He thought of heavenly bodies as perfect and unchanging and to explain their motion he said they were made of a fifth element, aether, that naturally moved in circles. Aristotle attempted to explain things by thought and logic, based on simple every-day experience, rather than conducting what today we would think of as scientific experiments.

armillary sphere A type of celestial globe that represents the sphere of the sky by a framework of intersecting rings, with the Earth at the center. The rings correspond to important circles on the celestial sphere, such as the

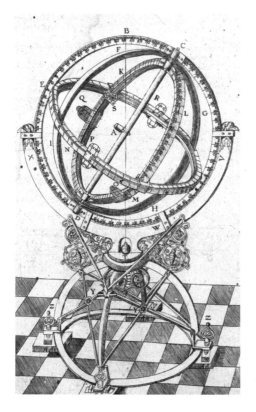

A sixteenth century engraving of an armillary sphere.

➤ *celestial equator* and the ➤ *ecliptic*. Some of the rings may be movable so that the sky's appearance at different times and at different latitudes can be reproduced. On some armillary spheres, the positions of the brighter stars are shown by small pointers attached to the fixed rings. The use of armillary spheres dates from at least the third century BC.

array An arrangement of linked radio ➤ *antennas* to make a ➤ *radio telescope*.

Arsia Mons A large ➤ *shield volcano* on Mars. It is about 350 km (220 miles) across at its base and rises to a height of 27 km (17 miles), 17 km above the level of the surrounding ridge.

Ascraeus Mons A prominent ➤ *shield volcano* in the ➤ *Tharsis Ridge* region of Mars. It is about 250 km (150 miles) across at its base and rises to a height of 27 km (17 miles), 17 km above the level of the surrounding ridge.

ashen light A dim glow that visual observers occasionally claim to see on the dark part of Venus when its phase is a very thin crescent. Its origin is unknown but, if it is a real physical effect in the atmosphere of Venus and not just an optical illusion, it may be similar to the ➤ *airglow* in Earth's atmosphere.

aspect The position of a planet or the Moon, relative to the Sun, as viewed from Earth.

association A loose grouping of young stars, typically with between 10 and 100 members. Stellar associations are found along the spiral arms of the ➤ *Galaxy*. They contain stars that were born together relatively recently in the same star-forming cloud and are always found along with interstellar matter. Associations are not held together very strongly and the stars will disperse within a few million years.

There are three main types of stellar associations. O or OB associations are made up of massive, luminous ➤ *O stars* and ➤ *B stars* scattered through a region up to several hundred light years across. T associations contain numerous, low-mass ➤ *T Tauri stars*. In R associations, the stars are embedded in a ➤ *reflection nebula*.

A star A star of ➤ *spectral type* A. A stars have surface temperatures in the range 7500–11 000 K and are white in color. The most prominent features in their spectra are the strong absorption lines due to hydrogen atoms. Lines of heavier elements, such as iron, are also noticeable at the cooler end of their temperature range. Examples of A stars are ➤ *Sirius* and ➤ *Vega*.

asterism A prominent pattern of stars, usually with a popular name, that is not a complete ➤ *constellation*. Well-known examples of asterisms are the ➤ *Big Dipper* (Plough) in Ursa Major and the ➤ *Sickle* in Leo.

asteroid (minor planet) A miniature planet composed of rock and/or metal. Asteroids range in size from the largest, ➤ *Ceres*, which is nearly 1000 km (600 miles) across, down to about 100 m (300 feet). Smaller objects are more often called *meteoroids*. Many thousands of asteroids have been individually identified and there could be half a million with diameters larger than 1.6 km

(1 mile). However, the total mass of all the asteroids put together is less than one-thousandth the mass of Earth. Most asteroid orbits are concentrated in the ➤ *asteroid belt* between Mars and Jupiter at distances ranging from 2.0 to 3.3 AU from the Sun. However, some asteroids follow orbits that bring them nearer to the Sun, such as the ➤ *Amor* group, the ➤ *Apollo* group and the ➤ *Aten* group, and some are more distant from the Sun, such as the ➤ *Centaurs*. The ➤ *Trojan asteroids* share Jupiter's orbit.

Asteroids are classified according to how they reflect sunlight: 75 percent are very dark, carbonaceous C-types, 15 percent are grayish, stony S-types and the remaining 10 percent are metallic M-types plus a number of very rare varieties. The darkest asteroids reflect only 3 or 4 percent of the sunlight falling on them, while the brightest reflect up to 40 percent. The brightness of many asteroids varies regularly as they rotate. Nearly all have irregular shapes; the smallest asteroids rotate the most rapidly and are the most irregular in shape.

The ➤ *Galileo* spacecraft, on its way to Jupiter, flew by two asteroids, ➤ *Gaspra* (on October 29, 1991) and ➤ *Ida* (on August 28, 1993). Detailed images showed their rocky surfaces to be pitted with numerous craters, and that Ida has a small satellite. The ➤ *NEAR Shoemaker* spacecraft flew by ➤ *Mathilde* in 1997 and went into orbit around ➤ *Eros* in 2000. The asteroid ➤ *Itokawa* was visited by the Japanese spacecraft ➤ *Hayabusa* in 2005. From the ground, it is possible to obtain information about the shape of asteroids using radar.

Asteroids are believed to be the remnants of the material from which the solar system formed. They are gradually disintegrating because of collisions between them. Most meteorites reaching Earth are small pieces broken off the asteroids. ➤➤ *dwarf planet, near-Earth asteroid, potentially hazardous asteroid.*

asteroid belt The region of the solar system, between 2.0 and 3.3 AU from the Sun, where the orbits of the vast majority of ➤ *asteroids* are. Within the belt, there are certain rings where the orbits of groups and families are concentrated and others, known as ➤ *Kirkwood gaps*, where there are very few asteroids,. The proportions of the different types of asteroid change markedly through the belt. At the inner edge, 60 percent are S-types and 10 percent C-types; at the outer edge the situation is reversed with 80 percent being C-types and 15 percent S-types.

asteroseismology The study of global oscillations of stars. Asteroseismology reveals details about the internal structure of stars in the same way that seismology uncovers information about Earth's interior. ➤ *helioseismology*.

5 Astraea An asteroid measuring about 120 km (75 miles) across When it was discovered in 1845 by the German amateur astronomer Karl L. Hencke, it was only the fifth asteroid to be found and was the first for 38 years.

astration The cyclic process in which interstellar matter is incorporated into newly formed stars, where its chemical composition is altered by nuclear processes, and is then expelled again into the interstellar medium to be used in the next generation of stars. Astration results in a steady increase in the proportion of heavier elements in a galaxy.

astrobiology The science concerned with the possibility of living organisms originating in space or on bodies other than Earth.

astrograph A historical astronomical telescope designed to take wide-angle photographs of the sky for measuring the positions of stars. In particular, the refracting telescopes constructed for the ➤ *Carte du Ciel* project were called astrographs.

astrolabe An ancient instrument for showing the positions of the Sun and bright stars at any time and date. Its invention is credited to Greek astronomers who worked in the second century BC.

A basic astrolabe consists of a circular star map (the "tablet" or "tympan") with a graticule (the "rete") over the top. These two parts are joined at their common center so that the rete can rotate over the tablet. Typically, it would be made of brass. Various engraved scales enable the user to display the positions of the stars and the Sun for any time and date, though any particular astrolabe is only useful within a narrow range of latitudes. Astrolabes were often fitted with a sight on a movable arm so that they could be used to estimate the ➤ *altitudes* of stars.

astrology An ancient tradition that claims to connect human traits and the course of events with the positions of the Sun, Moon and planets in relation to the stars. Before the seventeenth century, there was less of a clear distinction between astrology and the science of astronomy; many scientifically useful astronomical observations were originally made for astrological purposes.

A typical astrolabe.

astrometry The branch of astronomy concerned with the measurement of the positions and apparent motions of celestial objects in the sky and the factors that can affect them.

astronautics The science and technology concerned with all aspects of space travel.

Astronomer Royal Formerly the title of the director of the Royal Observatory in the UK but, since 1972, an honorary title bestowed on a distinguished astronomer in the UK.

Astronomers Royal

John Flamsteed	1675–1719
Edmond Halley	1720–42
James Bradley	1742–62
Nathaniel Bliss	1762–64
Nevil Maskelyne	1765–1811
John Pond	1811–35
Sir George Biddell Airy	1835–81
Sir William Christie	1881–1910
Sir Frank Watson Dyson	1910–33
Sir Harold Spencer Jones	1933–55
Sir Richard Woolley	1956–71
Sir Martin Ryle	1972–82
Sir Francis Graham-Smith	1982–90
Sir Arnold Wolfendale	1991–95
Sir Martin Rees	1995–

astronomical unit (AU or a.u.) A unit of measurement used mainly for distances within the solar system. It was originally based on the average distance between Earth and the Sun, though it now has a formal definition independent of Earth's orbit. Its value is 149 597 870 km (92 955 730 miles). There are about 63 240 astronomical units in a light year.

astronomy The study of the universe and everything in it beyond the bounds of the Earth's atmosphere. ➤ *astrophysics*.

astrophysics The physical theory of astronomical objects and phenomena. This term was introduced in the nineteenth century to distinguish between using physics to understand astronomical observations and the mere recording of positions, movements and phenomena. Though "astrophysics" keeps its original meaning, "astronomy" is generally considered to encompass all aspects of the study of the universe, including astrophysics.

Atacama Large Millimeter Array (ALMA) An array of 64 individual dish antennas, each 12 m in diameter, on an area 10 km square at Llano de

The Atacama Large Millimeter Array (ALMA). An artist's concept of the compact array.

Chajnantor in the Atacama desert in northern Chile. ALMA operates in the millimeter and submillimeter regions of the spectrum. Contruction of this joint US/European project started in 2003 and the telescope was expected to be fully completed and operational by 2009.

2062 Aten The prototype of the Aten group of Earth-approaching asteroids that have orbits lying mainly closer to the Sun than Earth. Aten was discovered in 1976 by Eleanor Helin and is only about 0.8 km (0.5 mile) across.

Atlas (1) The innermost small satellite of Saturn, discovered in 1980 by Richard Terrile during the ➤ *Voyager 1* mission. It measures $37 \times 34 \times 27$ km ($23 \times 21 \times 17$ miles) and orbits Saturn at a distance of 137 670 km (85 544 miles).

Atlas (2) A third-magnitude star in the ➤ *Pleiades* cluster.

atmosphere The gaseous outermost layer of a planet, moon or star. Since gas has a natural tendency to expand into space, only bodies with gravity strong enough can retain atmospheres. Mercury and the Moon, for example, are not massive enough to hold on to atmospheric gases. Earth, Venus, Mars and Titan are examples of rocky bodies with substantial atmospheres. In the giant planets, Jupiter, Saturn, Uranus and Neptune, there is no clear boundary between the gaseous layers on the outside and the liquid below; their "atmosphere" is the top layer of gas.

The more transparent outermost layers of a star are also described as an atmosphere.

atmospheric extinction A reduction in the brightness of an astronomical object caused by absorption and scattering in Earth's atmosphere. Extinction is worse the nearer the object is to the horizon and more blue light is cut out than red, which makes objects look redder than they really are.

atmospheric refraction A small deviation in the direction of light rays passing through Earth's atmosphere. Refraction makes objects seem slightly higher up than they really are. Atmospheric refraction is greatest near to the horizon.

atmospheric window A range of wavelengths of ➤ *electromagnetic radiation* that can pass through Earth's atmosphere without much absorption, scattering or reflection. There are two main windows: the optical window and the radio window.

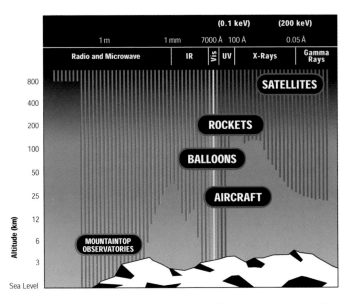

Atmospheric windows: the colored vertical lines show how far different kinds of radiation penetrate through Earth's atmosphere.

In the optical (or visible) region of the spectrum, wavelengths between about 300 and 900 nm can pass through the atmosphere. This range includes some near-ultraviolet and infrared radiation, invisible to the human eye. The radio window covers a range of wavelengths between a few millimeters and about 30 m, equivalent to frequencies from 100 GHz to 10 MHz. There are also some narrow windows in the infrared (micrometer wavelengths) and submillimeter parts of the spectrum where the atmosphere is moderately transparent to the radiation, particularly at locations such as deserts where the atmosphere is very dry.

AU (a.u.) Abbreviation for ➤ *astronomical unit.*

Auriga (The Charioteer) A large and prominent northerly constellation, described from ancient times as representing a charioteer. Its brightest star is ➤ *Capella,* associated by the Greeks with the mythological she-goat Amalthea, who nurtured the infant Zeus. The nearby triangle of fainter stars, Epsilon, Zeta and Eta, is called "the Kids." The star Elnath, formerly designated Gamma Aurigae and shared with the neighboring constellation of Taurus, now officially belongs to Taurus as Beta Tauri.

aurora (pl. auroras or aurorae) A display of luminous colors in the night sky. The lights often take the form of rays, moving curtains or a band in the east–west direction. Auroras are most often seen from high-latitude regions of Earth and the popular name for them is the northern or southern lights.

The usual cause of an aurora is a ➤ *magnetic substorm,* a disturbance in Earth's magnetic field triggered by particles from the Sun. At a height of

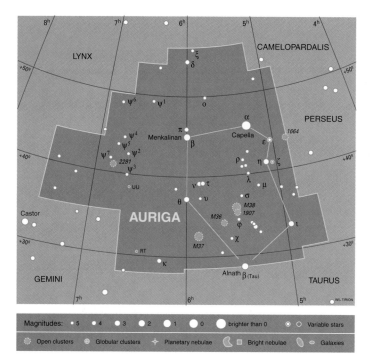

A map of the constellation Auriga.

around 100 km (60 miles), high-energy particles from Earth's ➤ *magnetosphere* collide with oxygen atoms and nitrogen molecules, which then emit mainly green and red light.

A large number of phenomena take place in the ➤ *ionosphere* during auroras, such as pulsations of the ➤ *geomagnetic field*, electric currents in the ionosphere and the emission of X-rays. Far more energy is emitted in the invisible parts of the spectrum than as visible light.

Strong auroras of ultraviolet light have been observed on Jupiter and Saturn.

auroral oval An oval-shaped belt on Earth where auroras are most likely to be seen. The two ovals lie asymmetrically around Earth's north and south magnetic poles. During the day, they are about 15° of latitude from the poles, increasing to 23° during the night. When Earth's magnetic field is disturbed, the ovals become wider and extend farther towards the equator.

auroral zone The zones on Earth's surface where most night-time auroras are seen. The zones are located at latitudes of about 67° north and south, and are about 6° wide.

Australia Telescope National Facility An Australian radio astronomy facility opened in 1988. It consists of a number of antennas at three separate sites in New South Wales and uses ➤ *aperture synthesis* to map astronomical radio sources.

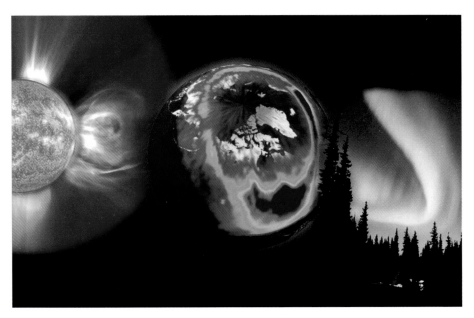

The aurora (right) is triggered by charged particles blasted from the Sun (left). The center image, taken from space, shows charged particles spreading down over the US during a large solar storm in July 2000.

Part of the Australia Telescope National Facility, showing five of the six antennas making up the compact array.

The Compact Array, located at the Paul Wild Observatory at Culgoora near Narrabri, consists of six antennas, each 22 m in diameter. Five can be moved along an east–west track, 3 km (2 miles) long. The sixth is on its own track 3 km farther to the west.

Greater resolving power is achieved by linking one or more of the antennas in the Compact Array with a 22-m (72-foot) dish, 100 km (60 miles) to the south at Mopra, near the Siding Spring optical observatory, and the 64-m (210-foot) dish at Parkes, which was completed in 1961 and is 200 km (120 miles) farther to the south. Together, these antennas form the Long Baseline Array.

Autonoe A small outer moon of Jupiter discovered in 2001. Its diameter is about 4 km (2.5 miles).

autoguider An electronic device for guiding a telescope automatically while it is making an observation. Even though a telescope may be driven by electric motors to follow the stars across the sky, further minor corrections are normally necessary to prevent the field of view drifting during a long observation. An autoguider detects drift and activates the drive motors to compensate.

axis The imaginary line through a body about which it rotates or is symmetrical.

azimuth The bearing of an object measured as an angle around the horizon eastwards, starting from north.

B

Baade, (Wilhelm Heinrich) Walter (1893–1960) Walter Baade was born in Germany and began his astronomical career at the Bergedorf Observatory of the University of Hamburg. While there, he discovered the unusual asteroid ➤ *Hidalgo*. In 1931 he moved to the ➤ *Mount Wilson Observatory* in California where he worked until his retirement. He discovered a total of 10 asteroids, including ➤ *Icarus*.

In the 1940s, he used the 100-inch telescope at Mount Wilson to resolve individual stars in the ➤ *Andromeda Galaxy* and two of its small companion galaxies. This led him to divide stars into two broad groups, ➤ *Population I* and ➤ *Population II*. He realized that Population I had the characteristics of young stars while population II was older. He also found that each population had its own kind of ➤ *Cepheid variable* star. Cepheids had been used incorrectly to judge the distance of the Andromeda Galaxy and Baade showed that it was twice as far away as previously thought. In the 1950s he worked on identifying radio sources, including ➤ *Cygnus A*.

Baade's Window An area of sky around the globular cluster NGC 6522 in the constellation Sagittarius, which is particularly rich in stars. The astronomer Walter ➤ *Baade* drew attention to it. He realized that very distant stars are visible in that direction because there is relatively little interstellar material to hide them.

Baikonur The Russian manned space-flight center, established by the former Soviet Union. It is situated north-east of the Aral Sea in Kazakhstan.

Baily's beads A phenomenon observed just before the Moon completely covers the Sun at a total solar eclipse and again just as the total phase of the eclipse ends. The very thin crescent of the Sun breaks up to look like a string of bright beads because the mountains and valleys on the Moon make its silhouette uneven rather than a perfect circle. The English astronomer Francis Baily (1774–1844) drew attention to the phenomenon at the solar eclipse of 1836.

324 Bamberga A large asteroid with a diameter of 252 km (156 miles), discovered by Johann Palisa in 1892.

Barlow lens An additional lens used in conjunction with a telescope ➤ *eyepiece* to produce a higher magnification.

Barnard, Edward Emerson (1857–1923) Barnard was one of the greatest and most prolific observers of his day. He was born in Nashville, Tennessee, and first developed his skills as an amateur astronomer before getting his first

An infrared image of Barnard's Galaxy.

professional job at Vanderbilt University in 1883 where he was also able to study. He then moved to ➤ *Lick Observatory* and in 1896 became Professor at Chicago. There he was able to use the newly built ➤ *Yerkes Observatory*.

In 1892, he discovered Amalthea, the first moon of Jupiter to be found since Galileo discovered the four largest in 1610. His many other discoveries included 16 comets and ➤ *Barnard's star*. He also realized that dark patches in the Milky Way are not due to the absence of stars but are dark nebulae hiding the stars behind them and he drew up a catalog of about 200 of them.

Barnard's Galaxy The galaxy NGC 6822 in Sagittarius, discovered by Edward ➤ *Barnard* in the 1880s. It is a small ➤ *irregular galaxy* belonging to the ➤ *Local Group* and is about 1.5 million light years away.

Barnard's Loop A faint ring of hot, glowing gas in the constellation Orion. It has the shape of an ellipse covering 14° by 10°. It is thought to be the result of the pressure of radiation from the hot stars in the region of Orion's belt and sword acting on interstellar material.

Barnard's Star A ninth-magnitude star in the constellation Ophiuchus that has the largest known ➤ *proper motion* of any star. Its rapid motion across the sky was discovered by Edward ➤ *Barnard* in 1916. Its position changes by 10.3 arc seconds each year as it moves through space relative to the Sun. It is the third-nearest star to the Sun at a distance of 5.88 light years. Possible "wobbles" in the motion of Barnard's star have raised suspicions that it may have unseen planets but this has never been confirmed.

barred spiral galaxy A common type of ➤ *spiral galaxy* that has a bright central bar of stars. The spiral arms seem to wind out from the end of the bar.

Barringer Crater ➤ *Meteor Crater*.

Barwell meteorite A 46-kilogram (101–lb) stony ➤ *meteorite* that fell near the village of Barwell, Leicestershire, UK, in 1965. Though it broke up, it is the largest stony meteorite known to have fallen in the UK.

The barred spiral galaxy NGC 1365, which lies about 60 million light years away in the Fornax cluster of galaxies.

barycenter The center of mass (balancing point) of a system of bodies moving under the their mutual gravitational attraction. The barycenter of the solar system lies about a million km (0.6 million miles) from the center of the Sun but is constantly moving as the relative positions of the planets change.

Bayer letters The letters of the Greek alphabet, used in conjunction with constellation names (Alpha Leonis, for example), as identifiers for brighter stars. Johann Bayer (1572–1625) was responsible for compiling the first complete star atlas, called *Uranometria*, which was published in 1603. In it he introduced the system of naming the brighter stars in each constellation by Greek letters, which he allocated approximately according to brightness or, in some instances, in order of position on the sky. The idea was soon taken up by others and Bayer's letters are still in use today.

Becklin-Neugebauer object One of the brightest of all astronomical sources of infrared radiation. It was discovered by Eric Becklin and Gerry Neugebauer in 1967 and is located in the ➤ *Kleinmann–Low Nebula*, within the ➤ *Orion Nebula*. It is thought to be a massive ➤ *B star*, hidden behind so much dust that very little visible light gets through. There are other infrared sources nearby in this star-formation region.

Beehive A popular English name for the open star cluster ➤ *Praesepe*.

Belinda One of the small satellites of Uranus discovered during the ➤ *Voyager 2* encounter with the planet in 1986. Its diameter is about 80 km (50 miles).

Bellatrix (Gamma Orionis) A giant ➤ *B star* of magnitude 1.6 in the constellation Orion. Its name comes from Latin and means "female warrior." Bellatrix is 240 light years away.

Belt of Orion The line made by the three second-magnitude stars ➤ *Mintaka*, ➤ *Alnilam* and ➤ *Alnitak* in Orion, regarded as the mythological figure's belt.

Bennett, Comet A spectacular comet discovered by Jack C. Bennett from South Africa on December 28, 1969. It reached magnitude zero in March 1970 and had a tail 30° long. Observations made from space revealed a vast hydrogen cloud surrounding the head and tail measuring 13 million km in the direction parallel to the tail.

BepiColombo A spacecraft that the European Space Agency proposes to place in a polar orbit around Mercury in order to map and survey the planet with a variety of remote sensing instruments. The projected launch date is 2012. The mission is named in honor of the Italian mathematician and engineer Giuseppe Colombo (1920–84) who suggested the trajectory ➤ *Mariner* 10 should follow past Mercury in 1974–75.

BeppoSAX An Italian/Dutch gamma- and X-ray satellite launched on April 30, 1996. Observations it made in 1997 led to the first optical identification of a ➤ *gamma-ray burst*. It operated until 2003.

Bessel, Friedrich Wilhelm (1784–1846) Bessel was both a mathematician and an observational astronomer. Born in Minden in what is now Germany, he was appointed Director of the Königsberg Observatory at the age of only 26. One of his major projects was to catalog accurate positions for many thousands of stars. He is most remembered for being the first to announce the distance to a star as a result of measuring its ➤ *parallax*. He gave the distance of 61 Cygni as 10.3 light years, which is very close to the modern value of 11.4 light years. Bessell also noted that the positions of ➤ *Sirius* and ➤ *Procyon* seemed to deviate back and forth slightly. He correctly deduced that both stars are orbiting dim companions that he could not see.

Beta Pictoris A fourth-magnitude ➤ *A star* in the constellation Pictor, surrounded by a disk of material that may be a planetary system in the process of forming. Strong infrared radiation from the disk drew astronomers' attention to it. The disk is about 10 times the size of Pluto's orbit around the Sun.

Betelgeuse (Betelgeux; Alpha Orionis) A massive, red ➤ *supergiant* star in the constellation Orion. With a diameter more than 1000 times the Sun's it is one of the largest stars known and one of the few to be detectable as a disk rather than just a point of light. It has bright "star spots" and is surrounded by a shell of dust and an extended ➤ *chromosphere*. The brightness of Betelgeuse

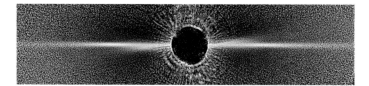

The disk around Beta Pictoris imaged by the Hubble Space Telescope. (The black disk at the center is to cut out the glare of the star itself. The colored effects are an artifact from the way the observation was made through filters.)

varies irregularly between magnitudes 0.4 and 0.9 with a rough period of around 5 years. Its distance is approximately 425 light years.

Bianca One of the small moons of Uranus discovered when ➤ *Voyager 2* flew past in 1986. It is 51 km (32 miles) across.

Biela, Comet A nineteenth century comet, famous because it split in two before disappearing completely. The comet was originally discovered in 1772 by Montaigne of Limoges. When it was recovered by Wilhelm von Biela in 1826, the orbit was calculated accurately enough to identify the two previous occasions when its appearance had been recorded. Its period was 6.6 years. At its 1846 return it was double. By 1852, the two components were separated by more than 2 million km though following the same orbit. Neither was ever seen again.

A November meteor shower, the ➤ *Andromedids*, is associated with Comet Biela. Brilliant displays have occasionally been seen both before and after the break-up of the comet.

Big Bang An explanation of the history of the universe that says it began in an infinitely compact state and has been expanding ever since. This theory is widely accepted because it explains both the ➤ *expanding universe* and the existence of ➤ *cosmic background radiation*. According to current estimates, the Big Bang took place about 13.7 billion years ago.

Big Bear Solar Observatory A solar observatory located at an altitude of 2000 m (6600 feet) on an island in Big Bear Lake in California. The site was chosen because the air is steadier over water than over land. There are four main telescopes on the same mount. The largest of them is a 65-cm (26-inch) reflector.

Big Crunch A hypothetical end to the universe in which it totally collapses in on itself. The universe could only suffer this fate if its present expansion were to slow down and be reversed. On current evidence, a Big Crunch is very unlikely because the expansion of the universe appears to be speeding up.

Big Dipper ("Plough" in Europe) The pattern formed by the seven stars Alpha, Beta, Gamma, Delta, Epsilon, Zeta and Eta in the constellation ➤ *Ursa Major*.

binary star A pair of stars in orbit around each other, held together by the gravitational attraction between them. About half of all stars have at least one companion, though many are so close that they cannot be separated even by a powerful telescope. In these cases, the evidence for more than one star comes from their combined spectrum.

The members of a binary system each move in an elliptical orbit around their center of mass. The further apart they are, the slower they move. Pairs far enough apart for the two stars to be distinguished, or "split," in a telescope often have orbital periods as long as 50 or 100 years. They are called visual binaries.

If one star is much fainter than the other, its presence may be revealed only by the obvious orbital motion of its brighter companion. Pairs of this type are called astrometric binaries.

As the members of a binary system move around their orbits, they move alternately towards and away from the Earth. Their spectra can reveal details about both the nature of the stars and their orbits. Binary stars recognized only by their spectra are called spectroscopic binaries. Their periods are usually between a day and a few weeks.

Some binaries are so close that the pull of gravity distorts the individual stars from their normal spherical shape. They are known as ➤ *contact binaries*. The energy released results in the emission of X-rays. ➤ *Novae* are another consequence of material being transferred between partners in certain binary stars.

In an eclipsing binary, the orbits of the stars are oriented so that one of them crosses in front of the other as seen from the Earth. Eclipsing binaries are also variables since one star periodically blots out light from the other.

binoculars An optical instrument consisting of two small telescopes, mounted side by side, one for each eye. The tubes are kept short and manageable by using prisms to reflect the light internally. The prisms also make the image upright, rather than inverted as it is in an astronomical telescope. The size and magnifying power of binoculars is usually given in the form $A \times B$, where A is the linear magnification and B is the diameter of each objective lens in millimeters (e.g. 10×40).

bipolar nebula A luminous nebula consisting of two lobes pointing in opposite directions. Any nebula of this shape may be called "bipolar," but this term is most often used for a group of nebulae that are intense sources of infrared radiation. They are thought to harbor a bright star that is completely concealed by a dense ring of dust and gas around its equator. The dust emits in the infrared because it is heated to a temperature of a few hundred degrees by the radiation from the star. The visible starlight is funneled along the star's poles and illuminates the more tenuous part of the nebula around the star. ➤➤ *bipolar outflow*.

bipolar outflow Gas streaming outwards in two opposing directions from a newly formed star. Because the star is surrounded by an ➤ *accretion disk*, the gas cannot escape around the equator and is forced to flow out from over the poles. This stellar wind sweeps up interstellar material before it and creates the

The bipolar outflow, nicknamed the Boomerang Nebula, coming from an old red giant star. Each lobe is nearly one light year long.

two lobes, which extend for a distance of about a light year. Bipolar outflows have been detected by the radio emission from the molecules they contain.

black body An object that absorbs all the radiation falling on it.

black body radiation The characteristic radiation emitted by a ➤ *black body*. How the radiation a black body emits varies with wavelength is affected only by its temperature and can be predicted by quantum theory. The graph of radiation intensity against wavelength is called the Planck curve after the physicist Max Planck, who predicted its shape theoretically. Planck curves are hill-shaped and peak at shorter wavelengths for hotter bodies. The total amount of energy emitted by a black body goes up steeply with temperature, as the fourth power (T^4).

black drop An effect observed during a ➤ *transit* of Venus or Mercury across the Sun, when the small dark disk of the planet is very near the limb of the Sun. When the limbs of the Sun and planet are not quite in contact, a small black spot, or drop, appears to join them.

black dwarf A dead star that has stopped shining. A star with less than about 1.4 times the Sun's mass spends the last stage of its life as a ➤ *white dwarf*. For a long time, white dwarfs keep shining because they are very hot to begin with but, since they cannot generate any new energy, they will gradually cool and fade into a dark stellar "corpse" or black dwarf. The universe is not yet old enough for any black dwarfs to have formed.

Black-eye Galaxy (M64; NGC 4826) A popular name for an unusual looking ➤ *spiral galaxy* in the constellation Coma Berenices that has very smooth spiral arms and a prominent dust cloud around its nucleus. It is about 65 000 light years in diameter.

black hole A region of space where the gravitational force is so strong that not even light can escape from it. Black holes are formed when matter collapses in on itself catastrophically, concentrating more than a critical quantity of mass into a particularly small region. Theory suggests that "mini" primordial

The Black-eye Galaxy.

black holes might have formed from large density fluctuations in the conditions prevailing in the early universe.

Stellar black holes are thought to form when massive stars explode, if the central relic is more than three solar masses, or is tipped over that mass when material cascades back onto it. To create a black hole, several solar masses of material would have to be packed into a diameter of just a few kilometers.

Matter falling into supermassive black holes is how most astronomers explain the exceptional power of ➤ *active galactic nuclei* and ➤ *quasars*. Direct observations of compact nuclei in galaxies, and motion of gas and stars near the centers of galaxies, appear to confirm that massive black holes do indeed exist at the centers of many galaxies. Typically, they have a billion times the Sun's mass.

Black holes cannot be observed directly. Their existence can only be inferred from their gravitational effects on their surroundings and the radiation emitted by material falling into them. A number of stellar X-ray sources, such as ➤ *Cygnus X-1*, are binary star systems in which one component appears to be a black hole. The black hole's orbit and mass can be computed from observations of its visible companion.

Black Widow pulsar A popular name for the ➤ *pulsar* PSR 1957 + 20. It is a member of a binary system and the action of its intense radiation is gradually evaporating its small stellar companion.

blazar A ➤ *BL Lac object* or a ➤ *quasar* that varies dramatically in brightness and emits strongly over the whole of the electromagnetic spectrum. Many blazars are powerful gamma-ray emitters.

Blaze Star A popular name for the recurrent ➤ *nova* T Coronae Borealis. It is the brightest recurrent nova ever recorded, having reached second magnitude in both 1866 and 1946.

blink comparator An instrument for comparing two photographs of a region of sky, usually a pair taken at different times. It shows up any images that do not have the same position or brightness on the two photographs. An optical system makes the two photographs appear to be exactly superimposed but illuminates them alternately. Any object whose brightness differs appears to blink on and off; one that is at a different position appears to jump between the two locations.

Blinking Nebula A popular name for NGC 6826, a ➤ *planetary nebula* in Cygnus. Observers using small telescopes say the nebula appears to blink on and off, leaving only the central star in view, if they switch between looking directly at it and looking to one side of it.

BL Lac object (BL Lacertae object; Lacertid) A type of ➤ *elliptical galaxy* with a bright, highly variable, compact nucleus. The first to be identified was BL Lacertae, which was thought to be a variable star when it was first spotted in 1929. Though it is now known to be a galaxy, it has kept its original name of the kind given to variable stars. BL Lac objects have dramatic short-term variations and a featureless spectrum. Their brightness can change by as much as a hundredfold over a period of a month, and day-to-day changes are sometimes observed. Many are also radio sources; BL Lacertae itself gives out intense radio bursts.

blue moon The origin of this expression, often used just to mean "a rare event," is not known. A suggestion that it refers to the second occurrence of a new Moon in one calendar month appears to be unfounded. An alternative explanation is that atmospheric effects may occasionally make the Moon appear blue; a possible cause would be dust in the upper atmosphere from volcanoes or forest fires.

blueshift A shift of a spectrum towards shorter wavelengths. Blueshifts are caused by the ➤ *Doppler effect* when the source of radiation and its observer are moving towards each other.

Blue Planetary A popular name for NGC 3918, a ➤ *planetary nebula* in Centaurus, which to visual observers looks like a blue featureless disk.

Blue Snowball A popular name for the ➤ *planetary nebula* NGC 7662 in the constellation Andromeda.

blue straggler A star that appears to belong to a ➤ *globular cluster* or an old ➤ *open cluster* but is excessively blue and bright compared with other cluster members. The explanation for the anomalous properties of blue stragglers is not known for certain. They might be ➤ *binary stars* where mass has been transferred from one star to the other, or the result of two very close stars merging.

Bode's law ➤ *Titius–Bode law.*

bolide A particularly bright ➤ *meteor* accompanied by an explosive sound or sonic boom.

bolometric magnitude The ➤ *magnitude* of a celestial object, taking into account all the energy it radiates at all wavelengths. The bolometric magnitude of an object that emits strongly in the ultraviolet or infrared, for example, differs greatly from its visual magnitude.

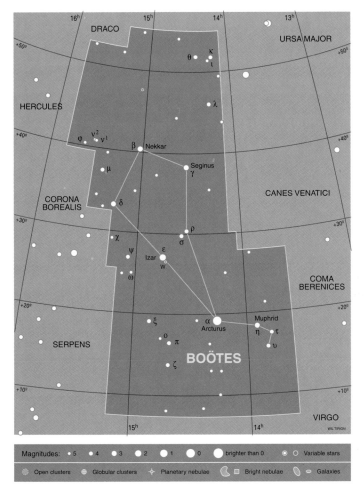

The constellation Boötes.

Boötes (The Herdsman) A constellation of the northern sky, dominated by the bright orange star ➤ *Arcturus*. It is usually said to represent a herdsman driving the bear, which is the neighboring constellation Ursa Major.

Borrelly, Comet A periodic comet that was visited by the ➤ *Deep Space 1* spacecraft in September 2001. Comet Borrelly was discovered by Alphonse L. N. Borrelly from Marseilles, France on December 28, 1904. Deep Space 1 discovered that its nucleus is about 8 km (5 miles) long by 4 km (2.5 miles) wide and very dark. Its orbital period is 6.8 years.

Bradley, James (1693–1762) The British astronomer James Bradley made the important discovery of the ➤ *aberration* of starlight, which he published in 1729. It was the first direct evidence that Earth is in motion around the Sun

41

Tycho Brahe.

and Bradley used his measurements to calculate the speed of light to within 2% of its correct value.

Bradley was introduced to astronomy by his uncle and studied at Oxford University. He became Professor of Astronomy at Oxford in 1721 and was appointed ➤ *Astronomer Royal* in 1742. He discovered the phenomenon of aberration while trying to measure stellar ➤ *parallax*. As a result of his accurate observations over many years he also discovered the ➤ *nutation* of Earth's axis, which he announced in 1748.

Brahe, Tycho (1546–1601) Tycho Brahe made the most accurate observations of the positions of stars and planets before the invention of the telescope. He was born into a noble Danish family and started to study astronomy after seeing a partial eclipse of the Sun in 1560. He realized the importance of making careful systematic observations of celestial bodies. When a "new star" blazed out in the constellation Cassiopeia in 1572 – what we now know to have been a ➤ *supernova* – Tycho recognized that it was something astronomical and not a phenomenon in the atmosphere and the short book he wrote about it made his name known.

The King of Denmark and Norway was impressed by Tycho and funded magnificently equipped observatories for him on the island of Hven. He worked there from 1577 but following a disagreement with a new king, left in 1596. By 1599, he had settled in Prague, where he found a new royal sponsor. In Prague, ➤ *Kepler* became his assistant and eventually inherited Tycho's observations, which he used to formulate his laws of planetary motion. Tycho himself was never persuaded that the Sun was the center of the planetary system. He thought the Earth could not be moving because he could not detect ➤ *parallax* in the positions of the stars he measured. (His instruments were not accurate enough to do so.) He proposed that the Sun and Moon revolved around Earth, while all the other planets orbited the Sun.

9969 Braille An asteroid that was imaged from a distance of 26 km (16 miles) by the NASA spacecraft ➤ *Deep Space 1* in 1999. It is an irregular, elongated body about 2.2×1.0 km (1.4×0.6 miles). The infrared spectrum of Braille proved to be nearly identical to that of ➤ *Vesta*, prompting speculation that Braille is a fragment broken off Vesta by an impact.

brown dwarf A gaseous body intermediate between a cool star and a giant planet. The mass of a brown dwarf is between 13 and 80 times the mass of Jupiter, which is too little for hydrogen-burning nuclear reactions to be sustained in its core. As it forms, a brown dwarf heats up because gravitational energy is

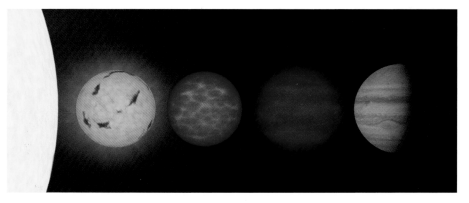

The size of brown dwarfs compared with Jupiter (right), the Sun (left) and a cool red star (to the right of the Sun).

released. Some nuclear processes may take place for a short time but, after peaking, the temperature steadily falls. Hundreds of brown dwarfs are now known to exist in the Sun's neighborhood of the Galaxy.

The system for classifying very cool stars and brown dwarfs builds on the long-standing system of ➤ *spectral types*. The new types introduced following the discovery of brown dwarfs are L and T.

Cool dwarf stars and the younger, warmer brown dwarfs look similar although their masses are different, and it is difficult to distinguish between them on the basis of their spectra alone. They both fall into types M and L. The M-type objects, with surface temperatures ranging down to 2100 K, have water and strong oxide features in their spectra.

The next cooler group, with temperatures of roughly 1500–2100 K are the L dwarfs (L0 to L8). Though the coolest, least massive stars fall into this range, their temperatures cannot be lower than 1800 K. Objects with temperatures between 1500 and 1800 K must be brown dwarfs.

The coolest objects so far detected are the methane brown dwarfs, so-called because their spectra show strong absorption by methane as well as water. These brown dwarfs, with surface temperatures ranging down from about 1000 K to 800 K, are allocated to type T.

B star A star of ➤ *spectral type* B. The surface temperatures of B stars are in the range 11 000–25 000 K and they look bluish white. The most prominent features in their spectra are ➤ *absorption lines* of neutral helium. Hydrogen lines are also present and are stronger the cooler the star. Examples of B stars are ➤ *Rigel* and ➤ *Spica*.

A Hubble Space Telescope image of the Bubble Nebula.

Bubble Nebula A popular name for a faint glowing nebula (NGC 7635) in the constellation Cassiopeia. The massive central star of the nebula is blowing off a fast ➤ *stellar wind*. The "bubble" marks where the stellar wind encounters the surrounding interstellar medium.

338 Budrosa An asteroid of the rare metallic type discovered in 1892 by Auguste Charlois. It is 63 km (39 miles) across and is the prototype of the Budrosa family of unusual asteroids, with six known members at 2.9 AU from the Sun.

Bug Nebula A name given to the ➤ *bipolar nebula* NGC 6302 in the constellation Scorpius. The star assumed to be at the center is not visible, but gas can be seen streaming out of the hot central region with speeds up to 400 km/s (over 2000 mph).

butterfly diagram The popular name for a diagram illustrating how the latitudes on the Sun at which ➤ *sunspots* appear vary through the ➤ *solar cycle*. It was first plotted in 1922 by E. Walter Maunder and its more formal name is the Maunder diagram. The diagram is a graph with solar latitude as the vertical axis and time (in years) as the horizontal axis. A short vertical line covering one degree in latitude is plotted for each sunspot group centered at that latitude within one rotation of the Sun. The result is a pattern reminiscent of pairs of butterfly wings.

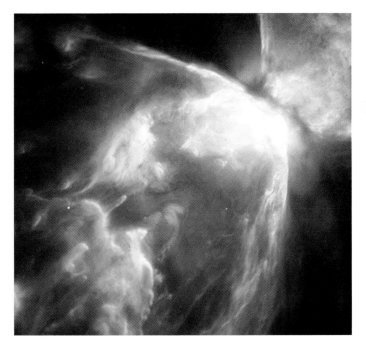

A Hubble Space Telescope image of the Bug Nebula.

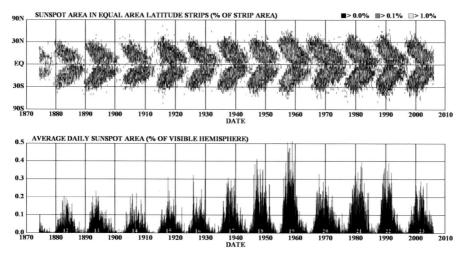

A butterfly diagram of the 11-year solar cycle.

C

Caelum (The Chisel) A small constellation with no star brighter than fourth magnitude. It was introduced into the southern sky in the mid-eighteenth century by Nicolas L. de Lacaille, who called it "Caela Sculptoris."

caldera A large volcanic crater.

Caliban One of two small moons of Uranus discovered in 1997 by Brett Gladman and others, using the ➤ *Hale Telescope*. It is reddish in color and thought to be a captured ➤ *Kuiper belt* object. Its diameter is estimated to be 98 km (61 miles).

California Nebula (NGC 1499) A bright ➤ *emission nebula* in the constellation Perseus, named for it resemblance to the shape of the US state. It forms the rim of a dark nebula of gas and dust illuminated by the star Xi Persei.

Callirrhoe A small outer moon of Jupiter discovered in 1999. Its diameter is about 9 km (6 miles).

Callisto The second largest moon of Jupiter and one of the four discovered in 1610 by Galileo. With a diameter of 4821 km (2996 miles), it is also the third largest moon in the solar system. Callisto is the darkest of the Galilean satellites and also the least dense, which suggests that it contains a high proportion of water, though detailed images returned by the ➤ *Galileo* spacecraft indicate that the surface has more rock and dust than previously supposed. Under its thick icy crust, Callisto may have a liquid ocean several kilometers deep. ➤ *Voyager* and Galileo images show a heavily cratered surface with no high hills or mountains. A multi-ringed impact basin, called Valhalla, is the most prominent feature. It has a bright central zone 600 km (375 miles) across, surrounded by numerous rings spaced between 20 and 100 km (12 and 60 miles) apart. There are at least seven other multi-ring features on Callisto.

Caloris Basin (Caloris Planitia) A large, multi-ringed, impact basin on Mercury. It is 1300 km (800 miles) in diameter and the most conspicuous feature on the planet.

Calypso A small satellite of Saturn discovered in 1980. It measures 30 × 16 × 16 km (19 × 10 × 10 miles) across and it shares the same orbit as ➤ *Tethys* and ➤ *Telesto* at a distance of 294 660 km (183 093 miles) from Saturn.

Camelopardalis (alternatively Camelopardus; The Giraffe) A large but not very conspicuous constellation near the north celestial pole. It was first mentioned in 1624 by the German mathematician Jakob Bartsch, who was a son-in-law of Johannes ➤ *Kepler*.

A close-up of Callisto showing ancient multi-ring impact basins.

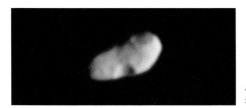

A Cassini image of Saturn's small moon Calypso.

107 Camilla An asteroid estimated to be 240 km (149 miles) across, discovered from Madras, India, by Norman R. Pogson in 1868. In 2001, astronomers observing Camilla with the Hubble Space Telescope reported that it has a small satellite.

Canada–France–Hawaii Telescope (CFHT) A 3.6-m (140-inch) telescope at the ➤ *Mauna Kea Observatories* in Hawaii. It was commissioned in 1979 and is used for both optical and infrared observations.

canals Imagined linear features ➤ *Percival Lowell* and others claimed to see on Mars. In the nineteenth century, the Italian astronomers Angelo Secchi and Giovanni Schiaparelli used the word *canale*, meaning "channel," to describe linear features they thought they could see on Mars. The word was translated into English as "canal." Percival Lowell was excited by these reports and built an observatory at Flagstaff, Arizona, with the main purpose of observing Mars. His drawings of the planet showed extensive networks of straight "canals," and he proposed that a civilization of intelligent beings on Mars was responsible for constructing them. Later observers found little evidence for Lowell's canals, and spacecraft images of Mars show no trace of them. The linear features that Lowell and others genuinely believed they had observed can probably be explained as optical effects.

Canary Islands Large Telescope ➤ *Gran Telescopio Canarias*.

Cancer (The Crab) A constellation in the traditional ➤ *zodiac*. It is said to represent the crab Hercules crushed under his foot when he was fighting the Hydra. None of its stars is brighter than fourth magnitude. The star cluster ➤ *Praesepe* at the center of Cancer can be seen with the naked eye.

Canes Venatici (The Hunting Dogs) A small constellation in the northern sky, lying between Boötes and Ursa Major. It was introduced by ➤ *Johannes Hevelius* in the late seventeenth century and is supposed to represent the dogs Asterion and Chara held on a leash by Boötes. Though small, Canes Venatici contains several interesting objects including the bright star ➤ *Cor Caroli*, the fine ➤ *globular cluster* M3 and the ➤ *Whirlpool Galaxy*.

Canis Major (The Greater Dog) A small constellation, just south of the celestial equator and next to Orion, containing the brightest star in the sky, ➤ *Sirius*. It is said to represent one of the dogs following the hunter, Orion.

Canis Minor (The Lesser Dog) A small constellation near to ➤ *Orion*. It is supposed to represent the smaller of two dogs following the hunter Orion. Canis Minor contains only two stars brighter than third magnitude. The brightest is ➤ *Procyon*.

Cannon, Annie Jump (1863–1941) The American astronomer Annie Jump Cannon is remembered for her phenomenal skill at classifying the spectra of stars. While working at ➤ *Harvard College Observatory* from 1896, she catalogued the ➤ *spectral types* of about 400 000 stars. She was born in Delaware, the daughter of a state senator, and studied physics and astronomy at Wellesley College and Radcliffe College. As a young woman, she became almost totally deaf after contracting scarlet fever. Cannon was responsible for reorganizing the way stellar spectra are classified into the system still used today. Her

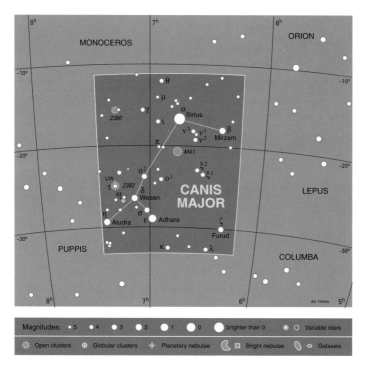

A map of the constellation Canis Major.

classifications of over 225 000 stars were published in several volumes, called the ➤ *Henry Draper Catalog*. She received a number of honors, including a medal from the National Academy of Sciences in the US and an honorary doctorate from Oxford University in the UK.

Canopus (Alpha Carinae) The brightest star in the constellation Carina and the second brightest star in the sky. Canopus is a supergiant ➤ *F star* of magnitude –0.7. It is 313 light years away. In Greek mythology, Canopus was the pilot of the fleet of King Menelaos.

Cape Canaveral The location in Florida, USA, of the Kennedy Space Center and of a US Air Force base from where many of NASA's space missions are launched.

Capella (Alpha Aurigae) The brightest star in the constellation Auriga, magnitude 0.1. It is a ➤ *spectroscopic binary* 42 light years away. Both members of the binary system are giant ➤ *G stars*. Capella means "little she-goat" in Latin.

Cape York Meteorite A large iron ➤ *meteorite* broken into three pieces that were found by Robert Peary in Greenland in 1894. Three years later he transported them to New York and they are now at the American Museum of Natural History in New York City. The heaviest piece, weighing 31 tonnes, was nick-named "Ahnighito," meaning "the tent," by the local people in Greenland. It is the largest meteorite on view in any museum.

Capricornus (The Sea Goat) One of the traditional constellations of the ➤ *zodiac*. Its brightest stars are third magnitude.

carbonaceous chondrite A rare type of stony ➤ *meteorite*. Because their average chemical composition (except for hydrogen and helium) is very similar to the Sun's, carbonaceous chondrites are thought to consist of primitive, unprocessed material dating from the time when the solar system formed. They are made up of carbon-rich minerals and have ➤ *chondrules* embedded in them. Their water content can be as high as 20 percent. The largest known example is the ➤ *Allende meteorite*.

carbon cycle (carbon–nitrogen (CN) cycle; carbon–nitrogen–oxygen (CNO) cycle; Bethe–Weizsäcker cycle) A series of nuclear reactions that takes place inside stars. The outcome of the carbon cycle is that hydrogen is converted to helium and large amounts of energy are released.

carbon star A peculiar, red giant star with unusually strong features in its spectrum caused by C_2, CN, CH or other carbon compounds. Carbon stars have temperatures similar to the more common ➤ *K stars* and ➤ *M stars* but contain more carbon and oxygen. Though carbon stars are rare in our own Galaxy, many thousands have been discovered in the Large and Small ➤ *Magellanic Clouds*.

The spectra of some carbon stars reveal the presence of the radioactive element technetium. Its longest-lived isotope has a half-life of only 210 000

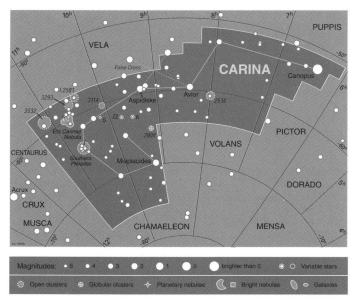

A map of the constellation Carina.

years, a short period on astronomical timescales. A few of the coolest carbon stars have in their spectra an extremely strong line of lithium, a chemical element that is easily destroyed by the nuclear processes in stars. Technetium and lithium in the outer layers of carbon stars show that material is being dredged up from deep inside them by the process of ➤ *convection*. All the evidence is consistent with carbon stars being in an advanced state of evolution.

Carina (The Keel) A large constellation in the southern Milky Way, formerly part of ➤ *Argo Navis*. It contains the second-brightest star in the sky, ➤ *Canopus*.

Carina Nebula (NGC 3372) A large cloud of glowing hydrogen gas in the southern Milky Way in the constellation Carina. The star ➤ *Eta Carinae* is embedded in it, near its center. The nebula is 3° across on the sky. It actual diameter is

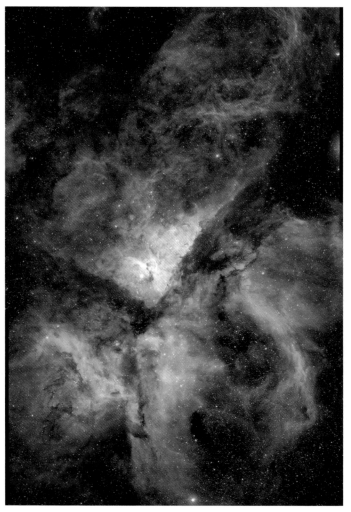

The Carina Nebula.

400 light years and it lies at a distance of 8000 light years. It is a region of star formation containing a number of young star clusters. ➤ *Keyhole Nebula*.

Carme A small outer moon of Jupiter, discovered in 1938 by Seth B. Nicholson. Its diameter is 46 km (29 miles).

Carpo A small outer moon of Jupiter discovered in 2003. Its diameter is about 3 km (2 miles).

Carrington rotation number A number that uniquely identifies each rotation of the Sun. The sequence began with rotation number one on November 9, 1853. The system was started by Richard C. Carrington, based on the average rotation rate of sunspots, which he had determined. In reality, the Sun does

An image of the Cartwheel Galaxy formed by combining X-ray data from the Chandra X-ray Observatory, ultraviolet light from the Galaxy Evolution Explorer satellite, a visible image by the Hubble Space Telescope and an infrared view from the Spitzer Space Telescope.

not rotate as if it were a solid body, its rotation rate varying with latitude. For the purpose of counting the rotation numbers, the period is taken as 25.38 days relative to the stars, which is equivalent to 27.28 days as viewed from Earth.

Carte du Ciel An ambitious project, begun in 1887, to make photographic charts of the entire sky and compile a star catalog. In the event, the charts were never completed because the standard methods laid down for the project were overtaken by technical advances in astrophotography. However, the catalog was finally completed and published in 1964.

Cartwheel Galaxy A ➤ *peculiar galaxy*, 500 million light years away, more formally known as A0035. It consists of a circular rim, 170 000 light years in diameter, inside which are a hub and spokes made up of old red stars. It is believed that the galaxy was once a large, ordinary ➤ *spiral galaxy*. A few hundred million years ago, a smaller galaxy passed right through it. That intruder is still nearby. The shock of the collision caused large numbers of massive stars to form in the "rim." When these stars reach the end of their relatively short lives, they explode as ➤ *supernovae*. As a result, the rate of supernova explosions in the Cartwheel Galaxy is now about a hundred times greater than in a normal galaxy.

Cassegrain telescope A reflecting telescope in which the image is focused just behind a central hole in the primary mirror. The design was proposed in about 1672 by Jacques Cassegrain (1652–1712), professor of physics at Chartres in France, some four years after Isaac ➤ *Newton* constructed the first reflecting telescope. Its secondary mirror is convex rather than flat as in Newton's design. Cassegrain did not build a telescope himself and it was some years before his idea was put into practice. Today, the Cassegrain focus is popular and widely used in both modest amateur instruments and large professional telescopes.

Cassini The orbiting spacecraft of the ➤ *Cassini–Huygens mission* to Saturn.

Cassini Division A conspicuous dark gap between the A and B rings of ➤ *Saturn*. It is 2600 km (1616 miles) wide.

Cassini, Giovanni Domenico (Jean-Dominique) (1625–1712) Cassini was born near Genoa in Italy and became a professor of astronomy in Bologna but in 1669 King Louis IV of France persuaded him to move to Paris to take charge of the newly established ➤ *Paris Observatory*. He became a French citizen in 1673 and adopted the French form of his first names. His astronomical work was concerned with the solar system and from his observations calculated that the distance from Earth to the Sun was about 140 million km (87 million miles). Though still 7 percent too small, his figure was much closer to the true value than previous estimates. He discovered four of Saturn's moons – Iapetus, Rhea, Dione and Tethys – and the gap in Saturn's rings called the Cassini division.

Both Cassini's son, Jacques, and his grandson, César François, subsequently became directors of the Paris Observatory.

The Cassini–Huygens mission. An artist's impression of Cassini releasing the Huygens probe.

Cassini–Huygens mission A joint NASA/ESA mission to explore Saturn, its rings, its magnetosphere and several of its moons. The spacecraft was launched in October 1997 and used ➤ *gravity assist* flybys of Venus (in April 1998 and June 1999), of Earth (in August 1999) and of Jupiter (in December 2000). It finally arrived at its destination in July 2004 and was placed in orbit around Saturn. The nominal length of its mission is four years.

One of the major objectives of Cassini–Huygens was a study of Saturn's moon ➤ *Titan*. Cassini carried the Huygens probe, an instrument package that successfully parachuted down through Titan's atmosphere in January 2005 and landed on the surface. The Huygens probe was the principal ESA contribution to the mission.

Cassiopeia A conspicuous W-shaped constellation near the north celestial pole. It is said to represent the seated figure of Queen Cassiopeia, a character from Greek mythology. Cassiopeia was the wife of ➤ *Cepheus* and the mother of ➤ *Andromeda*, both represented by their own constellations. Tycho ➤ *Brahe* observed a ➤ *supernova* in Cassiopeia in 1572. This constellation also contains the strongest radio source in the sky, known as ➤ *Cassiopeia A*.

Cassiopeia A The strongest radio source in the sky (other than the Sun). It is the remnant of a ➤ *supernova* that must have occurred around AD 1667 though no records exist of anyone seeing a supernova around that time. The light from

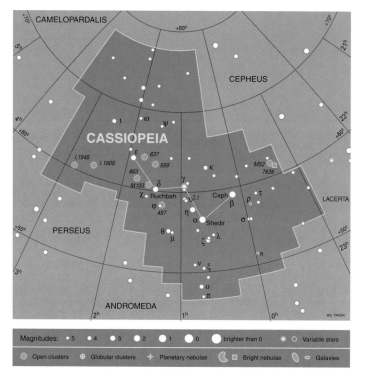

A map of the constellation Cassiopeia.

An image of the supernova remnant Cassiopeia A formed from infrared data from the Spitzer Space Telescope (colored red), optical data from the Hubble Space Telescope (yellow) and X-ray data from the Chandra X-ray Observatory (green and blue).

the explosion of the star, which took place 10 000 light years away, was obscured by the large quantities of dust that lie in the line of sight from Earth. The radio emission is concentrated in a ring 4 arc minutes across and there is X-ray emission from the ring too. Optical photographs of that region of the sky detect a faint nebula.

Castor (Alpha Geminorum) The second-brightest star in the constellation Gemini, after ➤ *Pollux*. Its magnitude as seen by the naked eye is 1.6, but this is the combined brightness of a multiple system with at least six components. There are two ➤ *A stars* of magnitudes 2.0 and 2.9 forming a close visual pair, each of which is a ➤ *spectroscopic binary*. There is also a more distant ninth-magnitude red star, which is an ➤ *eclipsing binary*. Castor lies at a distance of 52 light years.

cataclysmic variable A star that brightens dramatically and suddenly when some kind of explosive event takes place. The term is applied particularly to ➤ *novae*, ➤ *supernovae* and ➤ *dwarf novae*.

Cat's Eye Nebula The ➤ *planetary nebula* NGC 6543. It is about 3000 lights years away and lies in the constellation Draco. Estimated to be about 1000 years old, its intricate structure suggests that there is a binary star at its center.

CCD Abbreviation for charge-coupled device, an electronic imaging device widely used in astronomy.

A CCD is made of semiconducting silicon. Light falling on this material releases electrons. The number of electrons increases in proportion to the brightness of the light. To create an image, a CCD collects light over a matrix of small picture elements (pixels). The electric charge from each pixel is

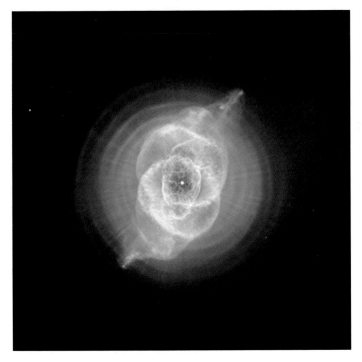

This Hubble Space Telescope image of the Cat's Eye Nebula shows at least eleven shells of gas around the central star, each blown off at a different time.

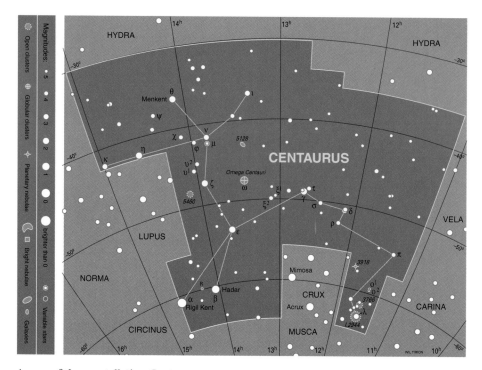

A map of the constellation Centaurus.

registered and then converted into a form that allows the whole image to be stored in a computer or displayed on a screen.

cD galaxy A type of giant elliptical galaxy that is very bright at the center and has an extended diffuse halo of stars. cD galaxies are found at the centers of rich galaxy clusters. Many are radio emitters and they are five to ten times more luminous than ordinary elliptical galaxies. Their masses are in the range 10^{13}–10^{14} solar masses and their diameters 1–3 million light years. A significant amount of their mass is in the form of ➤ *dark matter*.

    cD galaxies probably result from galaxy mergers. NGC 6616 is a typical example. It appears to have multiple nuclei and to be swallowing smaller nearby galaxies.

celestial equator The circle on the ➤ *celestial sphere* between the northern and southern celestial hemispheres. It is the projection into space of the Earth's equator.

celestial mechanics The branches of astronomy dealing with the movements and positions of astronomical objects.

celestial poles The two points on the celestial sphere about which the sky appears to rotate daily. Earth's rotation axis points towards them. The north celestial pole currently lies close to the star Polaris and the south pole is in the

The radio galaxy Centaurus A, imaged by the 8.2-m KUEYEN unit of the European Southern Observatory's Very Large Telescope.

constellation Octans, unmarked by any bright star. Because of the effects of ➤ *precession*, the positions of the poles are not stationary but sweep out circles with radii of about 23° over a period of 25 800 years.

celestial sphere The sky regarded as the inside of a distant hollow sphere. Though astronomical objects such as stars and planets really move through three-dimensional space, the idea of the celestial sphere is very useful for describing their positions and motion as seen from a particular location, such as Earth's surface. Every observer is located at the center of their own celestial sphere. Half the sky is always hidden from an observer on Earth's surface; the visible half varies according to the observer's latitude and with the date and time. Measurements on the celestial sphere are angles given in degrees, or minutes and seconds of arc, and take no account of how far away the objects actually are.

Centaurs A class of asteroids with orbits in the outer part of the solar system. Their orbits are within Neptune's but they approach no closer to the Sun than Jupiter. They are easily perturbed when they pass close to one of the giant planets. Two well-known examples are ➤ *Chiron* and ➤ *Pholus*.

Centaurus (The Centaur) A large southern constellation. The Milky Way runs through it so it is very rich in stars. Centaurus contains the nearest star to the solar system, ➤ *Proxima Centauri*, and the finest and brightest of all globular star clusters, ➤ *Omega Centauri*.

Centaurus A A ➤ *radio galaxy*, identified as the elliptical galaxy NGC 5128. At a distance of 15 million light years, it is the nearest radio galaxy and consequently one of the most studied. The visible galaxy is crossed by a thick dark lane of dust. The radio-emitting lobes are at right angles to the dust lane. They extend across 7° of sky, equivalent to almost two million light years. The galaxy is also a strong source of X-rays.

central meridian (CM) The imaginary north–south line bisecting the disk of an astronomical body as it is viewed by an observer.

Cepheid variable A type of pulsating ➤ *variable star*, named after the group's prototype, Delta Cephei, which varies between magnitudes 3.6 and 4.3 in a period of 5.37 days. Cepheid variables pulse in and out because their structure is unstable. Their size may change by as much as 10 percent and their temperature varies too. Pressure builds up inside the star and it expands until the pressure is released, rather like the action of a safety valve. The star then contracts and the cycle starts again.

Cepheids are luminous yellow giant stars. They radiate ten thousand times as much energy as the Sun so they can be seen at very great distances. In 1912, Henrietta ➤ *Leavitt*, who worked at Harvard College Observatory, noted a number of Cepheids in the Small Magellanic Cloud and plotted their light curves. She realized that there was a relationship between the periods over which the stars varied (typically between 3 and 50 days) and their average apparent brightness. The brighter the star, the longer it took to complete a cycle. This is called the period–luminosity relation. This important discovery meant that Cepheids can be used to determine the distances of nearby galaxies. Once the distance of a single Cepheid variable had been found by an independent method, the distances
of all others could be deduced simply by measuring their periods.

There are two distinct varieties of Cepheid variables: classical Cepheids and ➤ *Population II* Cepheids, also known as W Virginis stars. The period–luminosity relations of the two types are not the same. Classical Cepheids are about two magnitudes brighter than W Virginis stars with the same period because their masses and chemical composition are different.

Cepheus A constellation close to the north celestial pole, taking its name from the legendary King of Ethiopia, who was the husband of Cassiopeia and the father of Andromeda. It is one of the ancient constellations but is not conspicuous, having no star as bright as second magnitude.

1 Ceres The first ➤ *asteroid* to be discovered. It was found by Giuseppi Piazzi from Palermo, Sicily, on January 1, 1801. It is by far the largest asteroid, measuring 940 km (585 miles) across and according to criteria adopted by the

➤ *International Astronomical Union* in 2006, is also regarded as a ➤ *dwarf planet*. Ceres orbits in the main ➤ *asteroid belt* at a distance of 2.77 AU from the Sun. With a mass of 1.17×10^{21} kg, it accounts for about one-third the entire mass of the asteroid belt. However, it never gets brighter than magnitude 6.9 as its ➤ *albedo* is only 9 percent. Ceres rotates in just over 9 hours. Its color and brightness vary only slightly, suggesting that it is almost spherical and uniformly gray. Information from its spectrum indicates that the composition of the surface may be similar to that of ➤ *carbonaceous chondrite* meteorites.

Cerro Tololo Inter-American Observatory An observatory in Chile, forming part of the ➤ *National Optical Astronomy Observatories* of the USA. The headquarters are at La Serena, 480 km (300 miles) north of Santiago. The mountain site, 70 km (45 miles) inland, is at an altitude of 2200 m (7200 feet). The largest instrument is the 4-m (160-inch) Victor M. Blanco Telescope.

Cetus (The Whale or Sea Monster) A large constellation in the region of the celestial equator. It is said to represent the mythological sea monster that threatened Andromeda, though Cetus is frequently translated as the "whale." All but one of its stars are fainter than third magnitude. The most notable star in Cetus is the variable ➤ *Mira*.

Chaldene A small outer moon of Jupiter discovered in 2000. Its diameter is about 4 km (2.5 miles).

Chamaeleon (The Chameleon) A small, faint, southern constellation, probably invented by sixteenth century navigators and included by Johann Bayer in his 1603 star atlas *Uranometria*. None of its stars is brighter than fourth magnitude.

Chandler wobble Small variations in the position of Earth's geographical poles believed to result from seasonal changes in the distribution of mass on the Earth and movements of material within the Earth.

Chandrasekhar, Subrahmanyan (1910–1995) Chandrasekhar was one of the most distinguished astrophysicists of the twentieth century. He was awarded the Nobel Prize in 1983 for his theoretical work on the structure of stars and ➤ *stellar evolution*. He was born in Lahore (then in India, now in Pakistan) and studied physics in Madras and at Cambridge University in the UK. In 1937, he went to the USA to work at the University of Chicago and remained there for the rest of his life. He became an American citizen in 1953. He was known by everyone simply as "Chandra," and NASA named its orbiting ➤ *Chandra X-ray Observatory* in his honor. ➤➤ *Chandrasekhar limit*.

Chandrasekhar limit The maximum mass a ➤ *white dwarf* star can have. The astrophysicist Subrahmanyan ➤ *Chandrasekhar* proved theoretically that a white dwarf can be no more than 1.4 solar masses. When more massive stars exhaust their sources of nuclear energy, they must continue to collapse to a size much smaller than a white dwarf to form a ➤ *neutron star* or a ➤ *black hole*.

Subrahmanyan Chandrasekhar.

Chandra X-ray Observatory A NASA orbiting X-ray telescope launched from the ➤ *Space Shuttle* Columbia on July 3, 1999, then put into a very elliptical, high-altitude orbit. Its distance from Earth ranges between 10 000 and 140 000 km (6200 and 87 000 miles). It has instruments for making images and for recording spectra. It was named in honor of Subrahmanyan ("Chandra") ➤ *Chandrasekhar*. ➤➤ *X-ray astronomy*.

charge-coupled device ➤ *CCD*.

Charon The largest moon of ➤ *Pluto*. It was discovered by James Christy in 1978, when he noticed a slight elongation of the image of Pluto on a photograph taken at the US Naval Observatory. With a diameter of 1200 km (740 miles),

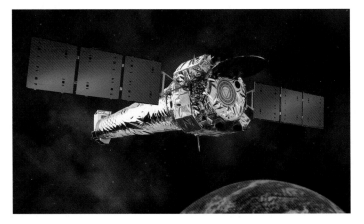

An artist's impression of the Chandra X-ray Observatory in space.

it is about half the size of Pluto. Charon orbits Pluto in 6.39 days, which is also the rotation period of both Pluto and Charon. There is evidence in the spectrum of Charon for water ice, but not methane, which is present on Pluto.

Chicxulub Crater A large terrestrial impact crater, 170 km (106 miles) across, located by the northern coast of the Yucatán Peninsula in Mexico. It has been identified as the crater excavated by an impact event 65 million years ago that appears to be linked with the mass extinction of living species, including the dinosaurs. The crater is now buried under a kilometer of sediment and was discovered in 1990 by a geophysicist looking for oil.

2060 Chiron An unusual asteroid discovered in 1977 by Charles Kowal. Its orbit lies between the orbits of the planets Jupiter and Uranus, well outside the main ➤ *asteroid belt*. It was the first of several asteroids found in orbits of this kind, which are now known collectively as ➤ *Centaurs*. Chiron's diameter is about 180 km (112 miles). According to infrared observations, it has a moderately dark, rocky or dusty surface and is nearly spherical. In 1989, a ➤ *coma* was discovered around Chiron, as a result of which it was also designated as a periodic comet (95P/Chiron).

chondrite A common kind of stony ➤ *meteorite*, characterized by the presence of ➤ *chondrules*. About 85 percent of meteorites are chondrites, as opposed to ➤ *achondrites*.

chondrule A small sphere of silicate minerals found in many stony ➤ *meteorites*. Chondrules range in size from less than 1 mm to more than 10 mm. ➤➤ *chondrite, achondrite*.

chromosphere The layer of the Sun (or of any star) above the ➤ *photosphere*. Chromosphere literally means "sphere of color." It is seen as a pinkish glow when the light of the photosphere is hidden at a total solar eclipse.

Circinus (The Compasses) A small, insignificant constellation of the southern sky. It was introduced by Nicolas L. de Lacaille in the mid-eighteenth century.

circumpolar star A star that never sets below the horizon as seen by a particular observer. For a star to be circumpolar, its ➤ *declination* must be greater than 90° minus the latitude of the observer's location. From the equator no stars are circumpolar; at the Earth's poles all stars are circumpolar.

circumstellar disk A disk of gas or dust surrounding a star. Newly forming stars develop disks as part of the normal process. In some cases at least, planetary systems form within circumstellar disks. ➤➤ *accretion disk, proplyd*.

Clavius A large lunar crater, 225 km (140 miles) in diameter, near the southern limb of the Moon.

Clementine A mission to the Moon undertaken by the US Department of Defense principally to test space hardware. Launched on January 25, 1994, it spent two months in orbit around the Moon. Using digital imaging with ultraviolet–visible and near-infrared cameras, it made better geological maps than any previous lunar mission and returned a large amount of scientific data.

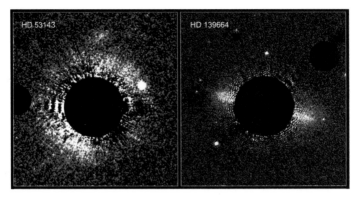

Circumstellar disks around two nearby stars. Data from the Hubble Space Telescope is in false color. Both stars are about 60 light years away. (The black circles are disks used to cut out the glare from the stars.)

The intention had been for Clementine to travel on to a flyby of the asteroid ➤ *Geographos*. However, that part of the mission was cancelled after the spacecraft malfunctioned on May 5.

Cloverleaf A ➤ *quasar* with a quadruple image cause by a ➤ *gravitational lens* effect. The quasar has a redshift of 2.55.

Cluster A European space mission to study the interaction between Earth's magnetic field and the ➤ *solar wind* and processes in Earth's ➤ *magnetosphere*. It consists of four separate satellites flying in formation. The distance between them varies between a few hundred kilometers and 20 000 kilometers (12 000 miles). The original four craft were destroyed when the first Ariane 5 launcher exploded on June 4, 1996. Four further craft were subsequently constructed for launch into highly elliptical polar orbits around Earth by two Soyuz rockets in 2000. The new mission was named Cluster II.

cluster of galaxies An assembly of galaxies, held together by gravity.

The distribution of galaxies in space is not uniform: they tend to be in clumps on distance scales of millions of light years. Clusters of galaxies take a variety of forms. They can be spherical and symmetrical, or ragged with no particular shape. They may contain a handful of galaxies or thousands. There may or may not be a concentration towards the center. Our own Milky Way Galaxy belongs to a small association known as the ➤ *Local Group*.

Clusters containing a concentration of many large galaxies are described as "rich." The nearest rich cluster is the ➤ *Virgo Cluster*, which has thousands of members. Even larger is the ➤ *Coma Cluster*, which is at least ten million light years across. The center of a rich cluster is usually dominated by a giant elliptical galaxy. The most massive galaxies known are at the centers of large rich clusters. Galaxies tend to merge at the centers of clusters – a process called "galactic cannibalism." The cannibal galaxies often look distended and some

The Fornax cluster of galaxies contains vast clouds of hot gas at 10 million degrees between and around the galaxies, which is revealed in this X-ray image.

seem to have more than one nucleus. They are usually strong sources of radio emission.

Tenuous hot gas fills the space between the galaxies in rich clusters. Its presence was revealed because of the X-rays it emits. In some clusters, there is as much matter in the gas between the galaxies as there is in the visible parts of the galaxies. ➤ *galaxy, intracluster medium*.

cluster of stars ➤ *star cluster*.

Coalsack An ➤ *absorption nebula* of interstellar dust in the ➤ *Milky Way*. It lies in the constellation Crux and is about 4° across.

COBE ➤ *Cosmic Background Explorer*.

Cocoon Nebula The diffuse nebula IC 5146 in the constellation Cygnus. It is a region of glowing hydrogen gas surrounding a sparse cluster of twelfth-magnitude stars, at an estimated distance of 3000 light years. It is thought to be a complex region of gas and dust where star formation is taking place.

Columba (The Dove) A small, faint constellation in the southern sky, introduced by Augustin Royer in 1679. It is sometimes said to represent the dove that followed Noah's Ark.

coma A cloud of gas and dust surrounding the nucleus of a ➤ *comet*. Typically, a coma reaches its maximum size of up to a million kilometers across when the comet has just passed ➤ *perihelion* in its orbit around the Sun.

Coma Berenices (Berenice's Hair) A small, faint constellation next to Boötes, which was introduced by Tycho ➤ *Brahe* in about 1602. It is supposed to represent the tresses of Queen Berenice of Egypt who cut off her hair and presented it to the gods in gratitude for the safe return of her husband from battle. The constellation is notable for the large number of galaxies it contains, members of both the Coma and Virgo clusters of galaxies.

Coma Cluster A rich cluster of galaxies in the constellation Coma Berenices. It covers several degrees of sky and contains more than a thousand bright galaxies. At an estimated distance of 300 million light years, it is one of the nearer rich clusters. Most of the galaxies are concentrated towards the center of the roughly spherical cluster, where the average distance between galaxies is three times smaller than the distance between the Milky Way and the Andromeda Galaxy in the ➤ *Local Group*.

comet An icy body orbiting in the solar system, which develops a ➤ *coma* and tails when it nears the Sun.

The nucleus of a comet is like a "dirty snowball" a few kilometers across. It is composed of water, carbon dioxide, methane and ammonia – all frozen – with dust and rocky material embedded in the ice. As a comet approaches the Sun, solar heating starts to vaporize the ice. A visible sphere of gas and dust,

The Coma Cluster of galaxies.

A comet in the process of disintegrating. The Hubble Space Telescope imaged these fragments of Comet Schwassmann–Wachmann 3, which broke up as it approached the Sun in 2006.

called the coma, develops around the nucleus. The coma may be up to a million kilometers wide. The coma is surrounded by a huge invisible cloud of hydrogen, many millions of kilometers in size. The hydrogen comes from the breakdown of water molecules by solar radiation. In 1996, X-ray emission was discovered from Comet ➤ *Hyakutake*, and it was subsequently found that other comets are also X-ray sources.

Dust and gas leave the comet nucleus from jets on the side facing the Sun, then stream away under the Sun's influence. Electrically charged atoms are swept away directly by the magnetic field of the ➤ *solar wind* and form a straight tail called an ion, Type I, plasma or gas tail. Variations in the solar wind cause structures to form in the ion tail. Sometimes the tail breaks off. Small dust particles are not carried along by the solar wind but get "blown" gently away from the Sun by radiation pressure. Dust tails (also called Type II tails) are often broad and flat. Comet ➤ *Hale-Bopp* was found to have a third distinct tail, made of sodium.

As a comet approaches the Sun, its tails grow. When it recedes from the Sun, the tails shrink again. Whatever the direction of the comet's motion, its tails are always directed away from the Sun. They can be as much as a hundred million kilometers long. Large dust particles get strewn along the comet's orbit and form ➤ *meteor streams*. Despite their often dramatic appearance, comets contain very little material, perhaps only a billionth the mass of the Earth. Their tails are so tenuous that only one five-hundredth of the mass of the nucleus may be lost in a passage around the Sun.

New comets are discovered regularly. Some are short-period comets, in elliptical orbits that take between about 6 and 200 years to complete. Most are long-period comets, in orbits so elongated that they do not return for many thousands of years.

It is now generally accepted that most comets come from a spherical cloud that surrounds the solar system at a distance of around 50 000 AU. This "reservoir" of comet nuclei is called the ➤ *Oort Cloud*. Other comets appear to come from the ➤ *Kuiper Belt*, located beyond the orbit of Neptune. The short-period comets have been captured within the planetary system by a close encounter with Jupiter.

When a new comet is discovered, or a known periodic comet is recovered, it is given a designation consisting of the year followed by a letter for the half-month period when it was found. The letters are A for January 1–15, B for January 16–31 and so on, up to Y for December 16–31. The prefix P/ is added for short-period comets or C/ for long-period comets. Periodic comets that have disappeared or been destroyed are prefixed by D/. New comets are named after their discoverers, though no more than three names are permitted when there are several independent claimants. A few comets have been named after individuals who calculated their orbits (Halley and Encke, for example) and after observatories or satellites, where discovery was essentially through the efforts of a team. When a short-period comet has been fully established as such, it is also allocated a number. This system for designating and naming comets was introduced in 1995.

cometary globule A small interstellar cloud resembling a comet. A cometary globule is what remains of a dense clump within a cloud of interstellar gas and dust after the more tenuous gas around it has been swept away by the action of strong ultraviolet radiation from nearby stars. The clump is the head of the globule. Its tail is like a shadow, where some of the material of the original cloud has been protected from the ultraviolet light. The famous ➤ *Horsehead Nebula* is a cometary globule in the process of formation.

comet group A class of comets that have orbits with similar characteristics. Members of a group (or family) of comets are not necessarily close to each other in space. Short-period comets are one example. These have come under the influence of Jupiter's gravity and have orbital periods typically between 6 and 8 years. The ➤ *sungrazers*, are another group. These long-period comets skim the Sun's outer layers and have other features of their orbits in common.

Compton Gamma Ray Observatory A NASA orbiting observatory with four ➤ *gamma-ray astronomy* experiments on board. It was launched from the ➤ *Space Shuttle* in April 1991 and operated for 9 years. Originally known simply as the "Gamma Ray Observatory," it was later named in honor of the American physicist Arthur Holly Compton (1892–1962). One of the things it achieved was a catalog of gamma-ray sources, including ➤ *supernova remnants*, stellar OB ➤ *associations* and ➤ *active galactic nuclei*.

Cone Nebula A dark, cone-shaped, dust nebula in NGC 2264, a complex region of nebulosity and stars near the fifth-magnitude star S Monocerotis.

conjunction An alignment of two bodies in the solar system with the Earth, so that they appear to be at the same place in the sky – or very close to each other – as seen from Earth.

A planet is at conjunction when it is approximately in line with the Sun and Earth. The planets Mercury and Venus have conjunctions when they are between the Sun and Earth and when they are behind the Sun as seen from Earth. These alignments are called inferior conjunction and superior conjunction, respectively. The planets farther from the Sun than Earth only have superior conjunctions.

The Cone Nebula imaged by the Hubble Space Telescope.

Conjunctions can also occur between planets or between the Moon and a planet.

constellation One of 88 designated areas in the sky or the pattern of stars in it.

From antiquity, different civilizations have given their own names to patterns of bright stars. Many of the constellations used by astronomers today originated in Mesopotamia and were further developed by the Greeks. Forty-eight were listed by ➤ *Ptolemy* in the second century AD and the rest have been added since about 1600. Some old star maps include constellations that are no longer used.

Originally, constellations were regarded simply as star patterns, but they gradually became a way for astronomers to specify stars and their positions. As the science of astronomy developed, the fact that the constellations were not precisely defined led to confusion. This was resolved in 1930 when there was international agreement among astronomers to define the boundaries of 88 constellations along lines of ➤ *right ascension* and ➤ *declination*.

Each of the 88 constellations has an entry in this dictionary under its official Latin name.

contact binary A pair of stars touching each other or surrounded by a common envelope of gas. As stars evolve, they expand and grow into giants. If an expanding star is close to another star in a binary system, the pair can come into contact. There may be ➤ *mass transfer* between the stars and their outer layers can even merge to make one sphere of gas around the two cores.

The brightness of close binary stars often varies as one star crosses in front of the other from our point of view on Earth. The shape of the ➤ *light curve* of some pairs, such as W Ursae Majoris, are best explained if the two stars are in contact. ➤➤ *Roche lobe*.

continuous spectrum A ➤ *spectrum* in which the strength of radiation varies only gradually with wavelength, in contrast with the sharp peaks of intensity in an ➤ *emission line* spectrum. Any object warmer than ➤ *absolute zero* emits a continuous spectrum. A continuous spectrum may be crossed by narrow ➤ *absorption lines*, as in the Sun's spectrum, for example. ➤➤ *black body radiation*.

convection The transport of heat energy by currents of gas or liquid. Convection is one of the ways in which heat reaches the surface of a star from its core where nuclear energy is generated. Huge circulating currents are created as gas rises because it is heated from below, then falls again after cooling.

coordinate system A way of specifying the position of something, either in space or on a surface, by giving a set of linear or angular distances from particular planes, lines or points. Latitude and longitude, for example, are the coordinates used to specify the location of a place on the surface of the Earth. Two coordinates are enough to identify a place on a surface but three are needed to fix a position in three-dimensional space.

In astronomy, several different systems of celestial coordinates are used to give positions on the ➤ *celestial sphere*. Each has its own special use. For example, ➤ *galactic coordinates* are most useful for studies of our own Galaxy.

Copernican system A description of the solar system in which the planets orbit around the central Sun, as proposed by Nicolaus ➤ *Copernicus* and published in his book *De Revolutionibus* in 1543. The theory Copernicus set out did not find favor at first since it predicted planetary positions no more accurately than the ➤ *Ptolemaic system*, which had been used for hundreds of years. Furthermore, it displaced the Earth from the center of the universe, which many people regarded as unacceptable from a religious point of view. Copernicus made the basic planetary orbits circles. He could not predict the movements of the planets correctly because their orbits are in fact ellipses. To reproduce the actual observations, Copernicus had to resort to ➤ *epicycles*. Later, Johannes ➤ *Kepler* developed a more accurate heliocentric theory after discovering that the planetary orbits are elliptical rather than circular.

Copernicus, Nicolaus (1473–1543) Copernicus's book, '*On the Revolutions of the Heavenly Spheres*' (usually called by its shortened Latin title, *De Revolutionibus*) is one of the most influential books of all time. In it, he set out his theory that Earth is in orbit around the Sun, along with the other planets. This was a truly revolutionary idea. Virtually everyone at that time believed that Earth was stationary at the center of the universe and that the Sun, Moon and planets circled around it. This view was deeply rooted in Christian beliefs as well as being the picture that had been accepted by astronomers for over a thousand years. Copernicus's work marked the start of a complete change of perspective on the universe.

Copernicus was born in the Polish city of Torun. He went to the university at Krakow, where he became interested in astronomy, and also studied law and medicine in Italy. His lifelong career was as a canon (an administrative position in the church) at Frauenburg. Between about 1510 and the early 1530s, Copernicus worked on his heliocentric theory, and produced the manuscript of *De Revolutionibus*. Though news of his ideas spread among scholars, he did not formally publish his book. Only years later was he persuaded to publish it. The first printed copies reached him just before he died in 1543. ➤➤ *Copernican system*.

Copernicus A large and conspicuous crater on the Moon at the center of a ray system that extends for more than 600 km (370 miles). Terraced walls and multiple central peaks are features of the crater, which is 93 km (58 miles) in diameter.

Cor Caroli (Alpha Canum Venaticorum) The brightest star in the constellation Canes Venatici. This Latin name means "Charles's heart" and is a reference to the execution of King Charles I of England in 1649. It is said to have been named by Charles Scarborough in 1660. Cor Caroli is 97 light years away and is

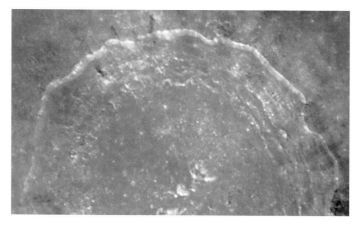

A close-up view of the terraced walls of the lunar crater Copernicus as seen by the Hubble Space Telescope.

a ➤ *visual binary* star. The components have magnitudes of 2.9 and 5.5. The brighter one is an ➤ *A star*.

Cordelia A small moon of Uranus discovered when ➤ *Voyager 2* flew by the planet in 1986. Cordelia is one of the two moons that act as "shepherds" of the planet's Epsilon ring (the other being Ophelia).

corona The outermost layers of the Sun, which become visible as a white halo during a total solar ➤ *eclipse*. The solar corona extends out to many times the Sun's radius, until it merges with the ➤ *interplanetary medium*. It consists of a number of components:

- The K corona (electron corona or continuum corona) is white light from the Sun's ➤ *photosphere* scattered by high-energy electrons, which are at a temperature of a million degrees. The structure of the K corona is variable, with streamers, condensations, plumes and rays.
- The F corona (Fraunhofer corona or dust corona) is light from the photosphere scattered by slower-moving dust particles around the Sun. ➤ *Fraunhofer lines* are seen in its spectrum. The extension of the F corona into interplanetary space is ➤ *zodiacal light*.
- The E corona (emission line corona) is light in the form of ➤ *emission lines* produced by highly ionized atoms, particularly iron and calcium. It is detected out to two solar radii only.

  The solar corona also emits extreme ultraviolet light and soft X-rays. The extent and shape of the corona changes during the course of the ➤ *solar cycle*, mainly due to the streamers produced in ➤ *active regions*.

Corona Australis (The Southern Crown) A small constellation on the southern border of Sagittarius. Though rather faint, it is quite conspicuous and is one of the ancient constellations listed by ➤ *Ptolemy*.

The Sun's corona as seen during a total eclipse in 1970.

Corona Borealis (The Northern Crown) A small but noticeable constellation of the northern sky. Its main stars form a rough semicircle.

coronagraph An instrument for observing the Sun's ➤ *corona*, which is normally seen only at a total solar eclipse. Invented by Bernard Lyot in 1930, the coronagraph is a special telescope containing a disk at the prime focus to block out the light from the Sun's face and create an artificial eclipse. However, even when a coronagraph is located where the sky is very clear, scattering of light by Earth's atmosphere is a problem. This is partly overcome by the use of special filters or by observing the light from the corona with a spectrograph.

coronal hole A region in the solar ➤ *corona* where the density and temperature are exceptionally low. They typically last for several rotations of the Sun, and they are sources of strong ➤ *solar wind*.

coronal mass ejection (CME) An eruption of material from the solar ➤ *corona*. CMEs are associated with features in the Sun's magnetic field. During periods of high ➤ *solar activity*, one or two occur each day and can happen anywhere in the corona. In periods when the Sun is quiet, there is one CME about every 3–10 days and they are confined to lower latitudes. The average ejection speed ranges from 200 km/s at minimum activity to twice that speed at solar maximum. Most occur without an accompanying ➤ *flare*. If a flare does occur, it usually follows the onset of the CME. CMEs are the most powerful kind of transient solar phenomena and have a significant influence on the ➤ *solar wind*.

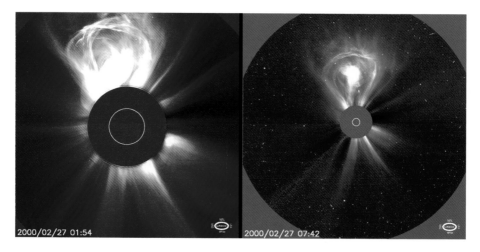

2000/02/27 01:54

2000/02/27 07:42

A coronal mass ejection imaged from space by SOHO. In each view, the white circle represents the size of the Sun, which is hidden behind the occulting disk at the center.

Large CMEs that eject material directly towards Earth provoke ➤ *magnetic storms*.

Corvus (The Crow) A small constellation on the southern border of Virgo. Its four main stars are third magnitude and arranged roughly in the shape of a kite.

Cosmic Background Explorer (COBE) A NASA astronomy satellite launched in 1989 to map far-infrared and millimeter-wave radiation from the whole sky. It measured the temperature of the ➤ *cosmic background radiation* precisely for the first time and found it to be 2.73 K. COBE confirmed that the cosmic background radiation is ➤ *thermal radiation* and consistent with the ➤ *Big Bang* theory of the origin of the universe. COBE's infrared detectors also mapped the distribution of dust in the Milky Way.

cosmic background radiation Diffuse electromagnetic radiation that appears to pervade the whole of the universe. Its discovery in 1964 by Arno Penzias and Robert Wilson (publicly announced in 1965) was immensely important for ➤ *cosmology* because it was strong evidence in favor of the ➤ *Big Bang* theory. The radiation received today is understood to be a relic of the radiation created when our universe came into existence. The spectrum of the background radiation is characteristic of a ➤ *black body* at a temperature of 2.73 degrees above ➤ *absolute zero* (2.73 K) and is most intense in the microwave region. For this reason, it is often called the cosmic microwave background (or CMB). The Milky Way Galaxy is traveling through space at 600 km/s relative to the background radiation.

The ➤ *Cosmic Background Explorer* satellite (COBE) in 1992 showed for the first time that the radiation is not totally even across the sky. It discovered ripple-like variations in temperature amounting to about ten millionths of a degree. These ripples show the first signs of structure emerging in the early universe.

cosmic rays Extremely energetic particles traveling through the universe at close to the speed of light. They were discovered by Victor F. Hess in 1912 during a balloon flight. Particles beyond Earth's atmosphere are known as primary cosmic rays. When they encounter the atmosphere, they collide with atomic nuclei and produce air showers of particles, called secondary cosmic rays.

The mix of atomic nuclei in cosmic rays largely mirrors the abundance of the different chemical elements making up stars like the Sun. Cosmic rays are the only particles we can detect that have traversed the Galaxy. The ones with most energy may even have come from remote ➤ *quasars* and ➤ *active galactic nuclei*. Lower-energy cosmic rays are generated within the Galaxy in ➤ *supernova* explosions, ➤ *supernova remnants* and ➤ *pulsars*. Solar ➤ *flares* are a source of the lowest-energy cosmic rays, which are more intense during maximum ➤ *solar activity*.

cosmic shear Distortion of the shape of a remote galaxy as observed from Earth, caused by the gravity of matter along the line of sight. This effect is predicted by ➤ *general relativity*. It has been used to map the distribution of ➤ *dark matter* in the universe. ➤➤ *gravitational lens*.

cosmic year (galactic year) A period of about 220 million years, which is the time taken by the Sun to complete one revolution about the center of the ➤ *Galaxy*.

cosmogony The study of how celestial systems and objects in general are formed and, in particular, the study of the origin of the solar system.

The modern view is that the solar system formed from a slowly rotating cloud of gas. As the cloud collapsed, a dense opaque core formed – ultimately to become the Sun – surrounded by a disk of gas and dust. The first such nebular theories were suggested by Immanuel Kant (1724–1804) in 1755 and Pierre Simon de ➤ *Laplace* in 1796. Ideas on how the planets actually formed within the disk have changed greatly over that time. Current thinking is that the smaller planets gradually accumulated by ➤ *accretion* while the giant planets may have formed from condensations within the disk, developing disks of their own which gave rise to their satellite systems. The fall-off in temperature with distance from the Sun accounts for the main differences between the inner rocky planets and the outer giants.

cosmology The branch of astronomy concerned with the origin, properties and evolution of the universe. Physical cosmology is about making observations to obtain information on the universe as a whole. Theoretical cosmologists create mathematical descriptions of the observed properties of the universe.

Cosmology in the broadest sense covers physics, astronomy, philosophy and theology because it aims to assemble a world picture explaining why the universe has the properties it has. Ancient cosmologies were simple pictorial models and myths. Greek cosmologists tried to describe the motion of the planets through mathematics.

Modern observational cosmology depends on collecting information about the universe as a whole, such as surveys of remote galaxies and studies of the ➤ *cosmic background radiation*. It is now generally accepted that the observational evidence strongly supports ➤ *Big Bang* cosmology. Theoretical cosmology is normally based on the theory of ➤ *general relativity*. Over large distances, gravity is the dominant force affecting matter and it controls the large-scale structure of the universe. General relativity describes the relationships between space, time, matter and gravity.

cosmos An alternative word for the universe as a whole or the realm of space beyond Earth.

coudé focus A place where the image from a telescope on an ➤ *equatorial mount* can be brought to a focus and remain stationary regardless of the orientation of the telescope. The coudé focus is at a point on the polar axis of the telescope mount. Light from the telescope is directed there along a zig-zag route by of a series of mirrors. The word "coudé" is French for "elbow." The coudé focus was often used in the past for bulky equipment. Modern, large, telescopes on computer-controlled ➤ *altazimuth mounts* use the ➤ *Nasmyth focus* instead.

Crab Nebula (M1; NGC 1952) The nebulous remains in the constellation Taurus of a ➤ *supernova* that exploded in AD 1054. In colored photographs it appears as a network of red filaments surrounding an elliptical area of pale white light. This white light is ➤ *synchrotron radiation* generated by hot ionized gas in a magnetic field. The filaments are the outer layers of the star that were blown off in the explosion and are traveling outwards at about 1500 km/s.

The core of the star that exploded remains at the center of the nebula. It is now a ➤ *pulsar*. Electrons emitted by the pulsar are responsible for the synchrotron radiation. The interval between flashes from the pulsar is 33 milliseconds; flashes are seen in visible light as well as the radio pulses.

The Crab Nebula is one of the strongest sources of radio waves in the sky. This source was called "Taurus A" by radio astronomers before it was identified with the known nebula. It is also a source of X-rays.

Crab pulsar The ➤ *pulsar* at the center of the ➤ *Crab Nebula*.

In 1942, astronomers speculated that a peculiar star in this nebula was a ➤ *neutron star*, created in the ➤ *supernova* of AD 1054. Radio astronomers discovered in 1968 that the central star is a pulsar spinning 30 times a second so that the period between pulses is 33 milliseconds. Shortly after discovery of the radio pulsar, the visible light from the neutron star was also shown to be coming in pulses. At the time of its discovery, this was the fastest pulsar by far and its spin rate proved that the object must be a neutron star: any

The Crab Nebula.

object with a typical star's mass but larger than a neutron star would fly apart if it were rotating so quickly.

Historically, this discovery was of great importance because theorists subsequently felt much more confident that highly condensed objects really do exist in the universe. This led directly to research on ➤ *black holes*.

The Crab pulsar is a source of electrons traveling at almost the speed of light. They give rise to strong X-ray and radio emission from the Crab Nebula surrounding the pulsar. The pulsar's rotation period is increasing at 36 nanoseconds per day. At that rate, its rotation speed will have halved within about 1200 years. The pulsar has random ➤ *glitches*, which may be caused by ➤ *starquakes*. Its mass is thought to be about half that of the Sun.

Craters on the Oceanus Procellarum on the Moon, including Schiaparelli C and Schiaparelli E, photographed from lunar orbit by the crew of Apollo 15.

crater A circular pit on the surface of a planetary body. The word literally means "cup," and craters are typically shaped like a shallow cup. They are lower in the center than the surrounding terrain but have raised walls. They may contain a central peak or depression. The vast majority of craters observed on planets and their moons are now known to have resulted from the impact of meteorites. However, volcanic craters (➤ *calderas*) also occur – on Mars, for example.

   Craters range in size from microscopic to hundreds of kilometers across. The largest are often called basins. The exact shape of a crater depends on many factors, including the composition of the planetary surface, the speed, mass and direction of the body that caused it, and weathering or geological activity that has modified it since it first formed.

Crater (The Cup) A small, faint constellation on the southern border of Virgo. It is one of the ancient constellations and is said to represent the goblet of Apollo. None of its stars is brighter than fourth magnitude.

Crepe ring One of the rings of ➤ *Saturn*, also called the C ring. It is fainter than the prominent A and B rings. It lies closer to Saturn than ring B, extending about halfway between the B ring's inner edge and the planet. The name comes from

The Crescent
Nebula.

a description of its appearance given by the observer William Lassell after
William Cranch Bond announced its discovery in 1850.

crescent One of the ➤ *phases* of the Moon, Venus or Mercury when less than half
the disk is illuminated.

Crescent Nebula A popular name for NGC 6888, a diffuse shell of gas surrounding
the ➤ *Wolf–Rayet star* HD 192163. One crescent-shaped segment of the spherical
shell is relatively bright.

Cressida One of the small moons of Uranus discovered during the encounter
➤ *Voyager 2* had with the planet in 1986. Its diameter is about 80 km (50 miles).

critical density In cosmology, the average density of matter and energy that
ensures the universe will not expand for ever and makes space flat rather than
curved. Its value is equivalent to about six atoms of hydrogen in a cubic meter,
but this is far larger than the amount of visible matter, such as stars and
galaxies, can account for. ➤ *Inflationary universe* theories predict that the
universe should have the critical density. Detailed studies of the ➤ *cosmic
background radiation* suggest that the average density of the universe is indeed
close to the critical value but ➤ *dark energy* and ➤ *dark matter* together account
for 98 percent of the density of the universe.

Crux (The (Southern) Cross) The best-known of all southern constellations. It used
to be part of Centaurus (which surrounds it on three sides) before Augustin
Royer introduced it as a separate constellation in about 1679. It is the smallest
constellation by area, yet one of the most distinctive and recognizable. It lies in
the ➤ *Milky Way* and contains a fine star cluster known as the ➤ *Jewel Box*.

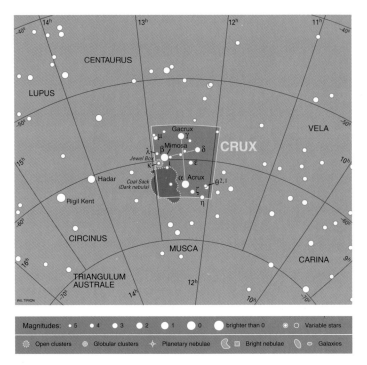

A map of the constellation Crux.

culmination The moment when a star or other celestial object reaches its maximum altitude above the horizon. Culmination occurs when the object crosses the observer's ➤ *meridian*. So, at culmination an object is either due south or due north of the observer. Circumpolar stars cross the meridian both below and above the pole. These events are called lower and upper culmination, respectively.

Cupid A small inner moon of Uranus, discovered in 2003. It is about 18 km (11 miles) across.

65 Cybele One of the larger asteroids, discovered by Ernst W. L. Tempel in 1861. Its diameter is 308 km (191 miles).

Cygnus (The Swan) A conspicuous constellation of the northern ➤ *Milky Way*. It has the shape of an elongated cross similar to a swan in flight and is sometimes called the Northern Cross. It contains 11 stars brighter than fourth magnitude, including the first-magnitude ➤ *Deneb* and a well-known double star, ➤ *Albireo*.

Cygnus A An elliptical galaxy that is one of the strongest radio sources in the sky.

Cygnus A, also known as 3C 405, is the strongest radio source in the constellation Cygnus and was detected by the first radio telescopes. The source consists of two similar clouds of radio emission, symmetrically located either

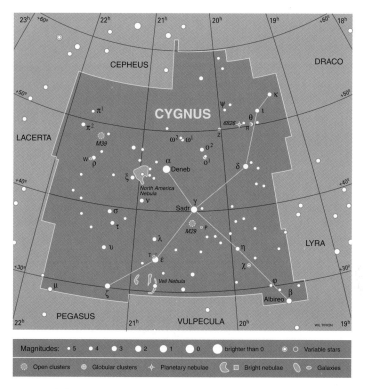

A map of the constellation Cygnus.

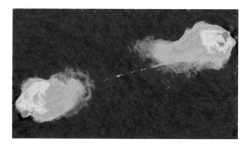

False color image of the radio jet and lobes of radio galaxy Cygnus A.

side of a fifteenth-magnitude galaxy at a redshift of 0.057. Cygnus A is the prototype of all powerful radio galaxies. It is one of the largest physical structures in the universe, its clouds spanning about 300 000 light years. The energy associated with the radio clouds is about ten million times greater than the output of normal galaxies such as the Andromeda Galaxy (M31).

The galaxy has strong emission lines in its optical spectrum, indicating that it contains an ➤ *active galactic nucleus*. It is widely accepted that the energy

81

Artist's impression of Cygnus X-1.

generated by Cygnus A can be explained only by a central ➤ *black hole* that is releasing very large amounts of energy as matter falls towards it.

Cygnus Loop A circular nebula 3° across in the constellation Cygnus. The ➤ *Veil Nebula* forms one section of the circular structure. Radio emission confirms that the nebula is a ➤ *supernova remnant*. It is estimated to be 30 000 years old and to lie at a distance of 2500 light years. What we see as a loop is in fact a spherical shell. It is currently expanding at a rate of 6 arc seconds every hundred years.

Cygnus X-1 An intense sources of X-rays in the constellation Cygnus, discovered in 1966. It is thought to be a ➤ *black hole* and has been the subject of much study and speculation.

Cygnus X-1 has been identified as a ➤ *binary star* system. The primary star is a hot supergiant of ➤ *spectral type* O or B. Its orbit around its unseen companion suggests that the invisible star is significantly more massive than the upper limit for neutron stars. If so, the invisible star can only be a black hole. The X-ray energy is generated by matter from the primary star streaming onto its compact companion.

Cygnus X-3 A source of X-rays in the constellation Cygnus. It is a close ➤ *binary star* system in which the two members orbit each other every 4.8 hours. The primary star's mass is about the same as the Sun's and it appears to be a ➤ *Wolf–Rayet star*. The secondary star is a pulsed source of gamma-rays with a period of 12.6 milliseconds. This star is likely to be a ➤ *pulsar* whose rate of rotation has been increased because of interactions with its partner.

➤ *Cosmic rays* have been detected at ground level on Earth from the direction of Cygnus X-3. This demonstrates that X-ray binary stars are important sources of high-energy cosmic rays. ➤➤ *X-ray astronomy*.

Cyllene A small outer moon of Jupiter discovered in 2003. Its diameter is about 2 km (1 mile).

D

Dactyl ➤ *Ida*.

5335 Damocles An asteroid following an unusual, highly elliptical orbit. Its distance from the Sun ranges between 1.6 and 22 AU. Damocles was discovered in 1991 by the Australian astronomer Rob McNaught. Its diameter is about 10 km (6 miles).

Daphnis A small moon of Saturn discovered in 2005 by the ➤ *Cassini* team. It orbits in the ➤ *Keeler Gap* in the ring system, 136 500 km (84 800 miles) from Saturn.

dark energy The unknown source of a repulsive force which appears to pervade the universe. It opposes the tendency of gravity to slow down the rate at which the universe is expanding. Because of its effects, the expansion of the universe has been speeding up for the last 500 million years. Dark energy accounts for about three-quarters of the total energy and mass in the universe. ➤➤ *accelerating universe*.

dark matter Matter in the universe that gives out no radiation and has so far been detected only by its gravity. Some of it is likely to be ordinary matter made from atoms containing neutrons and protons, but a large proportion is thought to be in the form of exotic elementary particles that hardly interact at all with radiation or with the neutrons and protons in ordinary atoms. About 23 percent of the total matter and energy in the universe is dark matter. Only 2 percent is ordinary matter. (The rest is ➤ *dark energy*.)

The motion of galaxies in clusters strongly suggests that many clusters are at least 10 times more massive than the luminous parts of the galaxies in them. It is also possible to estimate the masses of individual galaxies from the way their rotation varies between their center and their edge. Giant spiral galaxies studied in this way contain far more matter than can be accounted for by glowing stars and gas. Dark matter is also important for explaining how galaxies formed in the early universe.

dark nebula ➤ *absorption nebula*.

Darwin A project under study by the European Space Agency (ESA) for a flotilla of space telescopes designed to search for ➤ *extrasolar planets* and to analyze the composition of their atmospheres. The spacecraft would form an ➤ *interferometer* operating in the infrared and would be positioned 1.5 million km sunward of Earth, at the ➤ *Lagrangian point*. Darwin would be launched in 2015 at the earliest.

511 Davida An asteroid 324 km (201 miles) across, discovered by Raymond S. Dugan in 1903. It is one of the largest asteroids.

David Dunlap Observatory The observatory of the University of Toronto, Canada, located 25 km (15 miles) north of the university campus. It was presented to the university in 1935 by Mrs Dunlap as a memorial to her husband. The main instrument is a 1.88-m (74-inch) reflector, the largest in Canada.

Dawn A NASA mission to investigate the asteroids ➤ *Vesta* and ➤ *Ceres*, scheduled for launch in September 2007. The spacecraft is due to arrive at Vesta in October 2011 and orbit there until April 2012. It will then go on to orbit Ceres from February 2015. The mission will end in July 2015 unless extended with a visit to another asteroid.

day In astronomy, a unit of time defined as 86 400 seconds. This period is close to the average time Earth takes to rotate. However, Earth's rotation is not completely uniform. ➤ *solar day, sidereal day.*

De Chéseaux's Comet An exceptionally bright ➤ *comet* discovered independently by Dirk Klinkenberg from Haarlem on December 9 and by Jean Philippe Loys de Chéseaux from Lausanne on December 13, 1743. It reached magnitude –7 and developed a fan of 11 separate tails.

declination (Dec.) One of the coordinate used to define positions on the ➤ *celestial sphere* in the ➤ *equatorial coordinate* system. Declination is the equivalent of latitude on the Earth. It is the angular distance, measured in degrees, north or south of the ➤ *celestial equator*. Northerly declinations are positive and southerly ones negative.

Deep Impact A NASA space mission to Comet Tempel 1, launched on January 12, 2005. It released a 500-kg (1,110-pound) impactor that struck the comet in July 2005 and then studied the resulting crater and ejecta as it flew by.

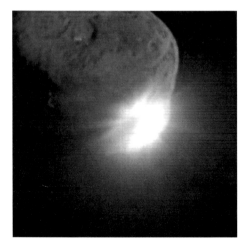

The Deep Impact spacecraft took this image of Comet Tempel 1 13 seconds after it had been struck by the impactor it had released.

deep sky object An astronomical object that does not belong to the solar system.

Deep Space Network A ground-based network of radio dishes used by NASA to communicate with space missions. The three 70-m (230-foot) antennas are located at ➤ *Goldstone* in California, near Madrid in Spain and near Canberra in Australia.

Deep Space 1 A NASA spacecraft launched in October 1998 with the main aim of testing 12 novel technologies. In July 1999 it flew close to the asteroid ➤ *Braille*. On September 22, 2001, it passed within 2200 km (1400 miles) of the nucleus of Comet ➤ *Borrelly* and returned the best images of a comet nucleus up to that time.

degenerate star A term that covers ➤ *white dwarfs* and ➤ *neutron stars*, both of which are made up of degenerate matter. These stars are in an advanced state of evolution and have suffered extreme gravitational collapse. Normal atoms cannot exist in them.

In white dwarfs, the electrons and atomic nuclei are squashed together so there is no space between them. A quantum effect, called degeneracy pressure, prevents further collapse. However, if the total mass of the star is over 1.4 times the mass of the Sun, even degeneracy pressure is insufficient to hold back more collapse. If the mass is up to about 3 solar masses, it becomes a neutron star. In a neutron star, the electrons and nuclei combine together to create matter consisting of tightly packed neutrons. If the star is more than about 3 solar masses, a ➤ *black hole* forms.

Deimos One of the two small moons of Mars, discovered in 1877 by Asaph Hall. Its shape is irregular and it measures $15 \times 12 \times 10$ km ($9 \times 7 \times 6$ miles). Images from the ➤ *Viking* 2 Mars mission revealed a heavily cratered surface. The properties of both moons suggest that they are captured ➤ *asteroids*.

Delphinus (The Dolphin) A small constellation lying in the ➤ *Milky Way* just north of the celestial equator. Though it is quite faint, its shape is very distinctive and it is one of the ancient constellations listed by ➤ *Ptolemy*.

Delta Aquarids A double ➤ *meteor shower* that radiates from the constellation Aquarius between July 15 and August 20 each year, peaking on July 29 and August 7.

Delta Cephei A yellow giant star in the constellation Cepheus that varies in brightness between magnitudes 3.6 and 4.3 over a period of 5.37 days. It is the prototype of the ➤ *Cepheid variable* stars and lies at a distance of 300 light years.

Demon Star ➤ *Algol*.

Deneb (Alpha Cygni) The brightest star in the constellation Cygnus. It is a supergiant ➤ *A star* of magnitude 1.3 and is 2600 light years away. Its Arabic name means "tail."

Denebola (Beta Leonis) An ➤ *A star* of magnitude 2.1, the third-brightest in the constellation Leo. It is 36 light years away. The name Denebola comes from Arabic and means "the lion's tail."

The diamond ring effect captured at a total solar eclipse in 1983.

Desdemona One of the small moons of Uranus discovered during the ➤ *Voyager 2* encounter with the planet in 1986. Its diameter is about 64 km (40 miles).

Despina A moon of Neptune discovered during the flyby of ➤ *Voyager 2* in August 1989. It measures 150 km (93 miles) across.

D galaxy A type of large ➤ *elliptical galaxy* with a bright nucleus surrounded by an extensive envelope. D galaxies are often radio galaxies. ➤ *cD galaxy*.

diagonal (star diagonal) An attachment for a small telescope used to turn the eyepiece tube through a right angle so the observer can look from the side. It contains a small plane mirror or prism. When a diagonal is used, the image is reversed right-to-left.

diamond ring effect A phenomenon observed at the very beginning and end of totality in a total solar ➤ *eclipse* when the last or first glimpse of the Sun shines through a valley on the limb of the Moon. ➤➤ *Baily's beads*.

dichotomy The time when the Moon, Mercury or Venus is exactly at half phase.

differential rotation The rotation of a gaseous body, such as the Sun or the planet Jupiter, at a rate that varies with latitude, or the rotation of a non-solid, disk-shaped structure, such as a galaxy, at a rate that varies with distance from the center.

diffraction The spreading of a beam of light as it passes by the edge of an obstacle. The effect is to bend some light into the obstacle's shadow. Interference

Dione imaged by Cassini with Saturn in the background.

between different parts of the diffracted light beam results in a pattern of light and dark areas called a diffraction pattern.

diffraction grating An optical device used to split light into a spectrum. It consists of a large number of narrow, closely spaced lines ruled either on glass to form a transmission grating or on polished metal to form a reflection grating. Typically there are several thousand rules per centimeter. Interference between the beams of light created by ➤ *diffraction* at each slit produces the spectrum. Diffraction gratings can produce very high dispersion spectra of good quality and are used in astronomical ➤ *spectrographs*.

diffuse nebula A gaseous ➤ *nebula*. The use of the word "diffuse" dates from a time when all objects of fuzzy appearance were classed as nebulae, and observers wanted to distinguish different types even though their true nature was unknown. Star clusters and galaxies are no longer called nebulae.

Dione A medium-sized moon of Saturn, discovered by Giovanni D. ➤ *Cassini* in 1684. Images from the ➤ *Voyager 1* mission show several different types of terrain on Dione: heavily cratered areas, plains with a lower density of craters, and smooth plains with few craters or other features. The largest craters are over 200 km (125 miles) across, and craters over 100 km are common in the heavily cratered areas. Another noticeable feature is an irregular network of light wispy streaks on a dark background, which may be frosty deposits. Dione is Saturn's fourth largest moon orbiting at a distance of 337 400 km (234 500 miles) and measures 1118 km (695 miles) across.

disconnection event A break in the ➤ *ion tail* of a ➤ *comet* that happens when the comet crosses a region where the direction of the magnetic field in the ➤ *solar wind* changes.

disk galaxy A ➤ *spiral galaxy* that has had most of its interstellar gas stripped away as it moves through the intergalactic medium in a ➤ *cluster of galaxies.*

D layer A region of the Earth's ➤ *ionosphere* at heights between about 50 and 90 km (30 and 55 miles). The lower D layer, between altitudes of 50 and 70 km, is also known as the C layer.

Dobsonian telescope A low-cost, large-aperture, reflecting telescope on a simple, ➤ *altazimuth mounting.* Dobsonians are used for visual observing by amateurs and are easily moved around. The design was pioneered in the 1960s and 1970s by John Dobson of the San Francisco Sidewalk Astronomers. Dobson also demonstrated that a large mirror could be made cheaply from ordinary plate glass.

Dog Star A popular name for the star ➤ *Sirius.*

Dominion Astrophysical Observatory The National Research Council of Canada's center for optical astronomy, located near Victoria, British Columbia. It is part of the Herzberg Institute of Astrophysics. It was founded by John S. Plaskett and its 1.85-m (72-inch) telescope began operating in 1918.

Dominion Radio Astrophysical Observatory The National Research Council of Canada's radio astronomy observatory, located 20 km (12 miles) south-west of Penticton in British Columbia. It forms part of the Herzberg Institute of Astrophysics and was founded in 1959. The main instrument is an ➤ *aperture synthesis* radio telescope consisting of seven 9-m (30-foot) dishes on an east–west baseline 600 m (2000 feet) long.

Donati, Comet A brilliant comet discovered by Giovanni B. Donati of Florence in 1858. Contemporary drawings show it with a broad, curved ➤ *dust tail* and two narrow, straight ➤ *ion tails.*

Doppler effect The change in the observed frequency and wavelength of sound waves or ➤ *electromagnetic radiation* when the source of waves and the observer are moving towards or away from each other. ➤➤ *redshift.*

Doppler shift The amount by which the ➤ *Doppler effect* changes the wavelength of light. A Doppler shift is a ➤ *redshift* when wavelengths are increased because the source of light is receding. When wavelengths are decreased because the source is approaching, the Doppler shift is a ➤ *blueshift.*

Dorado (The Goldfish or Swordfish) A southern constellation, probably invented by sixteenth century navigators and included by Johann Bayer in his 1603 star atlas ➤ *Uranometria.* Its stars are inconspicuous but the Large ➤ *Magellanic Cloud* lies on its southern border with Mensa.

Double Cluster in Perseus ➤ *h and chi Persei.*

double star ➤ *binary star.*

Draco (The Dragon) A large but rather faint constellation stretching halfway around the north celestial pole and enclosing Ursa Minor on three sides.

Draconids A ➤ *meteor shower*, associated with Comet Giacobini–Zinner, which is seen in some years around October 9 or 10. The meteors appear to radiate from near the "head" of Draco. The number of meteors varies from year to year. Spectacular displays were witnessed in 1933, when meteors briefly rained down at a rate of 350 a minute, and again in 1946. Moderate showers occurred in 1952 and 1985. The Draconids are also known as the Giacobinids.

Drake equation A mathematical expression formulated by Frank Drake in 1961 that gives the number of extraterrestrial civilizations we might detect as the product of several factors. Drake originally wrote his equation as:

$N = R f_p \, n_e \, f_l \, f_i \, f_c \, L$, where

N is the number of detectable civilizations

R is the rate of star formation

f_p is the fraction of stars forming planetary systems

n_e is the number of planets hospitable to life

f_l is the fraction of those planets where life actually emerges

f_i is the fraction of planets where life evolves into intelligent civilizations

f_c is the fraction of planets with intelligent beings capable of interstellar communication

L is the length of time that such a civilization remains detectable.

Draper, Henry (1837–1882) The American astronomer Henry Draper was a pioneer in astronomical spectroscopy. He trained as a doctor and became Dean of the Medical Faculty of New York University but he also established a private observatory, where he worked on photography and spectroscopy. In 1872 he

Henry Draper.

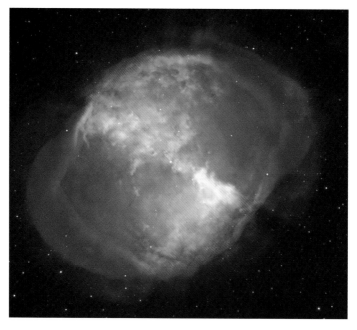

The Dumbbell Nebula.

took the first photograph of dark lines in the spectrum of a star (Vega) and in 1880 he took the first photograph of the ➤ *Orion Nebula*. However, he is most remembered because of the ➤ *Henry Draper Catalog* of stellar spectra, funded by his wife after his death as a memorial to him. "HD" numbers from the catalog are still widely used to identify stars.

Dubhe (Alpha Ursae Majoris) One of the two bright stars of the ➤ *Big Dipper* in Ursa Major forming the ➤ *Pointers*. It is a giant ➤ *K star* of magnitude 1.8 with a fifth-magnitude companion that orbits around it in 44 years. They are 124 light years away. "Dubhe" is a short version of an Arabic name meaning "the back of the greater bear."

Dumbbell Nebula (M27; NGC 6853) A large ➤ *planetary nebula* in the constellation Vulpecula, discovered by Charles Messier in 1764. It lies 1000 light years away and is a quarter of a degree across on the sky.

dust grains Small particles of matter, typically around 10–100 nm in diameter, which co-exist with atoms and molecules of gas in interstellar space. Interstellar dust grains are thought to be made mainly of silicates and/or carbon in the form of graphite. They form in the extended atmospheres of ➤ *red giant* stars. Dark clouds of dust show up when they obscure the light from stars and luminous gas clouds behind them. Though tenuous, dusty clouds absorb visible light very easily, but radiation at millimeter and longer wavelengths can pass through them unimpeded. The presence of dust is also revealed by infrared radiation, which the grains give off after they have absorbed energy in the form of visible and ultraviolet radiation. The temperature of interstellar dust is typically in the range 30–500 K.

Dust grains are thought to play an important role in the formation of ➤ *interstellar molecules* because gas atoms that cannot easily combine when moving freely in space can combine on the surfaces of grains. The molecules that form in this way can then leave the grains.

Dust clouds are also an important constituent of star-forming regions. The dust shields interstellar molecules from the destructive effects of high-energy radiation and helps ➤ *protostars* to radiate away surplus energy.

dust lane A dark layer of dust visible against a bright background of stars or glowing gas. ➤➤ *dust grains*.

dust tail (type II tail) One of the two principal types of tail ➤ *comets* develop as they approach the Sun. The dust tail is composed of particles about 1 μm in size that shine by reflecting sunlight. Dust tails can be as much as 10 million km (6 million miles) long. They curve away from the Sun under the influence of pressure exerted by the Sun's radiation.

dust trail A stream of interplanetary dust orbiting the Sun in the wake of a comet. Dust trails are made up of particles released from the surface of comets when warmth from the Sun evaporates the icy surface. After several revolution around the Sun, a dust trail tends to disperse, forming a continuous and more uniform stream of dust, often called a ➤ *meteor stream*. Dust streams and trails are responsible for ➤ *meteor showers*.

dwarf planet A body in the solar system, traveling in orbit around the Sun and with enough mass to have a more-or-less spherical shape due its own gravity, but not massive enough to dominate the vicinity of its orbit to the extent that its influence has cleared away all or most of the smaller bodies there. The ➤ *International Astronomical Union* adopted this definition in 2006 following the discovery in 2003 of ➤ *Eris*, a ➤ *Kuiper Belt object* larger than ➤ *Pluto*. Eris and Pluto, being located in the Kuiper Belt, fall into this category as does the largest asteroid, ➤ *Ceres*.

dwarf nova A type of cataclysmic variable star that suddenly brightens at intervals ranging from several days to a year. Dwarf novae are binary systems consisting of an ordinary star and a ➤ *white dwarf*. Material from its companion streams towards the white dwarf, building up an ➤ *accretion disk* around it. Outbursts happen when hot spots form on the accretion disk. Dwarf novae are also known as U Geminorum stars.

dwarf spheroidal galaxy (dSph galaxy) A dim, faintly glowing dwarf ➤ *elliptical galaxy*. Typically these galaxies are about 1000 light years across, which is less than one percent of the diameter of the Milky Way Galaxy. Several have been discovered in the ➤ *Local Group*.

dwarf star A star that is not especially large. All but the hottest, most massive stars on the main sequence of the ➤ *Hertzsprung–Russell diagram* may be described as dwarf stars. ➤➤ *giant star, stellar evolution, white dwarf.*

Dwingeloo Observatory A radio astronomy observatory in the Netherlands, established in 1956. It is also the headquarters of the Netherlands Foundation for Research in Astronomy (NFRA). The 25-m (82-foot) Dwingeloo radio telescope is owned and operated by the NFRA. ➤➤ *Westerbork Observatory.*

Dysnomia The moon of ➤ *Eris.*

E

Eagle Nebula (M16; NGC6611) A glowing nebula in the constellation Serpens shaped rather like a flying bird with broad wings. The nebula is about 7000 light years away and surrounds a small cluster of brilliant stars that are only about 2 million years old. Near the center, new stars are still being formed where columns of cool gas and dust, up to a light year long and known as "elephant trunks," protrude from a dark molecular cloud.

Earth The third planet from the Sun. Like Mercury, Venus and Mars, Earth is one of the rocky, smaller planets in the inner solar system known as the terrestrial planets. It is the only one of the four with a large natural satellite – the ➤ *Moon*. Earth's diameter is 12 756 km (7927 miles). It revolves around the Sun in a slightly elliptical orbit at an average distance of 149.6 million km (92.96 million miles), taking 365.256 days to complete one circuit. Earth is

The Eagle Nebula.

Earth imaged by the Galileo spacecraft in December 1990 from a distance of 2.1 million km (1.3 million miles).

closest to the Sun on January 3rd or 4th, when the Sun is 147.0 million km (91.35 million miles) away. Six months later Earth is at its maximum distance of 152.0 million km (94.45 million miles). Relative to the stars, Earth spins in 23 hours 56 minutes but, because it is simultaneously traveling around the Sun, it takes 4 minutes longer to turn once relative to the Sun giving us a day of 24 hours. Earth's rotation axis is tilted so that the equator makes an angle of 23° 27′ to Earth's orbit. The yearly cycle of ➤ *seasons* arises because of this tilt (and is not linked to the variation in Earth's distance from the Sun.)

Liquid water covers 71 percent of Earth's surface. Where there are continents, Earth's rocky crust is typically 30 km (20 miles) thick but under the oceans the crust is considerably thinner – 10 km (6 miles) or less. The crust lies over the mantle, a thick layer of silicate rock nearly 3000 km (1900 miles) deep. Many of the major features on Earth's surface result from the motion of crustal plates, which slide over the softer mantle below. Earthquakes and volcanoes occur particularly in areas along the boundaries between plates. Folded mountains rise up where plates collide. All this geological activity means that the surface of Earth is constantly changing and much of it is young, geologically speaking.

The ocean floors are some of the most recent rocks to have formed. On the oldest parts of Earth's surface, some traces remain of impact craters formed in the early days of the solar system but they have mostly been worn away by weathering.

The core of Earth is predominantly iron and nickel. The very center is a solid ball about 2800 km across but the outer part of the core is molten. The temperature in the core rises to over 5000 °C (9000 °F). Electric currents flowing in the molten metal generate Earth's magnetic field. The region of space around Earth where its own field is the strongest magnetic influence is called the ➤ *magnetosphere*.

Earth is surrounded by a shallow layer of gas – the atmosphere. It thins out gradually with height and most of the gas is concentrated within the 20 km (12 miles) immediately above the surface. The composition of the atmosphere has changed since Earth formed about 4.5 billion years ago. In particular, photosynthesis by plants increased the amount of oxygen significantly about 2 billion years ago. Molecular oxygen (O_2) now accounts for 21% of the atmosphere. The bulk of the gas (77%) is molecular nitrogen (N_2) with 0.9% argon, about 1% water vapor, and traces of other gases, the most important of which is carbon dioxide (CO_2). A layer of ozone (O_3) high in the atmosphere absorbs ultraviolet light from the Sun, which would be lethal to plants and animals if it were to reach the surface. Clouds typically cover half of the Earth at any time. Because of heat trapped by the atmosphere (the "greenhouse" effect) Earth's average surface temperature of 15 °C (59 °F) is 30 °C (54 °F) higher than it would be if there were no atmosphere.

earthshine Faint light illuminating the dark part of the Moon, seen when the Moon is at thin crescent phase close to new Moon. The popular name for it is seeing "the old Moon in the new Moon's arms." Earthshine is the result of sunlight reflected from Earth onto the Moon's surface.

eccentricity (symbol *e*) A number used to describe how elongated an ellipse is compared with a circle. The eccentricity of an ellipse is greater than 0 but less than 1. The larger its value, the longer and thinner the ellipse. A perfect circle has an eccentricity of zero, and a parabola, which is an open curve, has an eccentricity of one. Open curves with values of *e* greater than one are called hyperbolas.

eclipse When a celestial object is not visible for a while, either because another body passes directly in front of it or because it moves into a shadow in space.

On Earth, there are several eclipses of the Sun or Moon each year. When the Sun is eclipsed, the Moon passes in front of it. In a lunar eclipse, the Moon is not actually hidden. It passes through Earth's shadow where no sunlight falls directly on its surface to make it shine. Eclipses of the Sun and Moon happen regularly because the Moon's orbit around Earth is inclined to Earth's orbit around the Sun by only 5°. As a consequence, Sun, Moon and Earth line up in space frequently, though not every month because their relative movements are complex. At least two eclipses of the Sun happen every year. Most are partial, however. The maximum number of eclipses in any one year is seven, two or three of which are lunar. Theoretically, it is possible for two solar eclipses to occur a month apart and for there to be a lunar eclipse in between. However, there can never be lunar eclipses at two successive full Moons. Solar eclipses take place at (or very close to) new Moon.

An eclipse of the Moon in 2003 photographed from the Kennedy Space Center.

Though the Moon is much nearer than the Sun, both appear very nearly the same size in the sky. However, the apparent sizes of the Sun and Moon vary by a small amount because the orbits of the Moon and Earth are elliptical rather than circular. The ratio between the apparent diameters of the Moon and Sun at a solar eclipse is called the "magnitude" of the eclipse. If a solar eclipse that would otherwise have been total occurs when the Moon's diameter appears less than the Sun's, a ring of the Sun's disk remains visible. Solar eclipses of this kind are described as "annular" (from the Latin word "annulus," meaning "a ring").

The Moon's shadow is only a few hundred kilometers wide at Earth's surface. It traces out a curved path from west to east as the Moon and Earth move relative to each other. Observers in a wider region either side of the path of totality witness a partial eclipse. Partial eclipses may also occur when no part of the Earth experiences a total eclipse.

In the few minutes or seconds totality lasts, darkness falls. The faint outer parts of the Sun – the ➤ *chromosphere* and the ➤ *corona* – become visible. They cannot be seen from Earth except during an eclipse because the Sun's bright yellow disk is too dazzling. From a fixed place on Earth's surface, the longest a total solar eclipse can last is 7.5 minutes.

Lunar eclipses occur when the Moon passes through the shadow cast by Earth. They can take place only close to full Moon and can be seen from any location on Earth where the Moon has risen. The Moon does not normally become completely invisible. Illuminated by sunlight scattered from Earth's atmosphere, it often takes on a dark coppery color.

The full shadow cast by Earth – the umbra – is surrounded by a region of partial shadow, called the penumbra. Before and after entering the umbra, the Moon passes through the penumbra. This is often hardly noticeable to casual observers. Some lunar eclipses are only penumbral. The length of the Moon's path through the umbra, divided by the Moon's apparent diameter, defines the "magnitude" of a lunar eclipse. Lunar eclipses may last several hours.

The moon of another planet may also be eclipsed when the sunlight causing it to shine is cut off because it passes into the shadow cast by its parent planet. When a moon disappears behind its planet, the event is more usually called an occultation. However, a double star system in which one star periodically crosses in front of the other is known as an ➤ *eclipsing binary*.

eclipse year A period of 346.620 03 days, which is the time it takes for the Sun to move through the sky from one of the points where the Moon's orbit crosses its path until it reaches that same intersection point the next time. It is shorter than a normal year because the orientation of the Moon's orbit in space is gradually changing. ➤➤ *saros*.

eclipsing binary (eclipsing variable) A system of two stars that periodically hide each others' light. As a result, the combined brightness regularly dips.

In a binary system, two stars are held close together by gravity and follow elliptical orbits around their center of mass. The orbits of some pairs are oriented so that the two stars alternately cross in front of each other, partially or totally concealing their partner. If the stars are not identical, the drop in brightness when the fainter star is in front of its partner is greater than when it is behind. The greater drop is called the primary minimum and the lesser one the secondary minimum. The most well-known eclipsing binary is ➤ *Algol*. Some eclipsing star pairs belong to multiple systems with more than two members.

ecliptic The plane in space in which Earth's orbit around the Sun lies, or the circle around the sky traced out by the Sun in the course of a year. It is called the ecliptic because eclipses of the Sun or Moon can only take place on occasions when the Moon crosses it. On the sky, the ecliptic goes through the 12 traditional constellation of the ➤ *zodiac*. Because of the way astronomers have formally designated the boundaries of the constellations, it also goes through ➤ *Ophiuchus*.

The orbits of the other major planets in the solar system are tilted only slightly to the ecliptic plane. This means that their paths in the sky are all close to the ecliptic.

ecliptic coordinates A system of latitude and longitude for measuring positions on the ➤ *celestial sphere*, based on the ➤ *ecliptic*. These coordinates are used particularly for work on the positions and movements of the planets. Ecliptic latitude is measured in degrees north and south of the ecliptic. Northerly latitudes are positive and southerly ones negative. Ecliptic longitude is measured in degrees around the ecliptic from where the Sun crosses the ➤ *celestial equator* in March, a point called the vernal equinox.

Eddington, Arthur Stanley (1882–1944) Sir Arthur Eddington was one of the foremost astrophysicists of his time, remembered particularly for his work on the interiors of stars and in connection with Einstein's theory of ➤ *general relativity*. He was born in Kendal, England and from 1913 until his death was Plumian Professor of Astronomy at the University of Cambridge. He was knighted in 1930.

In 1919, Eddington traveled with collaborators to the island of Principe off the west coast of Africa for a total eclipse of the Sun. With observations of stars made during the eclipse, they showed that the Sun's gravity deflected the path of starlight slightly from a straight line, confirming a prediction Einstein had made three years earlier, Eddington's ground-breaking theoretical work on stars was published in 1926 in his book, *The Internal Constitution of the Stars*. His discoveries include the relationship between the masses of stars and their luminosity, and the fact that no star more massive than about 120 solar masses can exist. The "Eddington limit" is the greatest luminosity a star of a particular mass can attain without blowing itself apart.

Eddington also wrote books for general readers, such as *The Expanding Universe*, which became very popular.

Edgeworth–Kuiper Belt ➤ *Kuiper Belt*.

EGG Abbreviation for "evaporating gaseous globule." An EGG is essentially a dense sphere of gas in which a star is forming. It is denser than the cloud of gas in which it is embedded. The action of ultraviolet light from nearby hot stars strips away the thinner surrounding gas to expose EGGs so they can be seen. The name was first coined for globules identified in Hubble Space Telescope images of the ➤ *Eagle Nebula*.

Egg Nebula A popular name for the very young planetary nebula more formally known by its catalog number as CRL2688. Its central star, which was a red giant until a few hundred years ago, is hidden behind a ring of dust. This star illuminates a series of shells of gas that it has previously ejected.

Einstein, Albert (1879–1955) Albert Einstein is most famous for his special and general theories of relativity (➤ *special relativity, general relativity*), which revolutionized physics at the beginning of the twentieth century. He received the Nobel Prize in 1921 for his work on the photoelectric effect.

Einstein was born in Ulm, which is now in Germany. In 1901 he took Swiss citizenship. As a young man, he did not seem especially brilliant and failed to get an academic job. While working as a private tutor, and then for the Swiss

A Hubble Space Telescope image of EGGs – evaporating gaseous globules – in the Eagle Nebula.

The Egg Nebula imaged by the Hubble Space Telescope.

Patent Office, he used his spare time to think about a number of fundamental problems in physics and developed his special theory of relativity, which was published in 1905. In the same year he announced his famous equation, $E = mc^2$, in which E represents energy, m mass and c the velocity

Albert Einstein (1932).

of light. This followed from his conclusion that a body loses mass when it radiates energy.

From 1907, Einstein held a series of posts in universities before settling in Berlin in 1914. His greatest work was all done by the early 1920s. In 1933 he moved permanently to the United States where he was on the staff of Princeton University.

Einstein Cross The result of a galaxy called G2237+0305 acting as a ➤ *gravitational lens*, to produce four separate images of a quasar. The four images form a cross around the bright nucleus of the galaxy and vary in brightness and color on a timescale of about a day. The quasar happens to lie along the same line of site as the galaxy, but is much farther away. The quasar's redshift is 1.7 but that of the galaxy only 0.04.

Einstein Observatory An orbiting X-ray observatory that was launched on November 13, 1978 and operated until 1981. It was named to honor the centenary of the birth of Albert Einstein. It was the first observatory capable of making X-ray images of extended objects.

Einstein ring A circular image of a distant point-like source of light or radio waves, such as a quasar, formed when a massive galaxy along exactly the same line of sight acts as a ➤ *gravitational lens*. Einstein was the first person to set out the theory of how such a ring-shaped image would be formed.

The Einstein Cross imaged by the Hubble Space Telescope.

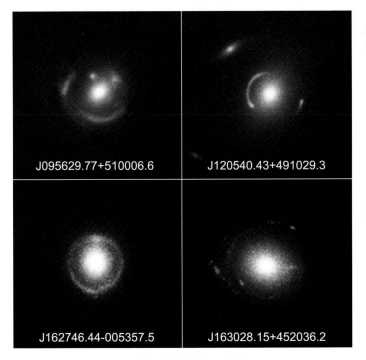

J095629.77+510006.6

J120540.43+491029.3

J162746.44-005357.5

J163028.15+452036.2

Einstein rings imaged by the Hubble Space Telescope.

ejecta Material thrown outwards when an impact forms a crater or by a volcanic eruption. It consists of fragments of rock and solidified droplets of gas or liquid. Ejecta often blankets a circular area around an impact or eruption, or it sometimes forms a pattern of rays. Some ejecta created by impacts may escape into space.

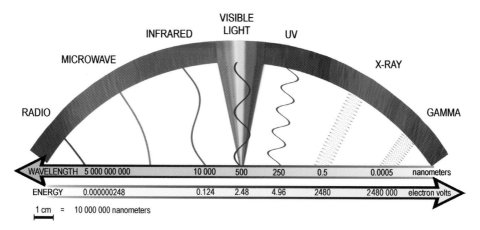

VISIBLE
INFRARED | LIGHT | UV

MICROWAVE | X-RAY

RADIO | GAMMA

| WAVELENGTH | 5 000 000 000 | 10 000 | 500 | 250 | 0.5 | 0.0005 | nanometers |
| ENERGY | 0.000000248 | 0.124 | 2.48 | 4.96 | 2480 | 2480 000 | electron volts |

1 cm = 10 000 000 nanometers

Electromagnetic radiation.

Elara A small moon of Jupiter discovered in 1905 by Charles D. Perrine. It is about 86 km (53 miles) across. With Leda, Himalia and Lysithea it belongs to a family of four moons with closely spaced orbits. Their average distances from Jupiter are between 11.1 and 11.7 million km (6.9 and 7.3 million miles).

Electra One of the brighter stars in the ➤ *Pleiades*. With a magnitude of 3.5 it is easily visible to the naked eye. Electra is also known as 17 Tauri.

electric propulsion ➤ *ion drive*.

electromagnetic radiation A form of energy with wave-like properties. It consists of linked electric and magnetic fields, which vary rapidly. Electromagnetic radiation travels through a vacuum at a speed of 3×10^8 m/s (represented by the symbol c), as discrete "wave packets," called photons. Its properties vary with wavelength. From the longest wavelengths to the shortest, the spectrum of electromagnetic radiation covers radio waves, microwaves, infrared radiation, visible light, ultraviolet radiation, X-rays and gamma rays.

Radio wavelengths cover a wide range from several meters down to millimeters. The shortest radio waves are usually called microwaves. Infrared radiation has even shorter wavelengths, ranging down to just under 1 μm. Visible light is a narrow band of wavelengths between about 700 and 400 nm. Ultraviolet radiation goes down to about 10 nm, and X-rays to 0.1 nm. The shortest waves of all are gamma rays. The shorter the wavelength of the radiation, the more energy carried by each photon.

element (chemical) One of the "building blocks" of matter in the universe. About 90 different elements are found naturally on Earth. Each one has a one- or two-letter symbol. Elements combine together chemically to make compound materials, like water (hydrogen and oxygen, H_2O) and common salt (sodium and chlorine, NaCl).

The elliptical galaxy M49 in the Virgo cluster of galaxies.

Elements consist of atoms. Atoms have a positively charged nucleus, which is usually surrounded by a cloud of negatively charged electrons. The nucleus of an atom contains protons, which carry the positive electric charge, and neutrons with no charge (except for hydrogen which has no neutron). Each distinct element has atoms with a unique number of protons, called the atomic number. Hydrogen has one, carbon six and iron 26, for instance. Different forms of an element may exist, called isotopes, each with a different numbers of neutrons. For example, most carbon has six neutrons but atoms can exist with seven or eight. Many isotopes are not stable and their nuclei break down sooner or later, emitting harmful particles and radiation. Unstable isotopes of an element are described as "radioactive."

elements (orbital) A set of quantities that define the shape, orientation and timing of an object's orbital motion.

elliptical galaxy A galaxy that looks smoothly elliptical in shape and has no structures such as a disk or spiral arms. In three dimensions elliptical galaxies are believed to be ellipsoidal – similar to a regularly shaped potato. The largest, most luminous galaxies in the universe are giant ellipticals up to 500 000 light years across. It is likely that many of them harbor massive black holes at their centers. It is thought that they formed through the merger of smaller galaxies. Large ellipticals tend to be where there are many galaxies relatively close together, such as in the center of a rich cluster of galaxies.

Some of the smallest galaxies are also ellipticals. Their typical size is about 1000–10 000 light years. These dwarf elliptical galaxies contain between a few hundred thousand and a few million stars. Small elliptical galaxies are most often found as companions to much larger galaxies.

All the evidence suggests that ellipticals are old galaxies. Almost all the stars in them are more than 10 billion years old – 5 billion years older than the Sun. With few exceptions, they contain very little gas and dust and hardly any newly forming stars or young star clusters.

Elnath (Beta Tauri) The second-brightest star in the constellation Taurus, of magnitude 1.7. The name Elnath (alternatively spelled Alnath) comes from the

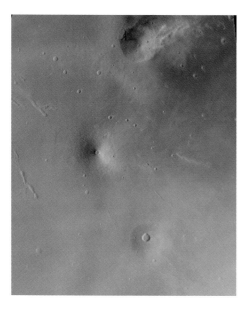

Elysium Planitia. This image from Mars Global Surveyor shows three volcanoes: Elysium Mons in the center, with Hecates Tholus to the north and Albor Tholus to the south.

Arabic for "the one butting with horns." In old star maps, this star was shared with the neighboring constellation Auriga, and was also designated Gamma Aurigae. Elnath is classed as a ➤ *B star.*

elongation The angle on the sky between the Sun and an astronomical body, such as the Moon or a planet. Elongations of 0°, 90° and 180° are called conjunction, quadrature, and opposition, respectively. Mercury and Venus both have maximum possible elongations because they are nearer to the Sun than Earth. The maximum distances they attain from the Sun during each orbit are called greatest elongation. Mercury's greatest elongation can be anywhere between 18° and 28° according to circumstances; the range for Venus is 45–47°. Any elongation is possible for the planets farther from the Sun than Earth.

Elysium Planitia The second largest volcanic region on Mars. The highest volcano, Elysium Mons, rises to a height of about 12 km.

emersion The reappearance of a body that has been hidden during an ➤ *occultation* or an ➤ *eclipse.*

emission line In a ➤ *spectrum*, a spike at a specific wavelength. Atoms and molecules produce emission lines when they lose energy by switching between two energy states.

emission nebula A cloud of glowing gas in interstellar space. Interstellar gas clouds give out light when they receive energy in the form of ultraviolet radiation from hot stars nearby. Hydrogen, the most common gas in space, glows pink.

Enceladus One of the medium-sized moons of Saturn, discovered by William Herschel in 1789. It measures $512 \times 494 \times 489$ km ($318 \times 307 \times 304$ miles) and it travels in an almost circular orbit 238 020 km (147 905 miles) from Saturn,

An emission nebula around a cluster of young stars (NGC 346) in the Small Magellanic Cloud, imaged by the Hubble Space Telescope.

which is within Saturn's ring system. The ➤ *Voyager 2* spacecraft returned images of Enceladus showing details as small as 2 km (just over 1 mile) and images have also been obtained more recently by the ➤ *Cassini* spacecraft. Large areas of the surface have no craters, and the density of craters in areas that do have them is relatively low. There is also widespread evidence of tectonic activity, such as troughs, grooves and ridges. The surface of Enceladus appears to have been completely remodeled by activity of some kind since it was first formed. In 2005, the Cassini spacecraft observed plumes of icy droplets jetting from a series of fissures, described as "tiger stripes," in the south polar region. These jets are believed to come from a body of liquid water below the surface. The orbit of Enceladus coincides with Saturn's faint E ring

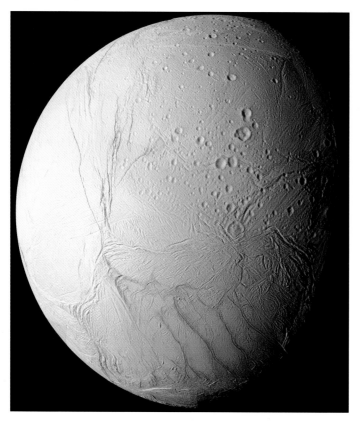

A false color image of Enceladus taken by the Cassini spacecraft.

and the eruptions on the satellite are likely to be the source of the ring material.

Encke, Comet 2/P A short-period comet named in honor of the German astronomer Johann Franz Encke (1791–1865). Encke did not actually discover it, but when it appeared in late 1818, he demonstrated that it was the same object as comets previously seen on three occasions. He also successfully predicted that it would be visible again in 1822. It was only the second comet (after Halley's) proved to return periodically. Pierre Méchain (1744–1804) of Paris was the first to record it in 1786. It was rediscovered by Caroline Herschel in England in 1795, by the French observer Jean Louis Pons and others in 1805 and by Pons again in 1818.

Comet Encke is in an unusual orbit for a periodic comet. On its very elliptical path, it takes only 3.2 years to make one revolution around the Sun. It has been in a very stable orbit for a long time and is probably one of the most evolved comets still actively producing a coma and tail. It may well be close to becoming a defunct core, about 2–3 km across. It has become fainter since first observed. In 1829 its magnitude was recorded as 3.5 but during the twentieth century if never became brighter than fifth magnitude. Streams of dust shed from Comet Encke causes the annual Taurid meteor shower around November 3 each year.

Encke Division (Encke Gap) A narrow gap toward the outer edge of the bright A ring surrounding ➤ *Saturn*. A ➤ *Voyager 2* image showed a narrow ringlet within the Encke Division. Saturn's small moon ➤ *Pan* orbits within the Encke Division. ➤➤ *planetary rings*.

Eos family A group of asteroids with similar composition orbiting the Sun in the ➤ *asteroid belt* at a distance of about 3.02 AU. The family was formed when a single parent asteroid broke up sometime in the past, and takes its name from its largest member, 221 Eos.

ephemeris (pl. ephemerides) A table of the celestial coordinates, magnitude and other data for the Moon, the Sun, a planet, a comet or other astronomical object. An annual handbook containing astronomical tables and data is also sometimes called an ephemeris.

epicycle A circular path around a center that is itself moving around a larger circle, which is called the deferent. The astronomer ➤ *Ptolemy*, who lived in the second century AD, made his Earth-centered model of the solar system fit the observations of the planets by having the planets move along epicycles. To make his predictions more accurate, he said that the center of the epicycle moved at a uniform angular rate about an off-center point called the equant and he also set Earth off-center on the side opposite the equant. Ptolemy's theory could predict the positions of planets to within a degree with the right choice of size for each epicycle and deferent.

Epimetheus A small moon orbiting Saturn at a distance of 151 422 km (94 094 miles), just beyond the far edge of the ring system. It is in the same orbit as another moon, Janus. The two satellites may be fragments of a larger object broken apart by an impact. Epimetheus is irregular in shape, measuring 138 × 110 × 110 km (86 × 68 × 68 miles). Little is known about it though an image returned by the ➤ *Voyager 1* spacecraft in 1980 shows numerous craters, including several 30 km or more across, and ridges and grooves on the surface.

equation of time The difference between mean ➤ *solar time*, as kept by a clock running at a steady rate, and the apparent solar time read off a sundial. It varies during the course of a year up to a maximum of 16.3 minutes.

equator The plane through the center of a rotating object that is perpendicular to the axis of rotation and, for a solid body such as a planet, the circle where that plane cuts its surface. In astronomy, equator is sometimes used as an abbreviation for ➤ *celestial equator*.

equatorial coordinates A system of celestial latitude and longitude for specifying positions on the sky, based on the ➤ *celestial equator*. The coordinate equivalent

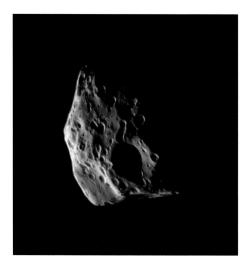

Epimetheus imaged by the Cassini spacecraft.

to latitude is declination (abbreviated to Dec. or δ (delta)). It is measured in degrees north and south of the celestial equator. Northerly declinations are positive and southerly ones negative.

The other coordinate, right ascension (RA or α (alpha)), is the equivalent of longitude but is measured in hours, minutes and seconds of time, because a telescope pointing at a fixed direction in the sky will "see" an angle of one hour of right ascension sweep by in one hour of ➤ *sidereal time*. There are 24 hours in a complete circle of 360° so 1 hour of right ascension is the same as 15°.

The zero point for right ascension is the ➤ *equinox* that falls in March. Because of ➤ *precession*, this point is very slowly moving along the equator. Equatorial coordinates, therefore, have to be specified with the date when they are valid, which is called the epoch.

equatorial mount A mount for a telescope which allows the instrument to track the east–west rotation of the sky by turning slowly around just one axis, called the polar axis. The mount is oriented so that the polar axis is parallel to Earth's rotation axis. The telescope can also rotate around a second axis, called the declination axis, which is at right angles to the polar axis. This arrangement makes it simple to point the telescope to particular ➤ *equatorial coordinate* (right ascension and declination).

Equatorial mounts are popular for amateur telescopes because they can easily follow the motion of the sky automatically if equipped with a small motor. Equatorial mounts were used for large professional telescopes until the late twentieth century when computerized controls made cheaper ➤ *altazimuth mounts* practicable.

equinox The time when the Sun is at one of the two points where the ➤ *celestial equator* and the ➤ *ecliptic* cross, or the actual position on the sky of either of these points. The Sun crosses the equator from south to north at the northern vernal (spring) equinox on March 20 or 21, and from north to south at the northern autumnal equinox, which falls between September 22 and 24. The dates vary because of leap years.

The position of the northern vernal equinox was traditionally known as "the first point of Aries" and is still often represented by the astrological symbol for Aries (γ). However, because of ➤ *precession*, the equinox points move around the ecliptic at a rate of about 1.4° per century. The vernal equinox point now lies in the constellation of Pisces, which neighbors Aries.

Equuleus (The Foal) The second-smallest constellation in the sky. Although it is small and its stars faint, Equuleus is an ancient constellation. It is located next to Pegasus near the celestial equator.

Eridanus (The River Eridanus) A large southern constellation, meandering from the celestial equator to declination –60°. ➤ *Ptolemy* named it in his list of 48 constellations, but its southern extension was added later. The most southerly point is marked by the first-magnitude star ➤ *Achernar*.

Erinome A small outer moon of Jupiter discovered in 2000. Its diameter is about 3 km (2 miles).

Eris A ➤ *dwarf planet* in the ➤ *Kuiper Belt*. Before receiving its official name in 2006, it was known informally as "Xena" and by its temporary designation 2003 UB313. With a diameter of about 2400 km (1490 miles), it is slightly larger than Pluto and when named in 2006 was the largest known dwarf planet.

Eris was discovered by Mike Brown, Chad Trujillo and David Rabinowitz on January 8, 2005, at ➤ *Palomar Observatory*. It had first been photographed in October 2003, but its nature was not detected until the subsequent observation in 2005. The orbit followed by Eris is a very elongated ellipse, which it takes 560 years to complete. Eris's distance from the Sun ranges between 38 and 97 astronomical units. At discovery, it was nearly as far away as it can ever be, but at closest it is nearer than Pluto. The spectrum of Eris suggests it is covered by a yellowish layer of frozen methane.

Eris has one known moon, Dysnomia, which is about 250 km (155 miles) across.

433 Eros The third largest of the known near-Earth asteroids, and the first asteroid to be discovered on an orbit that crosses inside the orbit of Mars. It was found on August 13, 1898 by Gustav Witt in Berlin and independently by Auguste Charlois at the Nice Observatory in France. Elongated and irregular in shape, it measures $13 \times 13 \times 33$ km ($8 \times 8 \times 21$ miles).

Eros's distance from the Sun ranges between 1.78 and 1.33 AU. In 1975, it came within 22 million km (14 million miles) of Earth on one of its nearest possible encounters. When Eros made a close approach in 1931, astronomers had never seen any other object in orbit round the Sun come so near to Earth.

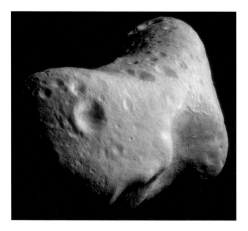

The northern hemisphere of Eros from a mosaic of images taken by the NEAR Shoemaker spacecraft orbiting at a height of 200 km (124 miles).

Observations of its path were used to make the best calculations possible at that time of the distance between Earth and the Sun. In February 2000, the spacecraft ➤ *NEAR Shoemaker* entered orbit around Eros to conduct the first ever long-term close-up study of an asteroid. During its year in orbit, it returned a wealth of images and information about Eros.

Eros is made entirely of primitive, rocky material which is not separated into layers with different compositions. Its average density is 2.7 g/cm^3, about the same as Earth's crust. It seems to be a single sold body, possibly porous or fractured inside, but not a collection of small fragments or "rubble pile." It was probably broken off a larger asteroid by an impact. The rough, heavily cratered surface is covered by a layer of ➤ *regolith* and scattered with numerous large boulders.

Erriapo A small outer moon of Saturn in a very elliptical orbit. It was discovered in 2000 and is 8.6 km (5.3 miles) across.

eruptive variable A star that changes brightness suddenly and unpredictably. There are many different kinds of eruptive variables, including ➤ *flare stars*, ➤ *luminous blue variables*, ➤ *R Coronae Borealis stars*, ➤ *T Tauri stars* and ➤ *Wolf–Rayet stars*.

ESA Abbreviation for ➤ *European Space Agency*.

escape velocity The least velocity a small body needs in order to escape from the gravitational attraction of a more massive one. The velocity of escape from Earth's surface is about 11.2 km/s.

Eskimo Nebula (NGC 2932) A ➤ *planetary nebula* in the constellation Gemini. Its popular name came from photographs of the circular nebula, which suggest the features of a face, surrounded by a fur-fringed hood. The "fur hood" consists of comet-shaped blobs of gas streaming away from the central star. The inner part of the nebula is two lobes of expanding gas seen end-on.

ESO Abbreviation for ➤ *European Southern Observatory*.

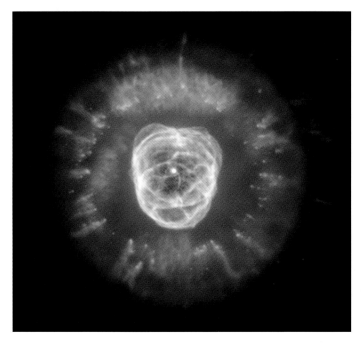

The Eskimo Nebula imaged by the Hubble Space Telescope.

Eta Aquarids A ➤ *meteor shower* caused by dust streams in the orbit of Comet ➤ *Halley*. Eta Aquarids can be seen between about April 24 and May 20 each year, peaking around May 4 or 5. The radiant of the shower lies in the constellation Aquarius.

Eta Carinae A ➤ *luminous blue variable* star inside the ➤ *Eta Carinae Nebula*. Estimates put its total mass at about 100 times greater than the Sun's but it is almost certainly a binary system rather than a single star. Eta Carinae is the strongest source of infrared radiation in the sky. Its total energy output, including infrared, is five million times that of the Sun.

Over the last 300 years, Eta Carinae has varied in brightness unpredictably. Halley recorded it as a fourth-magnitude star in 1677. By 1843 it was the second-brightest star in the sky. Subsequently, it faded irregularly and has been invisible to the naked eye for the last 100 years. Now it appears to be brightening again.

Eta Carinae is surrounded by the Homunculus Nebula, which is produced by gas flowing from the star. It loses gas at a rate of 0.07 solar masses a year, the fastest known rate for any star. This prodigious loss of material apparently started when the star was at its brightest but cannot be sustained for long – possibly only a few hundred years. The twin-lobed shape of the Homunculus Nebula is thought to be the result of a ring of gas restricting the outflow. The

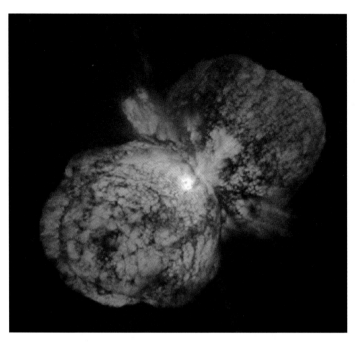

Eta Carinae. This image, made from views taken by the Hubble Space Telescope, shows huge blobs of material ejected by the supermassive star.

ring may have formed from the outer layers of the secondary star in the system during the 1843 outburst. A star as massive as Eta Carinae is not stable. It is likely to undergo more dramatic changes in the future and may possibly become a ➤ *supernova*.

Etched Hourglass A popular name for a ➤ *planetary nebula* with the formal catalog designation MyCn18. It is about 8000 light years away and lies in the southern constellation Musca. A ring of dust around the equator of its central star may be responsible for its hour-glass shape.

Euanthe A small moon of Jupiter discovered in 2001. Its diameter is about 3 km (2 miles).

Eudoxus of Cnidus (c. 408–c. 355 BC) Eudoxus was one of the greatest mathematicians of his age and is most famous for the theory of planetary motion he developed. He studied astronomy and made observations in Egypt but finally settled in his native Cnidus, in what is now Turkey. All the books he wrote himself are lost, but his work is known because later writers such as ➤ *Hipparchus* and ➤ *Aristotle* referred to it.

To predict the motion of the planets, Eudoxus proposed a complex system involving the rotation of 27 interlinked spheres centered on Earth. It was a remarkable achievement in geometry, but he still could not fully

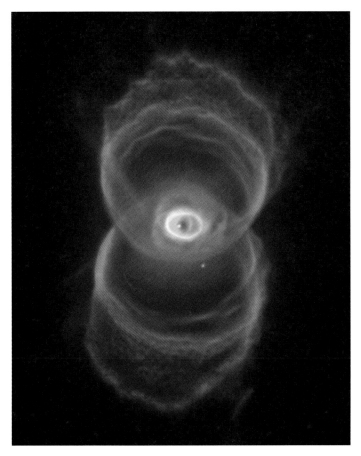

The Etched Hourglass. A false color image composed of views taken by the Hubble Space Telescope through three different filters.

reproduce the observed motion of the planets. Eudoxus probably did not think of his spheres as real objects in space, but simply as a mathematical method.

45 Eugenia One of the larger asteroids, discovered from Paris in 1857 by Hermann Goldschmidt. Its estimated diameter is 215 km (133 miles). In November 1998 astronomers working at the Keck Observatory found that it has a moon about 13 km (8 miles) in diameter, which was named Petit-Prince. It was the first asteroid moon to be discovered from Earth.

Eukelade A small outer moon of Jupiter discovered in 2003. Its diameter is about 4 km (2.5 miles).

15 Eunomia One of the larger asteroids, with a diameter of 260 km (162 miles), discovered from Italy in 1851 by Annibale de Gasparis.

Europa imaged by the Galileo spacecraft from a range of 677 000 km (417 900 miles).

31 Euphrosyne One of the larger asteroids, estimated to be 248 km (154 miles) across, and the first to be discovered from America. It was found 1854 by James Ferguson working at the US Naval Observatory.

Euporie A small outer moon of Jupiter discovered in 2001. Its diameter is about 2 km (1 mile).

Europa The smallest of Jupiter's four large Galilean moons, with a diameter of 3130 km (1945 miles). The ➤ *Voyager* spacecraft returned images showing a bright icy surface, criss-crossed by numerous dark lines but with relatively few obvious craters. Close-up images taken by the ➤ *Galileo* spacecraft revealed a complex network of grooves and bands, some straight and some curved or scalloped. These include long features several kilometers wide stretching for 1000 km or more, called triple bands, which are bright in the center with dark regions either side. Galileo's images also showed that there are numerous craters scattered over the surface.

It is not known for certain how Europa is structured inside but the evidence suggests there could be a core rich in iron, surrounded by rock. Europa's icy crust is at least several kilometers thick but underneath there could be an ocean of liquid water or a mixture of rock and icy slush. Tidal forces due to Jupiter's powerful gravity are strong enough to raise the temperature above the freezing point of water.

The appearance of Europa's crust shows that it must have changed greatly since it first formed. In some places it looks as if great rafts of ice have broken up and moved about, suggesting there is (or has been) water or slush beneath them. Some of Europa's features may have been created by liquid water welling up from below in a process similar to volcanism on Earth.

52 Europa One of the larger asteroids, with an estimated diameter of 312 km (194 miles). It was discovered in 1858 by Hermann Goldschmidt.

European Southern Observatory (ESO) A European research organization founded in 1962 to foster cooperation in astronomy and to provide European astronomers with a major modern observatory. The member-countries are Belgium, the Czech Republic, Denmark, Finland, France, Germany, Italy, the Netherlands, Portugal, Spain, Sweden, Switzerland and the UK. Its headquarters are at Garching bei München in Germany. Its observatory is split between two sites in Chile: the ➤ *La Silla Observatory* and the ➤ *Paranal Observatory*.

European Space Agency (ESA) An organization through which 16 European countries (Austria, Belgium, Denmark, Finland, France, Germany, Greece, Ireland, Italy, the Netherlands, Norway, Portugal, Spain, Sweden, Switzerland and the UK) collaborate on a joint space program. ESA builds and launches satellites and spacecraft for scientific research and commercial applications. The headquarters are in Paris. ESA's largest establishment is its Research and Technology Center (ESTEC) at Noordwijk in the Netherlands.

Eurydome A small outer moon of Jupiter discovered in 2001. Its diameter is about 3 km (2 miles).

EUV Abbreviation for extreme ultraviolet. There is no exact definition of where extreme ultraviolet begins and ends but it usually means the range of wavelengths between about 10 and 100 nm. ➤➤ *electromagnetic radiation, ultraviolet astronomy, XUV*.

evening star Venus or Mercury when it is visible in the western sky after sunset.

event horizon The nearest to a ➤ *black hole* it is possible to see anything. No light emitted inside the event horizon can escape because of the strength of the black hole's gravitational field. The radius of the event horizon, called the ➤ *Schwarzschild radius*, depends on the mass of the black hole. For a black hole with the minimum possible mass of 3 solar masses, it is 9 km (6 miles).

exit pupil The image of a telescope's field of view, formed by an eyepiece. The position of the exit pupil in front of an eyepiece is where an observer should place his or her eye to get the optimum view. If the system is well designed, the exit pupil should be at a place where it is comfortable to observe and should match the size of the pupil of the observer's eye.

exobiology Another word for ➤ *astrobiology*.

ExoMars A proposed mission to Mars, which would be the first to be led by the European Space Agency. The mission would deliver a "rover" carrying a package of experiments to the surface of Mars. The rover would communicate with Earth via the ➤ *Mars Reconnaissance Orbiter*. The target launch date is 2011.

exoplanet An ➤ *extrasolar planet*.

exosphere The outermost layers of a planetary atmosphere, where the gas is very thin and merges into the interplanetary medium, or the entire layer of gas around a planet if it is extremely tenuous. In an exosphere, the density is so low that gas atoms very rarely collide with each other, and atoms moving rapidly can escape from the gravitational pull of the planet. The Earth's exosphere starts at a height of about 400–500 km (250–300 miles).

expanding universe A model for the universe based on the idea that the scale of space has been getting larger since our universe came into being at the ➤ *Big Bang*. The discovery by Edwin Hubble in 1929 that the ➤ *redshifts* of galaxies increase with distance and the detection of the ➤ *cosmic background radiation* in 1964 are the main evidence that the universe is expanding.

extragalactic Outside the limits of our own Galaxy, the Milky Way.

extrasolar planet A planet orbiting a star other than the Sun.

The first extrasolar planets were discovered in 1991 when Alexander Wolszczan detected three planets orbiting the ➤ *pulsar* PSR1257+12 in the constellation Virgo. Later, evidence was found for a fourth. The planets are not directly visible but their presence affects the regularity of the radio pulses received from the pulsar.

The first evidence for extrasolar planets around ordinary stars was made public in 1995 by Michel Mayor and Didier Queloz of the Geneva Observatory who found a planet with at least half the mass of Jupiter orbiting the star 51 Pegasi. Since 1996, the count of extrasolar planets orbiting ordinary stars has risen steadily, passing the 200 mark in 2006. Several systems are known to contain two or more planets. Nearly all these planets have been found by the tiny regular changes they cause to the motion of their parent stars. This technique is not sensitive enough to detect planets as small as Earth. Other search methods include looking for the slight dimming of a star when a planet crosses in front of it, and ➤ *microlensing*. Several space missions are under consideration that would be capable of finding Earth-sized planets.

Extreme Ultraviolet Explorer (EUVE) An orbiting observatory launched by NASA in June 1992 to detect short-wavelength ultraviolet radiation. It was the first observatory dedicated to observing ultraviolet radiation in the wavelength range 7–76 nm. It operated until January 31, 2001. Results from EUVE included an all-sky catalog of 801 objects. ➤ *ultraviolet astronomy*.

eyepiece A combination of small lenses mounted in a tube, used to view the image formed by a telescope or other optical instrument. A telescope for visual use has a small tube where an eyepiece can be slotted. The field of view of an eyepiece depends on the combination of lenses used in its design. The magnification a particular eyepiece gives is the focal length of the telescope divided by the focal length of the eyepiece.

eye relief The distance from an ➤ *eyepiece* an observer's eye has to be in order to see clearly the whole ➤ *field of view*. Eyepieces of different designs and magnification have different eye relief. In general, the higher the power of the eyepiece, the lower the eye relief. Observers who wear spectacles need more eye relief than those who do not.

F

facula (pl. faculae) A bright region on the Sun's ➤ *photosphere*. Faculae are linked with ➤ *solar activity* and sunspots often appear in the same area of the Sun's surface.

fall A ➤ *meteorite* recovered after it has been observed to fall.

falling star A popular North American term for a ➤ *meteor*.

false color image A colored image in which the colors are not as they would be seen by a normal human eye in natural conditions. In astronomy, false color is used to enhance the contrast of an image and so bring out details that would otherwise be difficult to see. False color is also used to picture observations made in wavelength regions other than visible light.

False Cross The four stars Epsilon and Iota Carinae, and Delta and Kappa Velorum, in the southern constellations ➤ *Carina* and ➤ *Vela*. They are called the "False Cross" because they look similar to the nearby constellation ➤ *Crux*.

farside The hemisphere of the Moon that is permanently turned away from the Earth. ➤➤ *libration*.

Far Ultraviolet Spectroscopic Explorer (FUSE) A NASA astronomy satellite launched in June 1999 on a three-year mission for spectroscopy in the far ultraviolet region of the spectrum.

Ferdinand A small outer moon of Uranus, discovered in 2001. It is about 12 km (7 miles) across.

Fermi–Hart paradox The argument that extraterrestrial life does not exist because we see no evidence of it and have received no signals from extraterrestrials.

fibril A streak-like feature in an active region of the Sun. They are typically 725–2200 km (450–1400 miles) wide and 11 000 km (7000 miles) long. Individual fibrils have a lifetime of 10–20 minutes, though the overall fibril pattern of a region shows little change over several hours.

In the central parts of active regions, fibrils are arranged in patterns connecting spots and ➤ *plages* of opposite magnetic polarity. Ordinary sunspots are surrounded by a radial pattern of fibrils called the superpenumbra. This is material flowing down into the spot.

field galaxy A galaxy that appears in the same field of view as a cluster of galaxies but is not itself a member of the cluster. The alignment is purely coincidental and the field galaxy is nearer or farther away than the cluster.

field of view The angular size of the image formed by an optical instrument, such as a telescope. The field of view of a telescope becomes smaller the greater the magnification being used.

field star A star that appears in the same field of view as a ➤ *star cluster* but is not itself a member of the cluster. The alignment is purely coincidental and the field star is nearer or farther away than the cluster.

filament A solar ➤ *prominence* viewed against the bright disk of the Sun – that is, from above. Prominences can be seen in this way when the Sun is observed through a filter to isolate the light at the wavelength of particular ➤ *spectral lines* such as ➤ *hydrogen alpha*. Filaments then show up as dark streaks.

find A ➤ *meteorite* discovered accidentally or as the result of a search on the ground, as opposed to one actually seen to fall.

finder A small telescope attached to the tube of a larger instrument to help with pointing the main telescope correctly. A finder usually has a cross-wire in its field of view and is aligned so that an object located on the cross-wire appears centrally in the smaller field of view of the main telescope.

fireball A particularly bright ➤ *meteor*. There is no precise definition: magnitudes -3, -4 or -5 are variously quoted as the minimum for a meteor to be described as a fireball. Exceptional fireballs are sometimes a sign that a ➤ *meteorite* may have fallen, though the object more often burns up in the atmosphere.

first contact In an ➤ *eclipse* of the Sun, the point when the Moon first begins to move across the Sun's disk. In a lunar eclipse, first contact occurs when the Moon first enters the full shadow (umbra) of the Earth. The term also describes a similar stage in the progress of a ➤ *transit* or ➤ *occultation*.

First Point of Aries (γ) One of the two points on the celestial sphere where the ➤ *ecliptic* and the ➤ *celestial equator* cross. It is the point on the celestial equator where ➤ *right ascension* is zero. Because of the effects of ➤ *precession*, the celestial equator is slowly sliding around the ecliptic. It takes about 25 800 years to make one revolution. Though their crossing point was in the constellation Aries several thousand years ago, it is now in Pisces. However, its name has not been changed. The Sun is at the First Point of Aries on the northern vernal ➤ *equinox*, and "vernal equinox" is also used to mean this place on the celestial sphere.

first quarter The ➤ *phase* of the Moon when half its visible disk is illuminated and the Moon is waxing towards full. First quarter occurs when the celestial ➤ *longitude* of the Moon is 90° greater than the Sun's.

Flaming Star Nebula A popular name for IC 405, a star cluster surrounded by nebulosity in the constellation Auriga.

Flamsteed, John (1646–1719) The British astronomer Flamsteed was a skilled observer, who was appointed the first ➤ *Astronomer Royal* in 1675. He was born in Derby, where he attended school until aged 14. Thereafter he was self-taught. He studied astronomy from 1662 and began observations in 1671. His career developed under the patronage of the president of the Royal Society, which promoted the establishment of the Royal Observatory. The King appointed Flamsteed as the astronomer responsible for improving tables for determining longitude at sea, and built the Royal Observatory for him.

He designed the most accurate instruments of his day, to make a new star catalog and a star atlas, containing positions of about 3000 stars. Ever a perfectionist, he refused to release his data, then the most accurate in the world, to Isaac Newton and Edmond Halley, which led to bitter clashes. Halley published the catalog in 1712, against Flamsteed's wishes. Flamsteed later seized 300 of them and burnt them. His revised catalog was published in 1725.

Flamsteed numbers Identification numbers assigned to stars listed in *Historia coelestis Britannica* by John ➤ *Flamsteed*. In the official version of the catalog, which was published posthumously in 1725, the numbers are not included explicitly. However, they do appear in a preliminary version published by Edmond Halley and Isaac Newton in 1712 without Flamsteed's approval. Few copies of this version survive because Flamsteed was greatly angered at the action of Halley and Newton and destroyed many of them.

The numbers were allocated by constellation and in order of right ascension. They were subsequently used by other catalogers, including John Bevis (1750) and Joseph-Jérôme de Lalande (1783). The Flamsteed numbers are still commonly used for stars that do not also have a Greek ➤ *Bayer letter* designation.

flare A phenomenon in the solar ➤ *chromosphere* and ➤ *corona*, caused by a sudden release of energy that heats and accelerates gas in the Sun's atmosphere. Flares are explosions lasting typically for a few minutes. In a flare, the temperature reaches tens or hundreds of millions of degrees and particles are accelerated

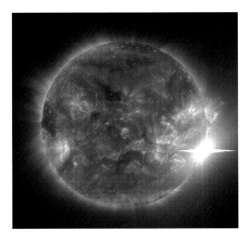

A powerful solar flare was recorded in this ultraviolet image by SOHO on November 4, 2003.

almost to the speed of light. Most of the radiation is emitted as X-rays but flares are also observed in visible light or radio waves. They are associated with active regions of the Sun. Charged particles ejected from the Sun by flares are potentially hazardous to humans and sensitive electronic equipment in space.

flare star A dwarf ➤ *M star* that unpredictably brightens by a magnitude or more for a few minutes. All flare stars have emission lines in their spectra and the flares are thought to be outbursts of energy in the star's ➤ *chromosphere*. Stellar flares are similar to solar ➤ *flares* but far more powerful. The nearest star to the Sun, Proxima Centauri, is a flare star. Flare stars are also known as UV Ceti stars.

flocculi A mottled pattern on the Sun visible when an image of the solar ➤ *chromosphere* is made that isolates the light emitted by calcium or hydrogen atoms.

Flora group A complex grouping of ➤ *asteroids* near the inner edge of the asteroid belt at a distance of 2.2 AU from the Sun. The group is separated from the main belt by one of the ➤ *Kirkwood gaps* and is not a true family with a common origin.

flyby A space mission in which a spacecraft passes close to a target without landing or going into orbit.

Fomalhaut (Alpha Piscis Austrini) The brightest star in the constellation Piscis Austrinus. The name comes from Arabic and means "the fish's mouth." Fomalhaut is an ➤ *A star* of magnitude 1.2 and is 25 light years away. Infrared observations show it to be surrounded by a vast ring of dust, 370 astronomical units in diameter, where planets may be forming.

Footprint Nebula ➤ *Minkowski's Footprint.*

The dusty disk around the star Fomalhaut imaged by the Spitzer Space Telescope.

forbidden lines ➤ *Spectral lines*, not normally observed in a laboratory situation. Under astrophysical conditions, however, spectral lines "forbidden" in a laboratory on Earth can sometimes be very strong.

fork mount A particular form of ➤ *equatorial mount*.

Fornax (The Furnace) A small, inconspicuous constellation of the southern sky. It was introduced in the mid-eighteenth century by Nicolas L. de Lacaille with the longer name, Fornax Chemica – the chemical furnace. None of its stars are brighter than fourth magnitude.

Fornax A A strong radio source in the constellation Fornax. It is associated with the spiral galaxy NGC 1316.

fourth contact In an ➤ *eclipse* of the Sun, the point when the Moon finally moves clear of the Sun's disk. In a lunar eclipse, fourth contact occurs when the Moon leaves the full shadow (umbra) of the Earth. The term also describes the similar stage in the progress of a ➤ *transit* or ➤ *occultation*.

Fra Mauro A lunar crater, 95 km (60 miles) in diameter, close to the Apollo 14 landing site. The geological formation sampled by the Apollo 14 astronauts was named after it. That formation is part of the ejecta or debris excavated when an impact formed the ➤ *Imbrium Basin*. ➤➤ *Apollo program*.

Francisco A small outer moon of Uranus, discovered in 2001. It is about 12 km (7 miles) across.

Fraunhofer, Joseph von (1787–1826) Fraunhofer was an instrument-maker who specialized in astronomical optics and investigated the solar spectrum. He was born in Straubing, Germany. After serving an apprenticeship as a glazier in Munich, he joined an optical institute specializing in optics of the highest quality. He made the world's finest optical glass for telescopes. In 1813, while researching the refractive properties of glass, he accidentally rediscovered the dark lines in the solar spectrum, first noted by William Wollaston in 1802. He measured the wavelengths of 574 ➤ *Fraunhofer lines*, introducing alphabetical names, such as sodium D lines, for the most intense. He also invented the spectroscope and the diffraction grating.

Fraunhofer lines The dark ➤ *absorption lines* in the spectrum of the Sun. Many of the stronger ones were first recorded by Joseph von ➤ *Fraunhofer*, who also labeled some of the most prominent ones with letters of the alphabet. Some of these identifying letters are still commonly used in physics and astronomy, notably the sodium D lines and the calcium H and K lines.

Fred Lawrence Whipple Observatory An observatory on Mount Hopkins in Arizona, operated by the ➤ *Harvard–Smithsonian Center for Astrophysics*. The instruments located there, at an altitude of 2600 m (8500 feet), include a 10-m (32-foot) optical reflector for gamma-ray astronomy and the reconfigured ➤ *Multiple Mirror Telescope* (MMT), which is operated jointly with the University of Arizona for optical and infrared observations. The observatory is named in honor of a former director of the ➤ *Smithsonian Astrophysical Observatory* who was particularly well known and distinguished for his work on comets.

F star A star of ➤ *spectral type* F. F stars have surface temperatures in the range 6000–7400 K. Their spectra feature strong absorption lines of ionized calcium. There are also many medium-strength absorption lines due to iron and other heavier elements. ➤ *Procyon* and ➤ *Polaris* are examples of F stars.

full Moon The ➤ *phase* of the Moon when its celestial ➤ *longitude* is 180° greater than the Sun's and its disk is fully illuminated.

G

Gaia hypothesis The idea that life on Earth regulates the composition of the lower atmosphere. Gaia was a Greek Earth goddess.

galactic center The central region of our Galaxy, which is hidden from view by dense concentrations of dust. A compact radio source called Sagittarius A* appears to mark the very center of the Galaxy and is the zero point for the system of ➤ *galactic coordinates*. It is believed to be a ➤ *black hole* of about three million solar masses. The ➤ *Chandra X-ray Observatory* has observed frequent X-ray flares from Sagittarius A*. Around 10 light years from the center there is a ring of gas and dust, rotating at about 110 km/s.

galactic coordinates A latitude and longitude coordinate system that takes the ➤ *galactic plane* as its equator and the ➤ *galactic center* (RA 17h 42.4m, Dec. –28° 55′) as the zero point of longitude measurement.

An infrared image of the galactic center from the Spitzer Space Telescope. The brightest white spot in the middle is the very center of the Galaxy, which also marks the site of a supermassive black hole.

galactic halo A spherical region around a spiral galaxy. The halo around our own Galaxy extends to about 50 000 light years from the center. The stars outside the disk of the galaxy but within the halo are the oldest and their distribution indicates the size and shape of the galaxy before much of it collapsed to a disk. Galactic halos contains very hot gas that emits X-rays.

galactic plane A circle around the sky, dividing it into two equal hemispheres, that passes through the ➤ *galactic center* and the densest parts of the ➤ *Milky Way*. It is inclined to the celestial equator by about 63°.

galactic poles The points on the sky at latitudes 90° north and south in ➤ *galactic coordinates*. The north galactic pole is in the constellation Coma Berenices and the south pole is in Sculptor.

galactic wind A flow of tenuous hot gas out of a galaxy. The temperature of the gas is millions of degrees and it comes from regions where there is a high rate of star formation.

galactic year (cosmic year) The time taken for the Sun to complete one orbit around the galactic center, roughly 220 million years.

Galatea A satellite of Neptune discovered during the flyby of ➤ *Voyager 2* in August 1989. Its diameter is 176 km (109 miles).

galaxy A large family of stars, often accompanied by clouds of gas and dust, kept together by gravity.

Galaxies cover a huge range of size and mass. The smallest known are relatively nearby dwarf galaxies containing only 100 000 stars. At the other end of the scale, the most massive galaxy known, the giant elliptical M87, contains 3000 billion solar masses, and is about 15 times more massive than our own ➤ *Galaxy*.

The shapes of most galaxies can be described as "spiral," "elliptical" or "irregular". Spiral galaxies are disk-shaped, with a central bulge, from which

An artist's impression of our Galaxy seen face on.

spiral arms appear to wind outwards. In barred spirals, a bar of stars extends out from the bulge and the arms come from the ends of the bar. Spiral galaxies contain very luminous young stars and have significant amounts of interstellar material in their arms.

Most of the conspicuous galaxies in the sky are spirals, but ellipticals are the most numerous. Both the smallest and largest galaxies are of this kind. Most consist entirely of old stars with relatively little interstellar material. In three dimensions they are spheroidal or sometimes nearly spherical.

Irregular galaxies form the third main group. These are neither spiral nor elliptical and account for up to a quarter of all known galaxies. At visible wavelengths, irregular galaxies show no particular circular symmetry and look chaotic.

A very small number of galaxies have very unusual structures, often because they have interacted with another galaxy. Active galaxies emit exceptionally large amounts of energy and show evidence of violent processes at work. Active galaxies include ➤ *Seyfert galaxies* and ➤ *radio galaxies*. ➤➤ *Hubble classification.*

Galaxy When written with a capital initial G, the galaxy to which the Sun and solar system belong, which is visible in the night sky as the ➤ *Milky Way*.

Our Galaxy is a barred ➤ *spiral galaxy*, containing of order 200 billion stars as well as much interstellar matter, both dark and luminous. It is disk-shaped with an almost spherical bulge in the middle, surrounding the ➤ *galactic center*. At the very heart of the bulge is a supermassive black hole. The disk is 100 000 light years across but much of its content is concentrated in a thin layer that is only 2000 light years thick towards its outer edges, though stars are distributed through a somewhat thicker disk. The central bulge has a radius of about 15 000 light years. The movements of the stars and interstellar material we can see suggest that they account for as little as 10 percent of the total mass of the Galaxy. The rest is ➤ *dark matter*, in a form not yet identified. This is distributed in a sphere around the Galaxy extending out as far as 300 000 light years.

The spiral arms are concentrations of stars and interstellar matter winding outwards from the edge of the bulge. Regions of star formation and ➤ *ionized hydrogen* are located in the arms. In the space between the arms, the average density of matter is a factor of 2 or 3 lower than within the arms. The Sun is about 28 000 light years from the galactic center, within the disk, near the inner edge of a spiral arm. The whole Galaxy is rotating, but not like a rigid object. The Sun takes about 220 million years to complete a circuit, but stars nearer the center take shorter times.

The disk of the Galaxy is surrounded by a sparsely populated, roughly spherical region, known as the galactic halo. Its radius is at least 50 000 light years. The halo contains the oldest stars in the Galaxy, including

The Galilean moons of Jupiter imaged by the Galileo spacecraft. From left to right, the moons shown are Ganymede, Callisto, Io, and Europa.

➤ *globular clusters*. There is very little luminous matter in the halo compared with the disk and central bulge.

galaxy cluster ➤ *cluster of galaxies*.

Galilean moons (Galilean satellites) The four largest moons of Jupiter, ➤ *Io*, ➤ *Europa*, ➤ *Callisto* and ➤ *Ganymede*, which were discovered by ➤ *Galileo* in 1610.

Galilean telescope A simple telescope of the kind used by ➤ *Galileo* in the first astronomical telescopes. It consists of two lenses. A long-focus convex (converging) lens acts as the objective lens and a single concave lens is used as the eyepiece. The resulting image is the right way up.

Galileo (spacecraft) A NASA spacecraft sent to explore ➤ *Jupiter*, its rings and satellites. It was launched in October 1989 from the ➤ *Space Shuttle* and arrived at Jupiter in December 1995. It went into orbit around Jupiter and returned high-resolution images and data for more than seven years. On the way to Jupiter, it made close flybys of the asteroids ➤ *Gaspra* and ➤ *Ida*.

A major disappointment was the failure of the craft's high-gain communications antenna to open properly. Relying only on a low-gain antenna limited the amount of data that could be transmitted back to Earth. Despite this, the mission was a great success overall. In 2003 it was deliberately crashed into Jupiter.

A probe carried by Galileo separated from the craft and entered Jupiter's atmosphere on December 7, 1995. It parachuted down, returning data on the composition and physical state of the atmosphere for 57 minutes.

Galileo Galilei (1564–1642) Galileo was an Italian physicist and astronomer who is regarded as the principal founder of modern scientific methods. Born in Pisa, Tuscany, the oldest of seven children, he received private tuition. From 1581 he studied for a medical career at the University of Pisa. His interest changed from medicine to mathematics and he left the university in 1585 without a degree. Nevertheless, four years later, his mathematical ability resulted in Pisa appointing him professor of mathematics. In 1592 he moved to a professorship in Padua.

Galileo Galilei.

In 1610 he constructed a simple refracting ➤ *telescope*, and became the first person to use such an instrument for astronomical observations. He was astonished by the sheer numbers of stars, by the rough surface of the Moon, and by ➤ *sunspots*. In January 1610 he discovered the ➤ *Galilean moons* of Jupiter. A few months later he observed the phases of Venus. All of these discoveries were at odds with the philosophy of ➤ *Aristotle* but were consistent with the ➤ *Copernican system*, which Galileo openly promoted in his writings and lectures.

High officers of the Church supported Aristotelian philosophy as a matter of faith. Pope Paul V summoned Galileo to Rome in 1616 to inform him that his public promotion of the Copernican system must cease. In 1632, after the election of a new Pope, Galileo wrote about his ideas in the form of a conversation between three men in his book, *Dialogue Concerning the Two Chief World Systems*. However, he presented the arguments in favor of a Sun-centered solar system so persuasively that he was tried before the Inquisition in Rome in 1633. Pope Urban VIII, a former friend of Galileo, threatened torture. Galileo was forced to recant his "heresies" and lived in enforced isolation for the rest of his life. In 1992 Pope John Paul II expressed regret for the unjust treatment of Galileo.

Galileo National Telescope ➤ *Telescopio Nazionale Galileo*.

gamma-ray astronomy The study of gamma rays from astronomical sources. Gamma rays are the most powerful form of ➤ *electromagnetic radiation*, with wavelengths shorter than X-rays (less than about 0.1 nanometers). Though the most powerful gamma rays can be detected at ground level, Earth's atmosphere absorbs the rest and nearly all astronomical gamma-ray observations are made from satellites.

Many spacecraft have carried detectors for ➤ *gamma-ray bursts* since 1969. Early sky surveys were carried out by the satellites SAS-2 and COS-B. SAS-2 was launched in 1972 and operated for seven months. COS-B was launched in 1975 and operated for over six years. In 1997, observations by the Italian/Dutch satellite BeppoSAX led to the first optical identification of an object that had emitted a gamma-ray burst.

A great advance in gamma-ray astronomy was achieved with the launch by NASA in April 1991 of the ➤ *Compton Gamma Ray Observatory*, which operated until June 2000. Many new sources were identified with greater positional accuracy than was possible previously. Its successor is the Gamma-Ray Large Area Space Telescope (GLAST), to be launched by NASA in 2007 or 2008. The European Space Agency's ➤ *International Gamma-Ray Astrophysics Laboratory* (INTEGRAL), with an expected lifetime of six years, started observations in 2002.

Astronomical sources of gamma rays include solar ➤ *flares*, ➤ *pulsars*, ➤ *X-ray binary* stars and ➤ *quasars*, but the nature of the most powerful sources, which are responsible for gamma-ray bursts, remains uncertain. Strong identifiable sources of gamma-rays include the ➤ *Vela pulsar*, the ➤ *Crab pulsar*, ➤ *SS433* and ➤ *Geminga*. There is also diffuse gamma radiation from the Milky Way.

gamma-ray burst A temporary intense burst of gamma rays and X-rays from a cosmic source. All bursts are short and exceedingly powerful but there two distinct varieties. Short bursts last a few tenths of a second while long burst go on for between 2 and 20 seconds. Gamma-ray bursts were first discovered by chance in the late 1960s by military satellites monitoring nuclear weapons tests and have since been observed by a variety of spacecraft. The ➤ *Compton Gamma-Ray Observatory* (CGRO) showed that bursts occur about twice a day, at random positions all over the sky.

Though the CGRO could determine the positions of the bursts with greater accuracy than previously possible, the positions were still not accurate enough to allow optical identification. From 1997, however, the ➤ *BeppoSAX* satellite was able to pinpoint the positions of gamma-ray bursts precisely enough for them to be identified optically, and for radio emission to be detected. The first one to have a spectrum recorded turned out to be about half way to the edge of the observable universe. This showed that the energy given out by a gamma-ray burst is immense – over a million times more energy than a whole galaxy.

The visible and X-ray afterglows of the longer gamma-ray bursts have similarities to ➤ *supernovae*. This has led astronomers to favor the idea that longer bursts result from the collapse of very massive stellar cores, leading to the formation of ➤ *black holes*. These events are also called "hypernovae." Short bursts are believed to come from the merger of two ➤ *neutron stars*, or a neutron star and a black hole.

Gamma-ray Large Area Space Telescope (GLAST) A NASA orbiting gamma-ray observatory, planned for launch in December 2007 or later. It will provide a larger field of view and better resolution than the ➤ *Compton Gamma-Ray Observatory* and is expected to operate for 10 years.

gamma-ray pulsar A ➤ *pulsar* that emits gamma rays.

Ganymede One of the four large moons of Jupiter. Measuring 5262 km (3270 miles) across, it is the largest natural satellite in the solar system.

The first high-resolution images of Ganymede were returned by ➤ *Voyagers 1 and 2*. Images showing even finer detail were obtained by the ➤ *Galileo* spacecraft. It has several different types of terrain. There are dark, heavily cratered areas and lighter grooved terrain covering around 60 percent of the surface area that has been seen. Galileo images of the dark areas suggest they have been changed by various episodes of shearing and furrowing. Galileo also revealed many small craters on the finely grooved areas.

One of the most significant discoveries made by the Galileo spacecraft was that Ganymede has a significant magnetic field, which is stronger at its surface than the fields of Mercury, Venus or Mars. Data from Galileo's trajectory, combined with the magnetic, field, suggest that Ganymede must have a molten iron-rich core. Overall, Ganymede's density is about twice that of

A close-up of the surface of Ganymede taken by Voyager 2.

Asteroid Gaspra imaged by the Galileo spacecraft. The colors are exaggerated in this view.

water. Its core is probably surrounded by a rocky mantle, overlain by a thick layer of ice.

Garnet Star The name William ➤ *Herschel* gave to the strikingly red star Mu Cephei. It is a red ➤ *supergiant* and a ➤ *semiregular variable* ranging in magnitude between 3.6 and 5.1.

gaseous nebula A glowing cloud of gas in interstellar space, which may be either an ➤ *emission nebula* or a ➤ *reflection nebula*.

951 Gaspra An asteroid in the main asteroid belt imaged by the ➤ *Galileo* spacecraft, which passed Gaspra at a distance of about 16 000 km (10 000 miles) on October 29, 1991. It was the first asteroid to be imaged from close by. It measures about $20 \times 12 \times 11$ km ($12 \times 7 \times 7$ miles) and the largest of many craters on its surface is 1.5 km (1 mile) across. Gaspra was discovered on July 30, 1916, by the Russian astronomer Grigory N. Neujmin and is a member of the ➤ *Flora group*. Galileo detected a magnetic field, suggesting that it has a metallic composition.

Gassendi A lunar crater, 100 km (60 miles) in diameter, on the northern border of the Mare Humorum. Clefts cross the crater floor and it has multiple peaks. Gassendi has been linked with ➤ *lunar transient phenomena*.

gegenschein ➤ *zodiacal light*.

Geminga A powerful gamma ray ➤ *pulsar* in the constellation Gemini. It is one of the most intense sources of gamma rays in the sky. It was discovered in 1972 by the orbiting observatory SAS 2 (➤ *Small Astronomy Satellite*). Weak X-rays from

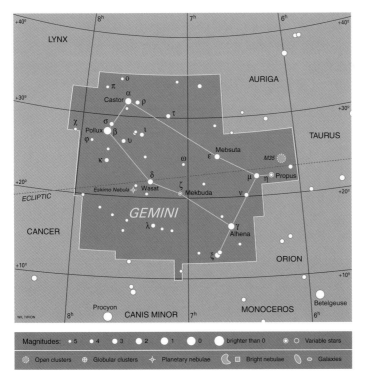

A map of the constellation Gemini.

Geminga were detected by the ➤ *Einstein Observatory* and its optical counterpart looks like a 25th-magnitude star. Geminga emits almost all its energy as gamma rays. Its X-ray emission is a thousand times weaker and its luminosity in visible light a thousand times weaker again. Observations made by ➤ *ROSAT* confirmed the X-ray emission and found that its radiation is pulsed, with a period of about a quarter of a second. Geminga is 350 000 years old and lies 500 light years away.

Gemini (1) (The Twins) One of the twelve constellations of the ➤ *zodiac*. The two brightest stars in Gemini, both first magnitude, have the names of the twins Castor and Pollux of classical mythology. Pollux, though the brighter of the pair, was given the designation Beta. Castor is Alpha.

Gemini (2) A series of manned, orbiting spacecraft, launched by the USA in the 1960s. They were an important part of the preparation for the ➤ *Apollo program* of Moon landings. Gemini 3 in 1965 was the first American flight with a crew of more than one astronaut, and Gemini 8 in March 1966 achieved the first successful docking in space. The last of the series was Gemini 12 in November 1966. Many of the astronauts who later took part in the Apollo Moon landings also flew in the Gemini program.

The enclosure of one of the Gemini Telescopes, located in Chile.

Gemini Telescopes Two 8-m telescopes for optical and infrared astronomy.
One is sited in the northern hemisphere, at the ➤ *Mauna Kea Observatories* in
Hawaii, while the other is in the southern hemisphere, on Cerro Pachón in
Chile, near to the ➤ *Cerro Tololo Inter-American Observatory*. They are operated
by an international consortium. The sites of the two telescopes were
chosen so that they can see the whole sky between them. The Hawaii
telescope was completed during 1998 and its more southerly twin
in 2000.

Geminids A major annual ➤ *meteor shower*, the radiant of which lies near the star
Castor in the constellation Gemini. The shower peaks around December 13,
and its normal limits are December 7–16. The meteor stream responsible has
an unusual orbit and in 1983 the ➤ *Infrared Astronomical Satellite* (IRAS)
discovered in the same orbit a cometary nucleus that is no longer active but
probably generated this stream in the past. It was named 3200 Phaethon.

Gemma An alternative name for the star ➤ *Alphekka*.

general relativity A theory of gravitation, published in its final form in 1916. It was
developed by Albert ➤ *Einstein* from his earlier (1905) ➤ *special relativity* theory.

Genesis A NASA space mission that collected samples of the ➤ *solar wind* and
returned them to Earth. It was launched on August 8, 2001 and spent 845 days
orbiting the Sun 1.5 million km (1 million miles) closer to the Sun than Earth.
Capsules containing samples of the solar wind amounting to about 0.4
milligrams were returned to Earth on August 9, 2004. The capsules crashed
when their parachutes failed to open but most of the samples were salvaged.

geomagnetic field The magnetic field in the vicinity of the Earth. At present,
Earth's magnetic field is roughly like that of a bar magnet displaced 451 km

(280 miles) from the center of the Earth towards the Pacific Ocean and tilted to Earth's axis by an angle of 11°. The strength and shape of the geomagnetic field varies gradually over a timescale of years.

geomagnetic storm ➤ *magnetic storm*.

geospace The region of space around Earth, inluding Earth's ➤ *magnetosphere* and ➤ *ionosphere*.

geostationary orbit ➤ *geosynchronous orbit*.

geosynchronous orbit An orbit around Earth in which a satellite's period of revolution is exactly a ➤ *sidereal day* (23 hours 56 minutes 4.1 seconds). If a satellite is in a circular geosynchronous orbit over the equator, it always stays very close to the same position in the sky and its orbit is described as geostationary. A geostationary orbit is at an altitude of 35 900 km (22 300 miles). A satellite in a geosynchronous orbit inclined to Earth's equator traces out a figure-of-eight shape over the course of a day.

German mount A type of ➤ *equatorial mount* for a telescope.

Ghost of Jupiter A popular name for NGC 3242, a ➤ *planetary nebula* in Hydra.

Giant Magellan Telescope A telescope to be sited in northern Chile consisting of seven 8.4-m (27-foot) mirrors arranged in a circle. It will be equivalent to a 22-m (72-foot) telescope. Work on the mirrors began in 2005 and the telescope is expected to be completed in about 2016.

Giant Metrewave Radio Telescope (GMRT) A ➤ *radio telescope* located near Poona in India. It consists of thirty 45-m (146-foot) dishes arranged in an array extending for 25 km (16 miles) and is the most powerful telescope for ➤ *radio astronomy* at meter wavelengths.

giant molecular cloud ➤ *molecular cloud*.

giant planet Jupiter, Saturn, Uranus or Neptune, or an ➤ *extrasolar planet* of similar size or larger.

giant star A star between 10 and 1000 times more luminous than the Sun, and between 10 and 100 times larger.

Stars that are not giants to begin with become giants when they run out of hydrogen fuel for nuclear fusion in their cores. Their outer layers expand greatly and their surface temperature drops, but their overall luminosity rises because they are so much larger. Massive hot stars, which are much larger than the Sun even when they first form, are also referred to as giants.

➤➤ *Hertzsprung–Russell diagram, red giant, stellar evolution*.

gibbous A ➤ *phase* of the Moon, or any other astronomical object, when it is between half and full.

Giotto A European Space Agency spacecraft that flew within 605 km (376 miles) of the nucleus of Comet ➤ *Halley* in March 1986. It was the first spacecraft to make a close encounter with a comet. Images, including close-ups of the nucleus, were returned but the camera was subsequently destroyed by impacts. In 1992, Giotto was reactivated for an encounter with Comet 26P/Grigg–Skjellerup.

ESA named the mission after the artist Giotto di Bondone, who is thought to have used the 1301 appearance of Halley's Comet as a model for the star of Bethlehem in his fresco, *The Adoration of the Magi*, which he painted in 1303 in the Scrovegni chapel in Padua.

GLAST ➤ *Gamma-ray Large Area Space Telescope.*

glitch A sudden change in the rotation rate of a ➤ *pulsar*. Glitches are thought to be caused by ➤ *starquakes*.

globular cluster A roughly spherical cluster of hundreds of thousands – or even millions – of stars. The globular clusters in our Galaxy contain some of its oldest stars and are distributed within the ➤ *galactic halo*. These old stars contain only small amounts of the elements heavier than helium because they formed from the original material of the Galaxy, before the interstellar medium had been enriched by heavier elements created inside stars. Globular clusters have also been identified in other galaxies.

The globular cluster M80 in the constellation Scorpius as seen by the Hubble Space Telescope. It is about 28 000 light years away.

Globules 5900 light years away in the constellation Centaurus imaged by the Hubble Space Telescope.

globule A small cloud of dark opaque gas. Globules often show up against a bright background of star clouds or a glowing nebula. Stars form in globules that become dense enough. The name of the Dutch–American astronomer, Bart Bok (1906–83), is associated with small globules, known as Bok globules, which may be only a few thousand astronomical units across.

gnomon A rod or plate mounted to form a shadow-stick, such as on a sundial. The altitude of the Sun can be calculated from the height of the rod and the length of its shadow. The direction of the shadow gives the ➤ *apparent solar time*.

Goddard Space Flight Center (GSFC) A large NASA establishment in Greenbelt, Maryland, 10 miles north-east of Washington, DC. The work done there includes astronomical research and the design, development and management of near-Earth orbiting spacecraft.

Gossamer ring The outermost of Jupiter's three known rings. ➤➤ *ring systems*.

Gould's Belt A formation of many of the brightest, most conspicuous stars, which appear to lie in a band around the sky, tilted at 16° to the plane of the ➤ *Milky Way*. The belt includes the bright stars of Orion and Taurus in the northern hemisphere and those of Lupus and Centaurus in the southern hemisphere. It was first noted in 1847 by Sir John ➤ *Herschel* and later studied by the American astronomer Benjamin A. Gould (1824–96). It is thought to be a spur branching off the nearest spiral arm (the Orion arm) of the ➤ *Galaxy*.

Gran Telescopio Canarias A Spanish 10-m telescope at the ➤ *Observatorio del Roque de los Muchachos* in the Canary Islands. Its mirror consists of 36 hexagonal parts (similar to the telescopes of the ➤ *Keck Observatory*) and it can be used for both optical and infrared observations. It began operation in 2005.

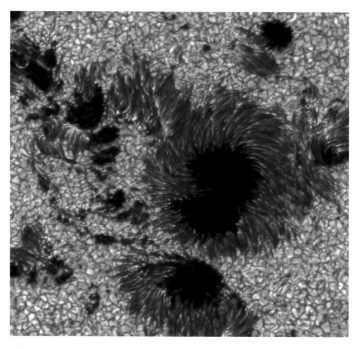

The granulation pattern on the Sun can be seen around these sunspots.

granulation A cell-like pattern observed in high-resolution images of the Sun's ➤ *photosphere*. It is caused by gas rising from deeper, hotter layers. Individual granules measure up to 1000 km (620 miles). ➤➤ *supergranulation*.

gravitation The attractive force between masses. According to the theory formulated by Isaac ➤ *Newton*, the force between two masses is proportional to their product divided by the square of the distance between them. In ➤ *general relativity*, gravitation is viewed as curvature of ➤ *spacetime*.

gravitational collapse The sudden collapse of a massive star when the internal pressure pushing outwards falls so that it cannot balance the weight of material pressing inward. The gravitational collapse of a massive star that has reached the end of its life is very sudden and catastrophic, perhaps taking less than a second. The enormous energy released triggers a ➤ *supernova* explosion, and the core of the collapsed star becomes a ➤ *neutron star* or ➤ *black hole*.

gravitational lens A massive object, such as a galaxy, that distorts the appearance of more distant galaxies behind it. A gravitational lens bends the path of light through space just as a glass lens bends light rays by refraction when they go through the lens. Gravitational lenses can produce multiple images of ➤ *quasars* and distort the appearance of a cluster of galaxies so that it looks like a pattern of arcs. Natural magnification by a gravitational lens sometimes

A cluster of galaxies (Abell 2218) 2 billion light years away acts as a gravitational lens, magnifying and distorting into arcs another galaxy cluster lying 5–10 times farther away but in the same line of site. This image was taken with the Hubble Space Telescope.

makes it possible to obtain detailed spectra of remote objects that would otherwise have been too faint.

    On a smaller scale, ➤ *microlensing* is sometimes seen when a small dark object, such as a planet, crosses our line of sight to a more distant star. ➤➤ *Einstein ring*.

gravitational redshift A ➤ *redshift* of the light from a massive object caused by the strong gravitational field.

gravitational waves Ripples in the structure of ➤ *spacetime* traveling at the speed of light. According to general relativity, gravitational radiation is emitted in certain extreme events, such as when the core of a star collapses and when matter falls into a black hole. It is very difficult to detect and up to 2005 had not been detected though astronomers believed they were close to doing so.

gravity ➤ *gravitation*.

gravity assist Using the gravity of a planet to change the speed and direction of a spacecraft without consuming any fuel.

grazing occultation An ➤ *occultation* in which the two objects involved appear just to touch.

Great Attractor A concentration of galaxies, containing perhaps 5×10^{16} solar masses of matter, that lies roughly 150–350 million light years from our Galaxy in the direction of the constellations Hydra and Centaurus. Its gravity is thought to be causing the motion of galaxies to deviate in its direction.

Great Dark Spot An oval feature on the planet ➤ *Neptune*, discovered on the images returned by the ➤ *Voyager 2* spacecraft in 1989. It was a storm system in

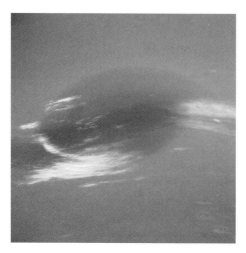

Neptune's Great Dark Spot as observed by the Voyager 2 spacecraft.

Neptune's cloud layers, similar to the ➤ *Great Red Spot* on Jupiter, but not so long lived. It had disappeared in 1994 when the Hubble Space Telescope began to take high-resolution pictures of Neptune. The spot's largest dimension was about the same as Earth's diameter (approximately 12 000 km or 7500 miles), making it about half the size of the Great Red Spot.

Great Nebula in Orion ➤ *Orion Nebula*.

Great Red Spot A large, red, oval spot on ➤ *Jupiter*, 24 000 km (15 000 miles) long and 11 000 km (7000 miles) wide. It was first reported by Robert Hooke in 1664 and has been observed ever since, through it has varied in size and color. The reason for its existence remains uncertain but it rotates like a giant anticyclone, with a westerly wind on its northern edge and an easterly wind to the south.

Great Rift A dark streak in the ➤ *Milky Way* in the constellations Cygnus and Aquila. It is due to dust, which hides the light of more distant stars.

Green Bank The location in West Virginia, USA, of a radio astronomy observatory belonging to the ➤ *National Radio Astronomy Observatory*. The 92-m (300-foot) dish built in 1962 collapsed in 1988. Its successor, the 100-m (325-foot) Robert C. Byrd Green Bank Telescope, was completed in 2000. It is the largest fully steerable dish in the world. The 43-m (140-foot) dish at Green Bank, completed in 1965, is the largest telescope in the world on an ➤ *equatorial mount*. There is also a radio interferometer on the site, consisting of three 26-m (85-foot) dishes, two of which can be moved along a track 1.6 km (1 mile) long.

green flash A phenomenon sometimes observed at the moment of sunset over a clear horizon, especially over the sea. Refraction by the Earth's atmosphere makes the last fragment of the Sun to sink below the horizon appear to break free and flash momentarily green before disappearing.

Changes to Jupiter's Great Red Spot observed by the Hubble Space Telescope between 1992 and 1999.

greenhouse effect Heating in the atmosphere of a planet because infrared radiation cannot escape. A greenhouse works in a very similar way, with glass playing the same role as the atmosphere.

The primary source of heat for a planet's surface and atmosphere is energy radiated from the Sun in the visible and near-infrared parts of the spectrum. However, longer-wavelength infrared radiation emitted by the warm planet cannot travel back out into space. It is trapped in the atmosphere, causing the planet's temperature to be higher than it would otherwise be. On Earth, the increase in temperature amounts to about 33 K. On Venus, a "runaway" greenhouse effect raises the temperature by 500 K. Mars is warmed by a modest 5 K.

The amount of heating the greenhouse effect causes depends on the composition of the atmosphere. Carbon dioxide is one of the main greenhouse gases, but water vapor and rarer gases also play a part.

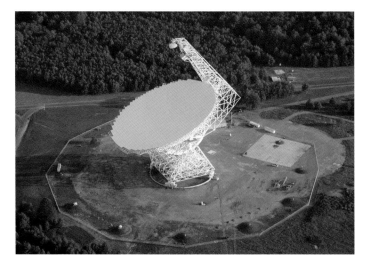

The Green Bank Telescope.

Greenwich Observatory ➤ *Royal Observatory, Greenwich.*

Gregorian calendar The civil calendar now in use in most countries It was introduced by Pope Gregory XIII in 1582 to replace the ➤ *Julian calendar.*

A workable civil calendar needs to be organized so that the seasons remain in step with the months of the year. This is a problem because the time taken by the Earth to orbit the Sun is not a whole number of days. The introduction of an extra day every fourth year improves matters, but more adjustments are necessary if the calendar is to stay synchronized with the seasons over centuries.

In the Gregorian system, years exactly divisible by four are leap years, except that century years must be exactly divisible by 400 to be leap years. Thus 2000 was a leap year, but 1900 was not. Averaged over 400 years, the rule gives an average year length of 365.2425, which is close to the true length of the ➤ *year*, 365.2422 days.

The Gregorian calendar came into effect in Roman Catholic countries in October 1582, when the seasons were brought back into step by eliminating 10 days from the calendar. Thursday October 4 was followed by Friday October 15. Also, on the introduction of the Gregorian system, the new year began on January 1 for the first time, instead of March 25. Britain and its colonies did not introduce the Gregorian calendar until September 1752, by which time an 11-day correction was needed.

Gregorian telescope A type of reflecting telescope proposed by James Gregory (1638–75) in 1663. The primary mirror is a paraboloid with a central hole and the secondary mirror is also curved. Gregory was unable to obtain mirrors shaped accurately enough to construct his telescope before Isaac ➤ *Newton* made the first working reflector to a simpler design with a flat secondary.

Grimaldi A large lunar crater, 222 km (138 miles) in diameter, situated near the western limb of the Moon on the border of the Oceanus Procellarum.

Grus (The Crane) A small southern constellation, introduced in the sixteenth century and included by Johann Bayer in his 1603 atlas ➤ *Uranometria*. It contains four stars brighter than fourth magnitude. Delta Gruis is a double star that can be seen as double by the unaided eye.

G star A star of ➤ *spectral type* G. G stars have temperatures in the range 4900–6000 K and are yellow in color. Their spectra contain many absorption lines. The Sun is a typical dwarf G star. ➤ *Capella* is an example of a giant G star.

Guardians The two stars Beta and Gamma in the constellation ➤ *Ursa Minor*.

guest star Used by Chinese astronomers in historical times to denote the appearance of a ➤ *nova*, ➤ *supernova* or ➤ *comet*.

guide star A star on which the manual or automatic guidance system of a telescope can be locked so that a fainter object being observed is correctly followed as the Earth rotates.

Gum Nebula A large, circular ➤ *emission nebula* in the southern constellations Vela and Puppis. It was discovered by an Australian astronomer, Colin Gum (1924–60). The nebula stretches across 36° of sky. Its diameter is 800 light years and it is 1300 light years away. It is thought to be the result of a ➤ *supernova* explosion about a million years ago.

H

Hadar (Beta Centauri; Agena) The second-brightest star in the constellation Centaurus. It is a giant ➤ *B star* of magnitude 0.6 and is 335 light years away.

Hadley Rille A sinuous channel on the Moon, running across Palus Putredinis. It is close to the landing site of the Apollo 15 mission and is believed to be a collapsed lava tube. ➤ *Apollo program*, *rille*.

Hale–Bopp, Comet One of the brighter comets of the twentieth century. Discovered by Alan Hale and Thomas Bopp on July 22, 1995, it reached perihelion on April 1, 1997 and at its brightest was magnitude –1. Its nucleus was estimated to be 40 km (25 miles) across, over twice the diameter of Comet ➤ *Halley*.

Hale, George Ellery (1868–1938) Hale is a key figure in twentieth century astronomy who foresaw that the development of astronomy required much larger telescopes. Born in Chicago, the son of a wealthy engineer, Hale took an early interest in instrument design. While an undergraduate at the Massachusetts Institute of Technology, he invented the ➤ *spectroheliograph*.

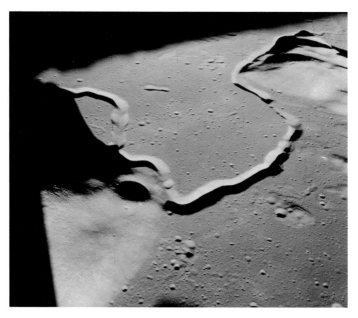

Hadley Rille photographed from orbit by the crew of Apollo 15.

Comet Hale–Bopp imaged in June 1999 by the European Southern Observatory's 3.5-m New Technology Telescope.

In 1892 he joined the astrophysics staff of the University of Chicago, where he established the ➤ *Yerkes Observatory* and its 40-inch refractor. To improve observational astronomy, he founded the ➤ *Mount Wilson Observatory* and equipped it with 60-inch and 100-inch reflectors. He resigned as director in 1923 and devoted the remainder of his life to raising funds for and planning the construction of the Palomar 200-inch telescope, now known as the ➤ *Hale Telescope*.

Hale Telescope The 5-m (200-inch) reflecting telescope at the ➤ *Palomar Observatory*. Work to construct the telescope began in the 1930s following the award of a grant to the California Institute of Technology from the Rockefeller Foundation. Completion was delayed by World War II. It was officially opened in 1948 and dedicated to the memory of George Ellery ➤ *Hale* (1868–1938) who had been the driving force behind the project.

Hall, Asaph (1829–1907) Hall was a self-taught America astronomer from Connecticut. He became an assistant at Harvard College Observatory in 1857 and moved to the US Naval Observatory in 1863. At the 1877 ➤ *opposition* of Mars he discovered and named ➤ *Phobos* and ➤ *Deimos*, the two moons of Mars.

Halley, Comet The most famous of all periodic comets. It travels in an elongated elliptical orbit around the Sun, returning to the inner solar system every 76 years. Historical records show that Comet Halley has been observed for over 2200 years.

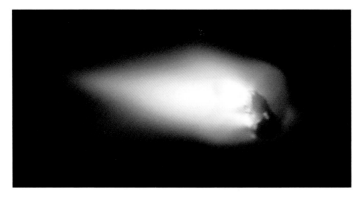

The nucleus of Comet Halley as seen by the Giotto spacecraft.

Edmond ➤ *Halley*, in whose honor the comet is named, did not discover it but he was the first person to realize the connection between the comet he saw in 1682 and certain other recorded appearances of comets separated by intervals of 76 years. He calculated the orbits of a number of comets, using Isaac Newton's newly published theory of gravitation. Noticing the similarity between the orbits of the comets seen in 1531, 1607 and 1682, he deduced that they were apparitions of one and the same comet and went on to predict that it would return in 1758–59. The comet duly appeared, as he had predicted.

Comet Halley's ➤ *perihelion* is 0.59 AU from the Sun, between the orbits of Mercury and Venus. At its most distant, the comet travels beyond the orbit of Neptune. Its orbit is inclined to the main plane of the solar system at an angle of 162° and it travels around its orbit in the direction opposite to the motion of the planets.

On its return in 1986, Comet Halley was never located where it could be observed well from Earth but several countries launched spacecraft to investigate it, with considerable success. The closest approach was made by the European craft ➤ *Giotto*.

The results showed that the comet has a solid nucleus made of ice and dust. Its shape is irregular and elongated and it measures 16 × 8 km (9 × 5 miles). It is very dark, reflecting only 4 percent of the sunlight falling upon it. The nucleus rotates slowly – once in 7.1 days. On the side facing the Sun, temperatures as high as 350 K were measured, enough to melt ice, and jets of escaping material were observed.

Two meteor showers, the ➤ *Eta Aquarids* and the ➤ *Orionids*, are associated with Comet Halley.

Halley, Edmond (1656–1742) Halley made enormous contributions to almost every branch of physics and astronomy. Born in London, he was the son of a wealthy merchant. He studied at Oxford, where he demonstrated outstanding

Edmond Halley

mathematical talent. His scientific career began in 1676 with two years on the island of St Helena, mapping the southern skies. His catalog of southern stars earned him election to the Royal Society.

Halley became interested in gravity, and particularly in a mathematical proof of ➤ *Kepler's laws*. When he discussed this with Isaac ➤ *Newton* in 1684, he found that Newton had solved the problem but had not published it. Halley personally financed the publication of Newton's *Principia Mathematica* (1687). His knowledge of geometry and historical astronomy enabled him to deduce that comets recorded in 1456, 1531, 1607, and 1682 were in fact one and the same comet. He correctly predicted that it would return in 1758. In 1720 he succeeded ➤ *Flamsteed* as ➤ *Astronomer Royal* and devoted himself to lunar observations.

The double star cluster h and chi Persei.

Hamal (Alpha Arietis) The brightest star in the constellation Aries. It is a giant
➤ *K star* of magnitude 2.0 and is 66 light years away. Its name comes from the
Arabic for "sheep."

h and chi (χ) Persei (Double Cluster in Perseus; NGC 869 and 884) A pair of open
star clusters in the constellation Perseus. They are visible to the naked eye as
faint hazy patches. Names of this kind were normally allocated to individual
stars, and were given to these clusters before their true nature was known. The
two clusters are very similar in appearance and are less than one degree apart
in the sky. They are 7100 light years away and estimated to be only 50 light
years apart. ➤➤ *open cluster*.

Harpalyke A small outer moon of Jupiter discovered in 2000. Its diameter is about
4 km (2.5 miles).

Harvard College Observatory The observatory of Harvard College, established in
1839. In 1847, it was equipped with a 0.38-m (15-inch) refracting telescope,
which is still in its original building, the Sears Tower on Observatory Hill. In
1973, the Harvard–Smithsonian Center for Astrophysics was formed by merging
the ➤ *Smithsonian Astrophysical Observatory* and Harvard College Observatory.

Harvard–Smithsonian Center for Astrophysics ➤ *Harvard College Observatory*.

harvest Moon The full Moon nearest the time of the autumnal ➤ *equinox*. At this
time of the year, the Moon's path is not inclined very steeply to the horizon and
the Moon rises at approximately the same time each evening for a short period.

Hayabusa A Japanese mission to explore the asteroid 25143 Itokawa and bring a sample of material from it back to Earth. Launched in May 2003, the spacecraft reached its target in November 2005 but a small lander called Minerva that should have investigated the surface was lost after being released from the main spacecraft. However, Hayabusa did collect samples despite some technical difficulties and they are due to be returned to Earth by June 2010.

6 Hebe The sixth asteroid in order of discovery. Its diameter is 204 km (127 miles) and it was first seen in 1847 by Karl L. Hencke.

Hegemone A small outer moon of Jupiter discovered in 2003. Its diameter is about 3 km (2 miles).

Heinrich Hertz Submillimeter Telescope A 10-m telescope at the ➤ *Mount Graham International Observatory*. It operates in the submillimeter waveband, between 0.3 and 1 mm. The first observations were made with it in 1994. It is operated jointly by the University of Arizona and the Max-Planck-Institut für Radioastronomie in Bonn, Germany.

624 Hektor The largest ➤ *Trojan asteroid* in the same orbit as Jupiter. It was discovered by August Kopff in 1907. It rotates in just under 7 hours and its brightness varies by a factor of 3 as it does so. The way its light varies indicates that Hektor is roughly cylindrical in shape, measuring 195 km (121 miles) wide by 370 km (230 miles) long. It has been suggested that Hektor may in fact be two asteroids in contact or a close binary.

Helene A small, irregularly shaped satellite of Saturn discovered in 1980. It measures about $36 \times 32 \times 30$ km ($22 \times 19 \times 19$ miles) and orbits at a distance of 377 400 km (234 505 miles).

heliacal rising The rising of a bright star just before sunrise. Since stars rise about 4 minutes earlier each day, a star's heliacal rising marks the start of a period of several months during which it can be seen for at least part of the night. The heliacal rising of Sirius signaled to the ancient Egyptians the season when the River Nile flooded.

Helike A small outer moon of Jupiter discovered in 2003. Its diameter is about 4 km (2.5 miles).

heliocentric model The concept of the solar system with the Sun at the center and the planets in orbit around it. It was suggested as early as *c.* 200 BC by ➤ *Aristarchus of Samos* but the idea that Earth is moving was philosophically unacceptable until centuries later. An Earth-centered system, refined by ➤ *Ptolemy* (*c.* AD 100–170), was in general use until the work of ➤ *Copernicus* (1473–1543). Over this time, the idea that the Earth was the center of the created universe became strongly rooted in religious dogma.

In his book *De Revolutionibus*, Copernicus argued in favor of considering the solar system as Sun-centered. However, the idea did not gain general acceptance until ➤ *Galileo* (1564–1642) and ➤ *Kepler* (1571–1630) made observations that could only be explained sensibly by a heliocentric system.

Copernicus presumed that the planetary orbits are circular. Because of this, he was no more successful than Ptolemy at predicting planetary positions, though his theory was more elegant and provided a natural explanation for the ➤ *retrograde* motion of the planets. Kepler solved this problem for the heliocentric theory when he discovered that the planetary orbits are elliptical rather than circular.

helioseismology The study of natural oscillations propagating through the Sun. These oscillations are detected by the ➤ *Doppler shifts* they cause in the ➤ *absorption line* spectrum of the Sun and they reveal information about the Sun's interior structure.

heliosphere The spherical region of space around the Sun, extending out to between 50 and 100 AU. Its outer boundary is where the ➤ *solar wind* merges with the ➤ *interstellar medium*. This boundary region is called the heliopause.

heliostat A movable flat mirror used to reflect sunlight continuously into a fixed solar telescope. A heliostat follows the motion of the Sun across the sky but produces an image that slowly rotates during the course of a day.

Helix Galaxy A popular name for the ➤ *spiral galaxy* NGC 2685 in the constellation Leo.

Helix Nebula (NGC 7293) A large, ring-shaped ➤ *planetary nebula* in the constellation Aquarius. Its apparent diameter is a quarter of a degree (half the size of the full Moon). At a distance of about 500 light years, it is the nearest planetary nebula.

Hellas Planitia An almost circular impact basin on the surface of Mars. Hellas Planitia is 1800 km (1100 miles) in diameter and is conspicuous even in a small telescope because it is lighter in color than the areas surrounding it. It was formerly known simply as Hellas.

An ultraviolet image of the Helix Nebula from the Galex spacecraft.

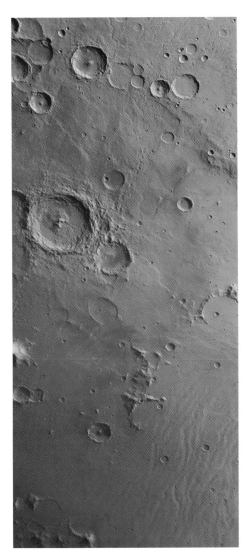

A view near the north-eastern edge of Hellas Planitia from Mars Express.

Henry Draper Catalog (HD Catalog) A catalog of stellar spectra, compiled at Harvard College Observatory. Its production was financed by a donation from the widow of Henry ➤ *Draper* (1837–82), a pioneering astrophysicist, and the catalog was named as a memorial to him. Under the direction of Edward C. Pickering (1846–1919), Annie Jump ➤ *Cannon* (1863–1941) classified the ➤ *spectral type* of most of the 225 300 stars in the nine-volume catalog between 1911 and 1915, though the first volume was not ready for publication until 1918 and the ninth did not appear until 1924.

The Herbig–Haro object HH 46/47, formed by a low-mass protostar ejecting a jet and creating a bipolar outflow. This infrared image is from the Spitzer Space Telescope.

Herbig–Haro object (HH object) A type of peculiar nebula associated with newly forming stars. The first three were discovered on images of the nebula NGC 1999 in Orion in 1946–47 by the American astronomer George Herbig and the Mexican Guillamero Haro. Many more similar objects have since been identified.

Herbig–Haro objects result when a powerful ➤ *bipolar outflow* from a newly forming star heats and compresses the surrounding interstellar gas. They are typically between 500 and 4000 AU in size but only 0.5–30 Earth masses, making them among the least massive objects detected outside the solar system. They move very quickly and, in many cases, their motion can be traced back to ➤ *T Tauri stars* or other young stars. ➤ *Hubble's Variable Nebula* is an example of an HH object.

Hercules A large constellation of the northern sky named after the hero of classical mythology. It contains no first magnitude stars but the brightest ➤ *globular cluster* in the northern hemisphere, M13, lies in Hercules.

Hercules A The strongest radio source in the constellation Hercules and the third most powerful in the northern hemisphere of the sky. It is also known as 3C 348 and is associated with a large, elliptical ➤ *cD galaxy* at the center of a cluster. Two long jets extend for half a million light years into space from the faint nucleus.

Hercules X-1 An X-ray ➤ *pulsar* in the constellation Hercules. The pulsar is a rotating neutron star in a binary star system that is drawing in gas from its companion. The rotation period of the neutron star is 1.2 seconds and the orbital period of the system 1.7 days.

Hermes An asteroid discovered by Karl Reinmuth in 1937 when it passed within 800 000 km (500 000 miles) of the Earth, in what was at the time the closest approach of an asteroid ever recorded. Hermes reached eighth magnitude and traveled across the sky at a rate of 5° per hour. However, after only five days, it was lost when its brightness fell to 23rd magnitude because its night side was facing Earth. It was recovered on October 15, 2003, by Brian Skiff of the Lowell Observatory.

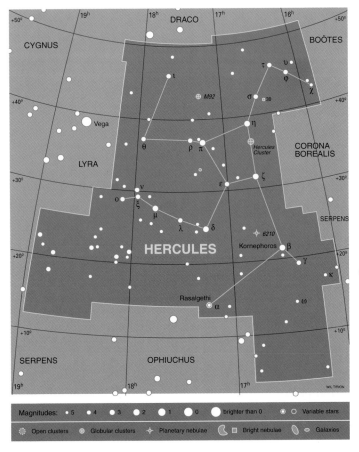

A map of the constellation Hercules.

Hermippe A small outer moon of Jupiter discovered in 2001. Its diameter is about 4 km (2.5 miles).

Herschel A spectacular impact crater on Saturn's moon ➤ Mimas. It is named in honor of William ➤ Herschel, who discovered Mimas. Herschel's diameter is 130 km (80 miles), one-third the diameter of Mimas. It is 10 km (6 miles) deep and has at its center a peak that rises 6 km (4 miles) above the crater floor. It is likely that the impact that created Herschel very nearly shattered Mimas, evidenced by fractures on the opposite side of the moon.

Herschel, Caroline Lucretia (1750–1848) Caroline was the sister of William ➤ Herschel. She was born in Hanover, but joined her brother in England in 1772. She became his able observing assistant, and then branched out into her own research, discovering eight comets and many nebulae. She prepared a catalog

151

of nebulae and star clusters for which she received the Gold Medal of the Royal Astronomical Society in 1828.

Herschel, Sir John Frederick William (1792–1871) John was the son of William ➤ *Herschel*. He studied mathematics at Cambridge University, and began to assist his father with observations from 1816. In 1834 he went to the observatory at the Cape of Good Hope to survey the southern skies where he discovered 2000 nebulae and 2000 double stars before returning to England in 1838. He later became interested in photography and pioneered its use in astronomy.

Herschel, Sir (Frederick) William (1738–1822) William Herschel, the discoverer of the planet Uranus, was the greatest observational astronomer of the eighteenth century. He was born in Hanover, then moved to England in 1757, where he worked initially as a musician in Bath. His interest in astronomy developed from 1773, when he started telescope making. His early observations came to the attention of King George III, a strong supporter of scientific enquiry, who employed him on a salary of £200 a year and financed the construction of large telescopes.

Herschel took full advantage of his royal pension, which enabled him to spend very long hours making astronomical observations. He discovered Uranus in 1781, which brought him worldwide fame, and two moons of Uranus in 1787. His sky surveys are notable for their precision and comprehensiveness. He cataloged 2000 nebulae and 800 double stars. Herschel made about 400 telescopes, culminating in the enormous 40-foot (12-m) reflector, with which he discovered two moons of Saturn, ➤ *Mimas* and ➤ *Enceladus*. His other scientific achievements include discovering infrared radiation, proving that some apparently double stars are true ➤ *binary stars*, demonstrating from ➤ *proper motion* studies that the solar system is moving through space. He also invented the term ➤ *asteroid*.

Herschel Space Observatory A European Space Agency 3.5-m space telescope for infrared and submillimeter-wave observations. It will detect radiation in the wavelength band from 100 μm to 1 mm. Launch is scheduled for May 2008 and it is expected to operate for about 3 years located about 1.5 million km (1 million miles) from Earth in the direction opposite to the Sun. Its mirror will be the largest ever deployed in space.

Hertzsprung, Ejnar (1873–1967) Hertzsprung discovered the main classes into which most stars can be grouped according to their overall luminosity: the bright and relatively scarce ➤ *giant stars* and ➤ *supergiants*, and the more common, fainter, "dwarfs." Born in Frederiksberg, Denmark, he trained as a photochemist, and used that experience to devise ways of determining the luminosity of stars. Unfortunately, he published his principal results, connecting the colors and luminosities of stars in obscure photographic

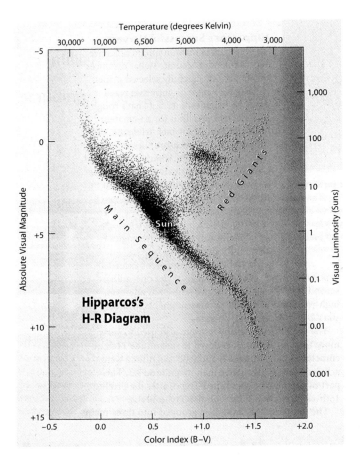

A Hertzsprung–Russell diagram made by plotting accurate data on 20 853 stars acquired by the ➤ *Hipparcos* satellite.

journals, so credit for the discovery initially went to Henry Norris ➤ *Russell*, who independently published essentially the same result in 1913. What later became known as the ➤ *Hertzsprung-Russell* diagram has been of immense importance in the study of stellar evolution.

Hertzsprung also made the first measurement of an extragalactic distance, using ➤ *Cepheid variable* stars to find the distance to the Small ➤ *Magellanic Cloud*.

Hertzsprung–Russell diagram (HR diagram) A graph of the relationship between the ➤ *spectral types* and luminosities of a selection of stars. Spectral type goes along the horizontal axis, with the hottest stars at the left. Luminosity is plotted along the vertical axis. Color, temperature or some other quantity related to spectral type is sometimes plotted along the

153

horizontal axis instead. Either ➤ *magnitude* or luminosity relative to the Sun are frequently used for the vertical scale. The plot may be called a color–magnitude diagram or color–luminosity diagram, depending on the actual quantities used.

What is now known as a Hertzsprung–Russell diagram was first plotted by Henry Norris ➤ *Russell* in 1913. It was later recognized that Ejnar ➤ *Hertzsprung* had independently put forward similar ideas at around the same time.

Any star whose spectral type and luminosity are known may be plotted as a single point on the HR diagram, but the diagram acquires particular significance when the data for a related group of stars, such as a star cluster, are plotted on it. Whatever sample of stars is chosen, the points are not distributed randomly. The points corresponding to most stars lie on a band running diagonally from the upper left to the lower right, which is known as the main sequence. The main sequence is essentially a mass sequence, with the most massive stars at the upper left and the least massive at the lower right.

Most stars spend 90 percent of their lives on the main sequence. Main-sequence stars generate their nuclear energy by converting hydrogen to helium in their cores. The effects of advancing age move stars away from the main sequence. Evolved stars are found in the giant and supergiant branches lying above the main sequence. The highly evolved ➤ *white dwarfs* form a group well below the main sequence. ➤➤ *stellar evolution.*

Hevelius, Johannes (1611–1687) Hevelius is remembered chiefly for his study of the Moon and the engravings he made of its features, and for the long unwieldy "aerial" telescopes of enormous focal length he constructed to make his observations. The son of a wealthy brewer, he was born in Danzig (now Gdansk) in Poland, where he established his observatory after studying in Leiden in the Netherlands. For his work in positional astronomy, which led to a star catalog (1690), he used a ➤ *quadrant* rather than a telescope. He was the last astronomer to do major observational work without a telescope.

944 Hidalgo An asteroid discovered in 1920 by Walter ➤ *Baade*. It follows a highly elliptical orbit that is inclined at 42° to the plane of the solar system and lies between 2 and 9.7 AU from the Sun, extending from the main ➤ *asteroid belt* to beyond the orbit of Saturn. Its unique orbit among asteroids has led some astronomers to speculate that Hidalgo may be a "dead" comet nucleus. It is estimated to be between 40 and 60 km (25 and 38 miles) across.

high-velocity clouds Clouds of hydrogen gas in and around our Galaxy moving at exceptionally high speeds. Several hundred have been located within the ➤ *Local Group* of galaxies. Each contains the same amount of mass as a small galaxy and they are typically 50 000 light years across. They appear to be composed of primordial material left over from the formation of the Local Group. Similar clouds have been observed in other galaxy groups.

high-velocity star A star traveling exceptionally fast (at more than about 65 km/s) relative to the Sun. High-velocity stars are very old stars that do not share the motion of the Sun and most stars in the solar neighborhood, which are in circular orbits around the center of the ➤ *Galaxy*. Rather, they travel in elliptical orbits, which often take them well outside the plane of the Galaxy. Although their orbital velocities in the Galaxy may be no faster than the Sun's, their different paths mean that they have high velocities relative to the Sun.

Hilda asteroids A group of asteroids at the outer edge of the main ➤ *asteroid belt*, 4.0 AU from the Sun. They are named after 153 Hilda, an asteroid about 180 km (110 miles) across, which was discovered by Johann Palisa in 1875. The ratio of their orbital periods to that of Jupiter is 3:2, and they are separated from the rest of the asteroid belt by a ➤ *Kirkwood gap*.

Himalia A moon of Jupiter discovered in 1904 by Charles Perrine. Its diameter is 170 km (106 miles). With Elara, Leda, and Lysithea it belongs to a family of four moons with closely spaced orbits. Their average distances from Jupiter are between 11.1 and 11.7 million km (6.9 and 7.3 million miles).

Hind's Nebula (NGC 1554/5) A variable ➤ *reflection nebula* surrounding the star T Tauri. ➤➤ *T Tauri star*.

Hinode A Japanese astronomy satellite, operated in collaboration with the USA and the UK, to study X-rays and extreme ultraviolet emissions from the Sun. It was launched in September 2006 and was a follow-up to the earlier ➤ *Yohkoh* mission. Hinode is Japanese for sunrise.

Hipparchus (*c.*170 BC – *c.*120 BC) Hipparchus, a Greek geometer and astronomer, worked in Rhodes and Alexandria. None of his works have survived, but we know of them through ➤ *Ptolemy*. After seeing a nova in 134 BC, Hipparchus constructed a catalog of the positions of 850 stars. By comparing his values with those in a catalog made 150 years earlier, Hipparchus discovered the ➤ *precession* of the equinoxes. He also established the basis of the ➤ *magnitude* system still in use today to describe the brightness of astronomical objects.

Hipparcos A European Space Agency orbiting telescope that surveyed the positions and brightnesses of over a million stars with unprecedented accuracy. It was launched in 1989 and observing ended on August 15, 1993. The name "Hipparcos" was an acronym for High Precision Parallax Collecting Satellite, chosen for its similarity to the name of the Greek astronomer ➤ *Hipparchus* (also spelt Hipparchos).

Hirayama family A group of ➤ *asteroids* with similar orbits located near each other in space. The existence of such groups was first noted by Kiyotsugo Hirayama in 1918. More than a hundred have been identified. In many cases the members of the group are asteroids of similar or related types, strongly suggesting that they were formed when a single parent body broke up. About half of all asteroids are thought to belong to Hirayama families.

Hoba meteorite The largest known ➤ *meteorite* in the world. It is of the iron type and weighs about 55 000 kilograms (60 tons). It remains at the place in Namibia, where it was discovered in 1928. A layer of rusty weathered material surrounds the meteorite. Taking this into account, the original meteorite must have weighed more than 73 000 kilograms (80 tons).

Hobby–Eberly Telescope (HET) A large telescope at the ➤ *McDonald Observatory* in Texas, designed specifically for ➤ *spectroscopy*. It became fully operational in 1999. Its 11-meter segmented mirror is permanently tipped at a 35° angle to the zenith and is mounted on a structure that can turn to point in any direction. The telescope tracks its targets with a movable secondary mirror. Though the tilt of the main mirror is fixed, the telescope is able to observe objects in about 70 percent of the sky at the site. ➤➤ *Southern African Large Telescope*.

Homunculus Nebula A small nebula surrounding the star ➤ *Eta Carinae*.

Hooker telescope The 2.54-m (100-inch) reflecting telescope at the ➤ *Mount Wilson Observatory*, near Pasadena in California. It was completed in 1917, having been financed by a gift from John D. Hooker. Until the opening of the 5-m (200-inch) ➤ *Hale Telescope* in 1948, it was the largest telescope in the world. It was temporarily closed in 1985 but subsequently renovated and brought back into use in the early 1990s.

horizon The boundary between the visible and hidden halves of the celestial sphere from the point of view of a ground-based observer. Horizon is also used more generally for any boundary between events that can in principle be observed and those that cannot. ➤➤ *event horizon*.

horizontal branch On a ➤ *Hertzsprung–Russell diagram*, the part of the diagram where low-mass stars that have lost material during the giant phase of their evolution are located.

horizontal coordinates A coordinate system in which points on the celestial sphere are identified by their altitude and azimuth. Altitude is angular distance above the horizon, and azimuth is angular distance around the horizon, measured eastwards from north. The altitude and azimuth of a celestial object vary according to the latitude and longitude of the observer and the time of observation.

Horologium (The Clock) A faint and inconspicuous constellation of the southern sky introduced by Nicolas L. de Lacaille in the mid-eighteenth century.

Horsehead Nebula (NGC 2024) A dark dust nebula in the constellation Orion, shaped similar to a horse's head. It is conspicuous because it protrudes into the bright emission nebula IC 434. It is about 1500 light years away.

Horseshoe Nebula An alternative name for the ➤ *Omega Nebula*.

The Horsehead Nebula, imaged by the 8.2-m KUEYEN Telescope of the European Southern Observatory.

Fred Hoyle.

hour angle (HA) For a celestial object, the ➤ *sidereal time* that has elapsed since it last crossed the meridian.

Hourglass Nebula A bright, luminous nebula within M8, the ➤ *Lagoon Nebula*. It was first noted by John ➤ *Herschel*, and its name describes its shape.

Hoyle, Sir Fred (1915–2001) Hoyle was one of the greatest theoretical astronomers of the twenteith century. He was born in West Yorkshire, to parents who strongly encouraged his interest in science. At Cambridge University, he graduated as the top applied mathematician. His postgraduate research was in nuclear physics but, in 1939, his interest changed to astrophysics, at which point his career was interrupted by five years of war service as a theorist working on radar. This secret research took him to the USA in 1944 where he visited Henry Norris ➤ *Russell* and Walter ➤ *Baade*, both of whom stimulated his interest in stellar evolution.

On his return to Cambridge in 1945, he applied his knowledge of nuclear physics to the origin of the chemical elements, showing how nuclear reactions in giant stars could create the elements from carbon to iron. In 1957, together with E. Margaret Burbidge, Geoffrey Burbidge, and William Fowler, he accounted for almost all the chemical elements and their isotopes by ➤ *nucleosynthesis* in stars. This was his most important and lasting contribution to astrophysics.

Famously, he proposed a ➤ *steady-state theory* as an alternative to ➤ *Big Bang* cosmology in 1948. The discovery of the ➤ *cosmic background radiation* in 1963 confirmed the Big Bang, much to his dismay.

HR diagram (H–R diagram) Abbreviation for ➤ *Hertzsprung–Russell diagram*.

HST Abbreviation for ➤ *Hubble Space Telescope*.

Hubble, Edwin Powell (1889-1953) Hubble was an American observational astronomer who discovered the expansion of the universe. He was born in Missouri and studied at the University of Chicago, where he was influenced by George Ellery ➤ *Hale*. Later, in 1919, he joined the staff of the Mount Wilson Observatory, where Hale was the Director. His early work demonstrated that spiral ''nebulae'' (what would now be called galaxies), such as the ➤ *Andromeda Galaxy*, lay far beyond our ➤ *Galaxy* (the Milky Way).

Using the 100-inch telescope at Mount Wilson, he determined the distances to 18 galaxies, an enormous achievement at that time. He found that the galaxies were all receding and that recession velocities increased in proportion with their distance. This work confirmed that the idea of an expanding universe, already proposed by theorists, was indeed correct.

Hubble also devised a scheme for classifying galaxies according to their shape, known as the ➤ *Hubble classification*, which is still used today.

Hubble classification A method devised in 1926 by ➤ *Edwin Hubble* for classifying ➤ *galaxies* according to their shape. The scheme sorts elliptical galaxies on a

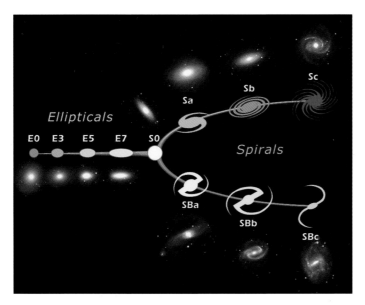

The Hubble classification scheme for galaxies.

scale from E0 for a circular disk, through E1, E2, and so on, to E7 in order of increasing elongation. Spirals are designated as Sa, Sb or Sc in order of increasing openness of the arms and decreasing size of the nuclear bulge in relation to the overall size of the galaxy. There is a parallel sequence for ➤ *barred spiral galaxies*, which are designated SBa, SBb or SBc. Galaxies that are neither elliptical nor spiral in form are designated Ir for "irregular." Hubble suggested in 1936 that ➤ *lenticular galaxies*, designated S0, provide a "missing link" in an evolutionary chain from E0 through to the open spirals Sc and SBc. This progression is no longer accepted as an evolutionary sequence, but Hubble's classification continues to be widely used as a simple way of describing the shapes of galaxies.

Hubble constant ➤ *Hubble parameter*.

Hubble diagram A graph on which the redshifts or recession velocities of galaxies are plotted against distance. The original diagram, plotted in 1929, gave the first strong evidence that the universe is expanding. It continues to be a crucial tool for cosmologists because any deviation from a straight line reveals information about the geometry of the universe and its rate of expansion in the past. ➤ *Hubble's law, Hubble parameter, expanding universe.*

Hubble parameter (symbol H) The ratio of a galaxy's velocity to its distance. It is related to the rate of expansion of the universe. Most theories of cosmology assume that its value changes over time as the universe evolves. The symbol H_0 is used for its present value, which is often called the Hubble "constant." Its

The Hubble Space Telescope after the release from Columbia's robot arm at the close of a successful servicing mission in March of 2002.

value was difficult to determine, because the distances of galaxies could not be measured accurately enough. However, two independent methods have given similar values with relatively small margins of error. A long-term project carried out using the ➤ *Hubble Space Telescope*, known as the Hubble Key Project, came up with the value 72 ± 8 km/s per megaparsec in 2001. The ➤ *Wilkinson Microwave Anisotropy Probe* (WMAP) results yielded 71 ± 4 km/s per megaparsec in 2003.

Hubble's law The observation that distant galaxies are receding at speeds that are proportional to their distance from us. This relationship was discovered by ➤ *Edwin Hubble* and was first announced by him in 1929. It is a consequence of the expansion of the universe. ➤➤ *Hubble parameter.*

Hubble Space Telescope (HST) An orbiting observatory built and operated jointly by NASA and ESA, and named in honor of Edwin ➤ *Hubble*. It has been used to make observations of virtually every kind of celestial object, from planets in the solar system to the most remote galaxies detectable. The science operations are conducted from the Space Telescope Science Institute (STScI) in Baltimore, Maryland.

The idea behind the HST was to put a telescope above the atmosphere, which degrades the quality of the images in ground-based telescopes. It was designed with a 15-year lifetime as an observatory that could be maintained and upgraded in orbit using the ➤ *Space Shuttle*. It can make observations in the ultraviolet, in visible light and in the near-infrared.

The HST was placed in orbit on April 25, 1990, by the Space Shuttle. However, a fault in the optical figuring of the 2.4-m (94.5-inch) main mirror meant that sharp focusing was impossible until the fault could be corrected. During the first servicing mission in December 1993, a Space Shuttle crew successfully installed a unit to correct the faulty optics. They also replaced the solar arrays. Since 1993, there have been two further servicing missions to replace instruments and worn-out parts. The second was in February 1997 and the third was partially carried out in December 1999 but the mission was cut short and the remainder of the work carried out in 2001.

Following an accident in 2003 in which a Space Shuttle was destroyed, Space Shuttle missions were suspended until 2005. NASA took the view that further maintenance of the HST could be too risky to undertake and the future of the HST beyond 2008 became uncertain. However, in 2006, NASA announced plans for a servicing mission in 2008. If successful, the HST's lifetime would be extended to 2013.

Hubble's Variable Nebula (NGC 2261) A luminous triangular-shaped nebula in the constellation Monoceros. From photographs taken between 1900 and 1916, Edwin ➤ *Hubble* discovered that the nebula varies in shape and brightness. An irregularly variable star, R Monocerotis, is embedded in the nebula. R Mon is a strong source of infrared radiation and is probably a very young star, surrounded by a circumstellar disk and ejecting a ➤ *bipolar outflow*. The nebula is now thought to be an example of a ➤ *Herbig–Haro object*.

Huggins, Sir William (1824–1910) Huggins was a British astrophysicist who was one of the principal founders of astronomical spectroscopy. He was the first person to make intensive investigations of stellar spectra. In 1863 he showed that the stars are composed of the same chemical elements as the Sun. His next great discovery was to show that ➤ *emission nebulae* are glowing clouds of gas. He was the first, in 1868, to measure a ➤ *Doppler shift*, which he did in the spectrum of the star Sirius. He subsequently determined the velocities of many stars from their spectra. Huggins' wife Margaret was his partner in research and she made significant contributions to their joint work.

Hungaria group A group of asteroids at the inner edge of the asteroid belt, 1.95 AU from the Sun, with orbits inclined at 24° to the plane of the solar system. The group is separated from the main belt by a ➤ *Kirkwood gap*, and is not a true family with a common origin. The group takes its name from 434 Hungaria, a small asteroid discovered by Max Wolf in 1898.

Huygens, Christiaan (1629–1695) Huygens was a Dutch physicist and astronomer, whose many contributions to science include the wave theory of light and the discovery of Saturn's largest moon, ➤ *Titan*. The importance of his work in dynamics and optics was second only to that of Isaac ➤ *Newton* in the seventeenth century.

Born in the Hague, and educated at the University of Leiden, he eventually settle back in the Hague. Working with his brother Constantin, he made significant improvements in the design of telescope optics. He used the telescopes he built to observe Saturn and found Titan in 1655. He was the first to give the correct description of Saturn's rings. In 1657 he invented the pendulum clock.

Huygens probe ➤ *Cassini–Huygens mission.*

Hyades An open star cluster in the constellation Taurus. Its members appear to be scattered over an area 8° in diameter around the star Aldebaran. However, Aldebaran is closer than the Hyades and does not belong to the cluster. The Hyades is the nearest star cluster, lying at a distance of about 150 light years. ➤➤ *open cluster.*

Hyakutake, Comet A bright comet that reached magnitude zero in March 1996 and developed a tail stretching at least 7° across the sky. It was bright mainly because it happened to come relatively close to Earth, passing within 15 million km (10 million miles).

Hydra (1) (Sea Monster) The largest constellation in the sky by area, but a difficult one to identify since it contains only one moderately bright star, the second-magnitude Alphard.

Hyperion imaged by the Cassini spacecraft.

Hydra (2) A small moon of Pluto discovered in 2005. It is estimated to be about 45–60 km (30–37 miles) across.

Hydra A The brightest radio source in the constellation Hydra. It is associated with a large elliptical galaxy at the center of a small cluster of galaxies about one billion light years away.

Hydrus (The Water Snake) An inconspicuous southern constellation, introduced by Johann Bayer in his 1603 star atlas. Its three brightest stars are third magnitude.

10 Hygeia The fourth largest asteroid, discovered by Anatole de Gasparis in 1849. Its diameter is 430 km (267 miles).

Hyperion A satellite of Saturn, discovered in 1848 by William C. Bond and William Lassell. It is an elongated irregular body measuring approximately $360 \times 280 \times 225$ km ($223 \times 174 \times 140$ miles). Its orbit is 1481 100 km (920 312 miles) from Saturn. There are large craters and a curving scarp-like feature, apparently 300 km (200 miles) long. The evidence suggests that it may be a remnant of a larger body shattered by an impact.

hypernova A cosmic explosion that releases a hundred times more energy than a ➤ *supernova*. A hypernova may be the result of the implosion of a massive, rapidly rotating stellar core, resulting in the formation of a ➤ *black hole*. The small number of hypernova remnants identified are exceptionally strong sources of X-rays. It is thought that hypernovae may be linked to ➤ *gamma-ray bursts*.

I

Iapetus The third largest moon of Saturn, discovered by ➤ *Giovanni Cassini* in 1671. It orbits Saturn at a distance of 3 561 300 km (2 212 888 miles) and measures 1436 km (892 miles) across. With a density only 1.1 times that of water, Iapetus must be mostly ice. The ➤ *Voyager* spacecraft confirmed a hypothesis, proposed by Cassini after he noticed variations in Iapetus's brightness, that one hemisphere is very much darker than the other. The satellite always keeps the same face towards Saturn and its dark and light hemispheres face Earth alternately as it travels around its orbit. The surface is cratered and the ➤ *Cassini* spacecraft showed a remarkable ridge of mountains up to 20 km (12 miles) high running almost parallel to the equator for about 1300 km (800 miles). The dark area is blanketed with a material as black as tar. Its nature and origin are unknown but Cassini images give the impression that the dark coating has fallen onto the surface.

IAU Abbreviation for ➤ *International Astronomical Union*.

IC Abbreviation for ➤ *Index Catalogue*.

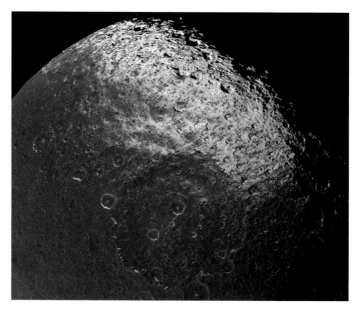

A Cassini spacecraft image of Iapetus.

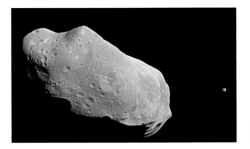

A Galileo spacecraft image of asteroid Ida with its moon Dactyl.

1566 Icarus A small asteroid, diameter 1.4 km (0.9 mile), discovered in 1949 by Walter ➤ *Baade*. It is a member of the ➤ *Apollo* group and its highly elliptical orbit takes it closer to the Sun than Mercury.

IceCube An experiment to detect cosmic neutrinos. built into the ice at the South Pole. It consists of 4200 light detectors, arranged in strings, buried in 70 holes, 2.4 km (1.5 miles) deep. The detectors record faint flashes of light emitted when neutrinos interact with particles in the ice. In all, the detectors occupy a cubic kilometer of ice. IceCube began operation in 2006 and was built around a smaller experiment called ➤ *AMANDA*. It is an international project led by the United States. ➤➤ *neutrino astronomy.*

ice dwarf A small planetary body composed of a mixture of ices and rock. Examples include ➤ *Pluto*, planetary moons such as ➤ *Triton*, and objects populating the ➤ *Kuiper Belt*.

243 Ida A member of the ➤ *Koronis family* of asteroids, imaged from close-up on August 28, 1993, by the ➤ *Galileo* spacecraft on its way to Jupiter. It measures 58×23 km (36×14 miles). The spacecraft observations revealed that Ida has a small satellite, subsequently named Dactyl, which measures about 1.6×1.2 km (1.0×0.75 mile). Observations of Dactyl's orbital motion put Ida's density in the range 2.2–2.9 g/cm^3. Ida and Dactyl do not have identical compositions, suggesting that the pair were created when larger asteroids collided and broke up to create the Koronis family. The surfaces of both Ida and Dactyl are heavily cratered.

Ijiraq A small outer moon of Saturn in a very elliptical orbit. It was discovered in 2000 and is about 10 km (6 miles) across.

Ikeya–Seki, Comet A particularly brilliant comet, discovered on September 18, 1965 by two Japanese amateur astronomers. It was especially conspicuous in the southern hemisphere after it had passed perihelion. It belonged to the group of comets known as ➤ *sungrazers*.

IKI The Space Research Institute of the Russian Academy of Sciences, which undertakes projects for the Russian Space Agency.

Imbrium Basin The largest and youngest of the very large circular impact basins on the Moon. The large crater created by the impact was subsequently flooded by lava to form the dark area known as the Mare Imbrium, which is 1300 km (800 miles) in diameter. The Imbrium Basin is surrounded by three concentric rings of mountains. The outer one is the most conspicuous and includes the Carpathian, Apennine and Caucasus mountains. The lunar Alps form part of the second ring.

immersion The disappearance of a star, planet, moon or other body at the beginning of an ➤ *occultation* or ➤ *eclipse*.

inclination (symbol *i*) The angle between the plane of an orbit and a reference plane. The reference plane for the orbits of planets and comets around the Sun is usually the ➤ *ecliptic*. For satellite orbits, the reference plane is normally the plane of the equator of the parent planet.

Index Catalogue (IC) Two supplements to the ➤ *New General Catalogue* (NGC) of nebulae and star clusters, compiled by John L. E. ➤ *Dreyer* and published in 1895 and 1908.

Indus (The Indian) An inconspicuous southern constellation, representing a native American. It was introduced in the 1603 star atlas of Johann Bayer and contains no stars brighter than the third magnitude.

inferior conjunction The position of Mercury or Venus when either of these planets lies directly between Earth and the Sun. Because the orbits of the planets are tilted to each other, ➤ *transits* of Mercury and Venus across the face of the Sun are rare. Normally they pass just north or south of the Sun at inferior conjunctions.

inferior planets Mercury or Venus, the two planets that have orbits closer to the Sun than Earth.

inflation A period in the very early history of the universe, soon after the ➤ *Big Bang*, when the universe expanded extremely rapidly. Inflation can account for the immense size of the universe today, and its uniformity. According to the theory, inflation converted tiny quantum-scale energy fluctuations into the uneven density of matter that seeded stars and galaxies. Observations made by the ➤ *Wilkinson Microwave Anisotropy Probe (WMAP)* support the theory of inflation.

Infrared Astronomical Satellite (IRAS) An orbiting infrared telescope with a primary mirror 57 cm (22.5 inches) across, which was launched on January 25, 1983, and operated until November 23, 1983. It was a collaborative mission between NASA, the Netherlands Aerospace Agency, and the UK. During its 10-month mission, it scanned 96 percent of the sky twice and detected a quarter of a million individual sources. It also discovered five comets. The first and brightest, Comet IRAS–Araki–Alcock discovered in May 1983, passed within five million km (3 million miles) of the Earth – the closest approach of any comet for 200 years.

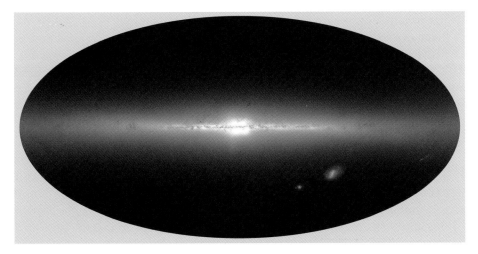

Infrared astronomy. An image of infrared radiation from the whole sky shows the plane of the Galaxy and the "bulge" at the galactic center.

infrared astronomy The study of infrared radiation from astronomical sources. Infrared is ➤ *electromagnetic radiation* in the wavelength range between visible red light and radio waves, which extends from about 0.1 to about 100 μm. It is invisible to the human eye and is absorbed almost completely in the lower layers of the Earth's atmosphere, primarily by water vapor. For this reason, infrared astronomy observations have to be conducted from the highest mountain sites, or from aircraft or satellites.

The first infrared observation was made accidentally by William ➤ *Herschel* in 1800 when a thermometer he placed just to one side of the red end of a visible solar spectrum recorded a rise in temperature. Infrared images predominantly show the distribution of heat. Since all warm objects radiate infrared, infrared telescopes must be cooled to a few degrees above absolute zero so they are not blinded by the radiation they are emitting themselves.

Systematic infrared astronomy began in the 1960s, when suitable detectors became available. The first infrared survey of the sky was carried out by Gerry Neugebauer and Robert Leighton of the California Institute of Technology (Caltech). They published a list of 5612 sources in 1969. Infrared astronomy has made important advances with the development since the 1980s of two-dimensional arrays of infrared detectors, capable of making a complete image in a single exposure.

Infrared astronomy was boosted by the successful operation of IRAS, the ➤ *Infrared Astronomical Satellite*, in 1983. Its successor, the ➤ *Infrared Space Observatory* (ISO), was launched in November 1995. The largest, most capable infrared telescope ever placed in orbit was NASA's ➤ *Spitzer Space Telescope*, launched in 2003. NASA's proposed James Webb Space Telescope and the European Space Agency's orbiting ➤ *Herschel* telescope will also operate in the infrared.

The best ground-based site for infrared astronomy is the ➤ *Mauna Kea Observatories* in Hawaii. Three infrared telescopes started operation there in 1979: the ➤ *United Kingdom Infrared Telescope* (UKIRT), NASA's ➤ *Infrared Telescope Facility* (IRTF) and the ➤ *Canada–France–Hawaii Telescope* (CFHT), which also functions as an optical telescope. The telescopes of the ➤ *Keck Observatory* and many other recently constructed large telescopes can detect the near-infrared as well as visible light.

Infrared radiation is detected from stars and galaxies, and from dust clouds within the solar system and in the interstellar medium. Strong infrared emission is particularly characteristic of dust that has been heated by visible and ultraviolet radiation from stars. Protostars in the process of formation and evolved red giant stars are surrounded by shells of dust that emit infrared. Unlike visible light, infrared radiation passes relatively easily through dust clouds. So, for example, the ➤ *galactic center*, which is largely obscured by dust as far as visible light is concerned, can be explored by means of infrared and radio astronomy. The way in which infrared radiation is scattered from the surfaces of objects in the solar system provides important clues to their composition. Infrared observations are also important for remote objects with large ➤ *redshifts*.

infrared galaxy A galaxy that emits most of its energy (typically more than 90 percent) as infrared radiation. Such galaxies are thought to have unusually high rates of star formation and so are also called ➤ *starburst galaxies*.

Infrared Space Observatory (ISO) An orbiting infrared telescope launched by the ➤ *European Space Agency* on November 17, 1995. It operated until May 1998 and made observations in the waveband between 2.5 and 200 μm with a sensitivity much greater than that of its predecessor, the ➤ *Infrared Astronomical Satellite*, IRAS.

Infrared Telescope Facility (IRTF) A NASA infrared telescope located at the ➤ *Mauna Kea Observator*ies in Hawaii, where it has been in operation since 1979 as a national facility for the USA. The main mirror is 3 m (120 inches) in diameter.

inner planets The planets Mercury, Venus, Earth and Mars.

Institut de Radio Astronomie Millimétrique (IRAM) A collaborative project between France, Germany and Spain for studies in ➤ *millimeter-wave astronomy*. The institute operates a 30-m dish in the Sierra Nevada, Spain, and a four-dish interferometer located in France, south of Grenoble.

INTEGRAL Abbreviation for ➤ *International Gamma-Ray Astrophysics Laboratory*.

interacting galaxies Galaxies so close to each other that they are distorted by the effects the gravity of one has on another. Most galaxies are in clusters, and

Interacting galaxies. The large barred spiral galaxy, NGC 6872, is interacting with a smaller lenticular galaxy, IC 4970 (just above the centre). The bright object to the lower right of the galaxies is a star in the Milky Way.

interactions between pairs are not uncommon. They often result in long wisps or filaments of stars and gas that form bridges between them.

704 Interamnia The sixth-largest asteroid known. It was discovered in 1910 by Vincenzo Cerulli, who gave it the Latin name for Teramo, the Italian city where he worked. Its diameter is 338 km (210 miles).

interferometer An instrument that brings together two or more separate beams of electromagnetic radiation from a celestial object so that they interfere and produce a characteristic pattern. In basic interferometry, telescopes are used in pairs and the resulting interference patterns are analyzed by computer. By using more than two mirrors, or radio antennas, it is possible to produce high-resolution maps or images. This technique is often known as ➤ *aperture synthesis*.

Interferometry has been used in radio astronomy for decades. More recently it has been applied to the infrared and to visible light. Instruments for optical interferometry have been constructed, such as the Cambridge Optical Aperture Synthesis Telescope in the UK and the Navy Prototype Optical Interferometer in the USA. In addition, several very large telescopes have been designed so they can be used for optical interferometry. They include the

An artist's impression of the International Gamma-Ray Astrophysics Laboratory (INTEGRAL).

➤ *Keck Observatory*, the ➤ *Very Large Telescope* and the ➤ *Large Binocular Telescope*. ➤➤ *radio interferometer*, *very-long-baseline interferometry*.

International Astronomical Union (IAU) An organization formed in 1919 for fostering international cooperation in astronomy. Its members include both national organizations and individuals.

The history of the IAU goes back to the international cooperation established for the ➤ *Carte du Ciel* project. Today, the IAU is recognized as the international authority on astronomical matters requiring cooperation and standardization, such as the official naming of astronomical bodies and features on them. The Central Bureau for Astronomical Telegrams and the Minor Planet Center located at the ➤ *Smithsonian Astrophysical Observatory* operate under its auspices. The IAU is also concerned with the promotion of astronomy in developing countries.

International Gamma-Ray Astrophysics Laboratory (INTEGRAL) An international orbiting observatory for ➤ *gamma-ray astronomy*. The project is led by ESA, in cooperation with Russia and the USA. The observatory was launched into Earth orbit on October 17, 2002. There are four instruments – an imager, a spectrometer, an X-ray monitor and a camera for visible light – all of which observe the same region of sky simultaneously. The nominal mission of two years was extended until 2008.

International Space Station (ISS) An orbiting space station, which is a joint project between the space agencies of the USA, Russia, Japan, Canada and Europe. Construction began in 1998 and is expected to be complete in 2010.

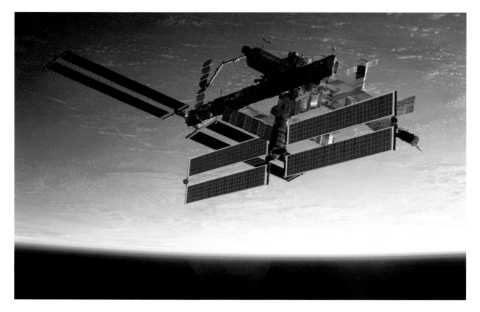

The International Space Station under construction in orbit in 2006.

The ISS is being assembled in orbit from a variety of modules and elements transported by the Space Shuttle or by Russian launch vehicles. The size and scope of the ISS has been reduced compared with the original plan but the finished space station will still have 10 main pressurized modules. It is in low Earth orbit, at an altitude of 360 km (220 miles) and makes one circuit of Earth in 92 minutes. The first crew arrived at the ISS in November 2000 and there has never been fewer than two astronauts or cosmonauts on board since.

The completion date was to have been 2005, but the project has been subject to delay for several reasons, including budget problems and the destruction of the ➤ *Space Shuttle* Columbia in an accident in 2003, which led to the grounding of all Space Shuttles for more than two years.

International Ultraviolet Explorer (IUE) An ultraviolet astronomical telescope with a 45-cm (18-inch) primary mirror, which was launched into Earth orbit in 1978. A joint NASA–ESA–UK project, it continued to observe successfully for 18 years, finally ceasing operations in September 1996.

interplanetary dust Small dust particles in the space between the planets. The particles are thought to come from collisions between asteroids in the ➤ *asteroid belt* and from the gradual break-up of comets. Fragments from comets initially form ➤ *meteor streams*, but disperse over long periods. The dust cloud in the plane of the solar system extends out from the Sun for at least 600 million km (370 million miles). It shows up in the sky as the ➤ *zodiacal light*.

Its density is very low, equivalent to one grain in a cube hundreds of meters across at the Earth's distance from the Sun.

interplanetary magnetic field (IMF) The general magnetic field in the solar system. Its origin is the ➤ *solar wind*, which carries the solar magnetic field with it as it blows outwards. The Sun's rotation makes its shape a spiral. The field varies considerably from place to place and over time, depending on the behavior of the solar wind. Its strength declines with distance from the Sun.

interplanetary medium The material in space between the planets of the solar system, which is composed of ➤ *interplanetary dust*, gas from the ➤ *interstellar medium*, and electrons, protons and helium nuclei (alpha particles) streaming outwards from the Sun.

interstellar dust Small particles in the ➤ *interstellar medium*. Interstellar dust grains range in size from 0.005 μm to 1 μm and are generally mixed in with gas. Though accounting for less than one percent of the mass of the typical interstellar medium, the dust absorbs far more light and emits far more infrared radiation than the gas. It causes ➤ *interstellar extinction* and ➤ *interstellar reddening*. Starlight scattered from dust particles creates ➤ *reflection nebulae*.

Energy the dust absorbs from starlight raises its temperature to a few tens of degrees above absolute zero. Even at such a low temperature, the dust emits some radiation, mainly in the infrared. If dust is heated above about 1500 K, it is destroyed.

It is unlikely that all interstellar dust has the same composition. Graphite (a common form of carbon) and silicates of iron, aluminum, calcium and magnesium are thought to be some of the commonest materials. Some of the particles, if not all, are not spherical.

Most interstellar dust is thought to come from dust shells that form around cool ➤ *red giant* stars. Solid particles can condense in the cool gas of their outer layers. Grains of material may also condense within ➤ *molecular clouds*.

interstellar extinction The dimming of light from distant stars because of absorption and scattering by interstellar dust. Red light is dimmed less than blue light, which results in ➤ *interstellar reddening*. The blue light from a star near the center of the Galaxy is reduced in brightness by 25 magnitudes by the interstellar material along our line of sight.

interstellar medium (ISM) The material in the space between stars in a galaxy. The interstellar medium in our own Galaxy has at least one-tenth the total mass of the stars. In general, spiral galaxies have substantial amounts of interstellar material and elliptical galaxies little or none.

The interstellar medium is not uniform and is made up of a number of components: dark clouds of gas and dust, regions of ➤ *ionized hydrogen* and ➤ *neutral hydrogen*, ➤ *molecular clouds*, ➤ *globules*, very hot tenuous gas and high-energy ➤ *cosmic ray* particles. Its chemical composition is enriched by material blown off ➤ *supernovae* and other stars. On a scale of thousands of light years, the structure of the interstellar medium is probably dominated by merging supernova remnants. ➤ *astration, Local Hot Bubble.*

interstellar molecules Molecules in the ➤ *interstellar medium*, especially in ➤ *molecular clouds*. They are destroyed by the ultraviolet radiation from hot stars and so are found chiefly in dense clouds where they are shielded. Before 1963, CH (methylidyne), CH^+ and CN (cyanogen) were the only interstellar molecules known. In 1963, radio emission at a wavelength of 18 cm was recognized as being from hydroxyl (OH). Around 100 different molecules have been identified since 1968, mainly by their spectra at millimeter wavelengths. Most are simple organic molecules.

interstellar reddening The apparent reddening of light from distant stars by ➤ *interstellar dust*. The degree to which light is scattered and absorbed in the ➤ *interstellar medium* depends strongly on its color: blue light is dimmed more than red light. As a result, the colors of stars viewed through interstellar material are altered and appear redder.

intracluster medium (ICM) The material between the galaxies in a ➤ *cluster of galaxies*. The ICM is made up of several components. Tenuous hot gas emits X-rays. Diffuse radio emission is produced by high-energy particles. Stars that have been torn from interacting galaxies are also found in intergalactic space. These known components of the ICM, and the galaxies themselves, only account for 20 percent of the total mass of a typical cluster. Some of the unseen mass is almost certainly associated with the galaxies but the motion of galaxies in clusters suggests that there is also ➤ *dark matter* in the ICM.

Io The third largest of the moons of Jupiter and one of the four discovered by ➤ *Galileo* in 1610. Its diameter is 3643 km (2264 miles). It is the most volcanically active body in the solar system and its surface is brightly colored – much of it a greenish yellow dappled with patches of orange and white.

Volcanism had been predicted on the basis that strong tides caused by Jupiter would stir up and melt rock in the interior of Io. Eight active eruptions were identified in images returned by the ➤ *Voyager 1* spacecraft. Monitoring by ground-based observatories, by the Hubble Space Telescope, and the ➤ *Galileo* spacecraft have confirmed the continuous high level of eruption activity on Io. Over 100 centers of volcanic eruptions are now known. Many are surrounded by roughly circular halos of ejected material.

There are a variety of features associated with the volcanism. These include calderas, lava flows and lava lakes where molten material gushes out from beneath the surface, and immense fountain-like plumes of gas and dust. The colored crust is made of sulfur and solid sulfur dioxide. No impact craters are seen; any that were formed in Io's early history have long since been

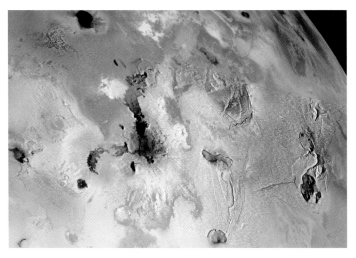

A close-up of the volcanic surface of Io from the Galileo spacecraft.

covered. Mountains rise up to 16 km (52 000 feet) high, but they are not volcanoes. Io has a thin atmosphere of sulfur dioxide and a ring of electrically charged particles. A ring of such particles also surrounds Jupiter, enclosing the orbit of Io. Data from the Galileo spacecraft suggest that Io has a metallic core.

Iocaste A small outer moon of Jupiter discovered in 2000. Its diameter is about 5 km (3 miles).

ion An atom that has gained or lost one or more electrons and so is electrically charged.

ion drive A popular name for electric propulsion, a method of propelling a spacecraft which uses an electrically generated jet rather than the gases produced by burning chemical fuel.

Ion drive. An artist's impression of the ion engine powering the SMART-1 spacecraft.

ionization The removal of electrons from atoms or molecules. Atoms and molecules can be ionized by collisions or by absorbing radiation.

ionized hydrogen Hydrogen in which the atoms have been split to make separate electrons and protons. (A neutral hydrogen atom has a single proton as its nucleus and one electron.)

 The main way hydrogen clouds in interstellar space are ionized is by absorbing ultraviolet radiation. Ionized hydrogen is the chief constituent of hot clouds known as H II (or H^+) regions. They are roughly spherical and can be up to 600 light years across. Intense ultraviolet radiation from young hot stars embedded in the clouds ionizes the gas. There is also ionized hydrogen in ➤ *supernova remnants* and the shells of ➤ *planetary nebulae*.

ionosphere An ionized layer in the atmosphere of a planet. Earth's ionosphere lies between heights of about 50 and 600 km (30 and 350 miles), though its extent varies considerably with time, season and ➤ *solar activity*. It is created by the effects of ultraviolet radiation and X-rays from the Sun. There are four different layers with different characteristics known as the D, E, F_1 and F_2 layers in order of height. The E and F_1 regions, between 90 and 230 km up, form the main part of the ionosphere.

ion tail (type I tail) One of the two types of tail ➤ *comets* develop when they get near the Sun. Ion tails, also known as the gas tails or plasma tails, consist of ionized atoms and molecules that glow after absorbing sunlight. The ion tail forms in the plane of a comet's orbit. It streams nearly straight away from the

The irregular galaxy NGC 1427A.

Sun but curves by a few degrees. It is "blown" by the ➤ *solar wind* and affected by the Sun's magnetic field. ➤➤ *disconnection event*.

IRAS Abbreviation for ➤ *Infrared Astronomical Satellite*.

7 Iris One of the larger asteroids in the main asteroid belt, with a diameter of 208 km (129 miles). It was discovered from London by John Russell Hind in 1847.

iron meteorite A type of ➤ *meteorite* composed almost entirely of iron and nickel.

irregular galaxy Any galaxy that is not obviously an ➤ *elliptical galaxy* or a ➤ *spiral galaxy*. About a quarter of known galaxies are irregular. Many appear to have star-forming regions, and are dominated by regions of luminous gas and bright young stars. Radio observations of the hydrogen gas in irregular galaxies often reveal that there is a rotating disk of gas; in this respect, and in their star content, they resemble spiral galaxies.

irregular variable A pulsating variable star that changes in brightness slowly and in an irregular way.

Isaac Newton Telescope (INT) A British 2.5-m (98-inch) reflecting telescope at the ➤ *Observatorio del Roque de los Muchachos*, La Palma, Canary Islands. The telescope was originally in the UK. It was rebuilt and provided with a new primary mirror on removal to La Palma, where it came into operation in 1984.

Ishtar Terra One of the major upland areas on the planet ➤ *Venus*, comparable in size with the continent of Australia. It includes the highest mountain peaks on Venus, Maxwell Montes.

ISO Abbreviation for ➤ *Infrared Space Observatory*.

Isonoe A small outer moon of Jupiter discovered in 2000. Its diameter is about 4 km (2.5 miles).

ISS Abbreviation for ➤ *International Space Station*.

An image of the asteroid Itokawa returned by the spacecraft Hayabusa.

25143 Itokawa A small asteroid visited by the Japanese spacecraft ➤ *Hayabusa*. Images returned by Hayabusa show it to be elongated, measuring 535×209 meters (1755×685 feet). It seems to be a loose collection of fragments, probably debris left over when a larger asteroid was shattered by an impact. The surface is littered with many rocks and boulders but there are no impact craters. Itokawa's path around the Sun crosses the orbits of both Earth and Mars.

IUE Abbreviation for ➤ *International Ultraviolet Explorer*.

J

James Clerk Maxwell Telescope A submillimeter-wave telescope at the ➤ *Mauna Kea Observatories* on the island of Hawaii operated by the UK, the Netherlands and Canada. Its 15-m (49-foot) reflector is on an ➤ *altazimuth mount*. The dish is constructed from 276 lightweight panels and is designed so its shape does not change wherever it is pointed.

James Webb Space Telescope (JWST) The space telescope under development as the successor to the ➤ *Hubble Space Telescope*. The projected launch date is no earlier than 2013 and the telescope's anticipated lifetime is 5–10 years. The JWST will have a primary mirror about 6.5 m in diameter and will operate in the wavelength region between the red end of the visible spectrum and the mid-infrared.

It will be placed in orbit around the Sun at a point 1.5 million km (1 million miles) from Earth in the direction opposite the Sun, rather than in orbit around Earth. This has the advantage of simplifying operations by minimizing stray light and temperature changes.

The project was initially known as the Next Generation Space Telescope but was later named in honor of a former NASA administrator.

The James Clerk Maxwell Telescope.

An artist's impression of the James Webb Space Telescope.

Jansky, Karl (1905–1950) Jansky was an American radio engineer, who in 1931, while he was investigating interference with long-distance communications, became the first person to detect cosmic radio emission. He surmized, correctly, that the radio waves came from interstellar gas in the ➤ *Milky Way*. The unit in which radio astronomers measure the strength of radio emission from cosmic sources is called the jansky in his honor.

Janus A small moon of Saturn, discovered by Audouin Dollfus in 1966 when the rings were edge-on as seen from Earth. Janus orbits just beyond the outer edge of the ring system at a distance of 151 422 km (94 089 miles) and its orbit is shared by another small moon, Epimetheus. The two may be fragments of a larger object that was shattered by an impact. Janus is irregular in shape, measuring 194 × 190 × 154 km (121 × 118 × 96 miles).

JD Abbreviation for ➤ *Julian date*.

Jeans, Sir James Hopwood (1877–1946) The British astronomer and physicist Sir James Jeans is remembered as a prolific author of popular books and textbooks that did much to promote astronomy to the general public. He read applied mathematics at Cambridge University, graduating in 1900. He applied his talent to a range of theoretical problems, including the dynamical theory of gases and the origin of the solar system. From 1923 to 1944 he was on the staff of the ➤ *Mount Wilson Observatory* in California.

Jet Propulsion Laboratory (JPL) An institution in Pasadena, California, operated by the California Institute of Technology (Caltech) in support of programs of

Karl Jansky.

A Cassini spacecraft image of Janus.

The Jewel Box star cluster.

the US ➤ *National Aeronautics and Space Administration* (NASA) and of other agencies. It is the principal center in the USA for the development and operation of interplanetary spacecraft.

Jewel Box (NGC 4755) An open star cluster in the constellation Crux. Its brightest member is a sixth-magnitude, blue, supergiant star, Kappa Crucis. It includes several blue and red supergiants and its appearance in a small telescope is impressive. The contrasting colors of the stars is said to be why John ➤ *Herschel* described the cluster as the "Jewel Box." The cluster's distance is 7800 light years.

Jodrell Bank Observatory The astronomy facility of the University of Manchester in the United Kingdom. It incorporates a radio astronomy observatory where the main instrument is a 76-m (250-foot) fully steerable dish, which was first opened in 1957. It has been known as the Lovell Telescope since 1987 in honor of the astronomer Sir Bernard Lovell, who was the driving force behind its conception and construction.

Johnson Space Center (JSC) A large space technology complex at Houston Texas, which is the headquarters for the astronauts of the US National Aeronautics and Space Administration (NASA). It was named after the late US President Lyndon B. Johnson, who took a strong interest in the space program.

jovian Pertaining to the planet Jupiter.

jovian planets The giant planets Jupiter, Saturn, Uranus and Neptune.

JPL Abbreviation for ➤ *Jet Propulsion Laboratory*.

Jodrell Bank Observatory. The control room for the Lovell Telescope (visible through the window) and MERLIN.

JSC Abbreviation for ➤ *Johnson Space Center*.

Julian calendar A calendar brought into use in the Roman Empire by Julius Caesar from 46 BC. Three years of 365 days were followed by one of 366 days, making the average length of a year 365.25 days. Since this is 11 minutes 14 seconds longer than a true ➤ *tropical year*, the seasons gradually shifted relative to the civil year. To correct for this, the ➤ *Gregorian calendar* was introduced from 1582.

Julian date (JD) The interval of time in days since noon at Greenwich on January 1, 4713 BC.

Juliet A small moon of Uranus, about 94 km (58 miles) in diameter, discovered by ➤ *Voyager 2* in 1986.

3 Juno The third asteroid to be discovered. Its diameter is 248 km (154 miles) and it was found by Karl L. Harding in 1804.

Jupiter The largest planet in the solar system and the fifth in order from the Sun. After Venus, it is the second-brightest planet as seen from Earth. It travels around the Sun in a slightly elliptical orbit at an average distance of 5.20 AU, taking 11.86 years to complete one revolution.

   With a diameter of 143 082 km (88 911 miles) at its equator, Jupiter is over 11 times larger than Earth and about one-tenth the size of the Sun. Its mass is 318 times Earth's, and about 0.1 percent that of the Sun, making it more massive than all the other planets put together.

   Jupiter's composition (by number of molecules) is very similar to the Sun's: 90 percent hydrogen and 10 percent helium. The most significant trace gases in its atmosphere are water vapor, methane and ammonia. There is no solid surface beneath the cloud layer. Instead, there is a gradual transition from gas

Jupiter. An image taken by the Cassini spacecraft.

to liquid as the pressure increases with depth below the outermost layers. At a particular depth, the nature of the liquid changes abruptly. Below that level, the hydrogen has the properties of a metallic liquid, in which the atoms are stripped of their electrons. At the very center of Jupiter there may be a small core of rock and perhaps ice. Some heat remains inside Jupiter from when it first formed, causing it to radiate between 1.5 times and twice as much heat as it currently absorbs from the Sun.

Observed by eye, the disk of Jupiter appears crossed by alternating light zones and dark belts. There are five or six in each hemisphere, corresponding to wind currents. Results from four spacecraft that passed by Jupiter between 1973 and 1981 (➤ *Pioneers 10 and 11*, ➤ *Voyagers 1 and 2*), and from the more recent ➤ *Galileo* mission, have revealed how complex the patterns created by the flow of gases are within these bands.

White and colored ovals appear and are relatively long-lived features. The best-known and most conspicuous of all is the Great Red Spot, which has been observed for around 300 years and is as wide as Earth. Its origin is uncertain but it is essentially a huge anticyclone.

The colored clouds are in the highest layers, which take up only 0.1–0.3 percent of Jupiter's radius. The origin of their coloration remains a mystery, though it must stem from trace constituents of the atmosphere. The cloud colors correlate with depth in the atmosphere. Blue features are the deepest, followed by brown, then white, with red being the highest.

A probe released by the Galileo spacecraft in 1995 parachuted through Jupiter's upper atmosphere and returned data on the composition and physical conditions. Ground-based observations of the entry site indicated that it may have been a relatively cloud-free spot, explaining why hardly any evidence was found for the expected three layers of cloud consisting of ammonia crystals at the highest level, ammonium hydrosulfide in the middle, with water and ice crystals below. Winds up to 530 km/h (330 mph) were even faster than anticipated. The abundance of helium was only about half that expected. A likely explanation is that the helium is more concentrated towards the center of the planet. The probe also discovered an intense radiation belt.

The existence of a faint ring around Jupiter was first suggested by results from Pioneer 11 in 1974 and was confirmed by Voyager images. The main part lies between 1.72 and 1.81 Jupiter radii from the center of the planet. The ring consists of tiny particles only micrometers in size, which must constantly be replenished.

There are at least 62 moons orbiting Jupiter, most of them small. The four large Galilean satellites – Io, Europa, Ganymede and Callisto – are easily visible with a small telescope or binoculars.

Radio emission from Jupiter was discovered in 1955. It was the first indication of Jupiter's strong magnetic field, which is 4000 times greater than Earth's. The radio emission is caused by electrons spiraling around in the magnetic field.

K

K Abbreviation for ➤ *kelvin*, a unit in which temperature is measured. One kelvin corresponds to an interval of one degree on the Celsius scale. The kelvin temperature scale starts from absolute zero, which is −273.16 °C.

Kale A small outer moon of Jupiter discovered in 2001. Its diameter is about 2 km (1 mile).

Kallichore A small outer moon of Jupiter discovered in 2003. Its diameter is about 2 km (1 mile).

Kalyke A small outer moon of Jupiter discovered in 2000. Its diameter is about 5 km (3 miles).

Kamiokande ➤ *Super-Kamiokande*.

Kappa Crucis The brightest star in the open cluster NGC 3324, popularly known as the ➤ *Jewel Box*.

Kaus Australis (Epsilon Sagittarii) The brightest star in the constellation Sagittarius. With Kaus Meridionalis (Delta) and Kaus Borealis (Lambda), it marks the Archer's bow. It is a ➤ *B star* of magnitude 1.9 and 145 light years away.

Keck Observatory Two 10-m (400-inch) reflecting telescopes belonging to the California Institute of Technology (Caltech) and the University of California. They are at the ➤ *Mauna Kea Observatories* on the island of Hawaii and have been

The twin domes of the Keck Observatory.

funded by the W. M. Keck Foundation. The first telescope was completed in 1992 and the second in 1996. The primary mirrors each consist of 36 hexagonal segments. The precise shape of the mirrors is maintained by specially designed supports under computer control. The use of ➤ *adaptive optics* makes it possible to produce images with a resolution of 0.04 arcseconds at a wavelength of 2 μm.

The two telescopes can be used together as an ➤ *interferometer*. Because Keck I and Keck II are about 85 m (nearly 280 feet) apart, the interferometer has a resolution equivalent to a telescope with an 85-m mirror, which is only 0.005 arc seconds at a wavelength of 2 μm.

Keeler, James Edward (1857–1900) Keeler was an American pioneer in astronomical spectroscopy. On graduation from Johns Hopkins University in 1881 he became an assistant at the Allegheny Observatory, Pittsburgh. In 1891 he was appointed the director there. Using spectroscopy he found that Saturn's rings are not solid. He worked at the Lick Observatory, California in 1898 and 1900, where he photographed 120 000 galaxies. This achievement established the superiority of large reflecting telescopes over refractors.

Keeler Gap A narrow gap towards the outer edge of the bright A ring of ➤ *Saturn*. ➤➤ *ring system*.

Kennedy Space Center (KSC) A NASA space launch center at Cape Canaveral on the Atlantic coast of Florida. It is the only place from which the ➤ *Space Shuttle* has been launched, and is one of the few facilities where the Shuttle can land. In the past, all the manned launches in the Mercury, Gemini, Apollo and Skylab programs were from KSC. It was named after the late US President, John F. Kennedy.

Kepler (1) A lunar crater, 32 km (20 miles) in diameter, in the Oceanus Procellarum. It has terraced walls and a central peak, and is the center of a large, bright ray system.

Kepler (2) A NASA space observatory designed to search for ➤ *extrasolar planets*. It is scheduled for launch in June 2008 and will be placed in an orbit around the Sun, trailing behind Earth. The 0.95-m telescope will monitor the brightness of thousands of stars, looking for minute dips when a planet passes in front of a star. It will be able to detect planets as small as Earth.

Kepler, Johannes (1571–1630) Kepler was one of the most important scientists of his time and made contributions in a number of areas. He is remembered especially for the laws of planetary motion he formulated (➤ *Kepler's Laws*) and he believed in a mathematical harmony underlying the universe.

Kepler was born in Würtemburg, Germany, the son of poor parents. He graduated from the University of Tübingen in 1591, where he studied for the Lutheran ministry. However, he took up mathematics instead and acquired a deep interest in the ➤ *Copernican system*.

The endless religious turmoil and numerous epidemics of his times considerably affected his career. In 1598, all Lutherans were expelled from

A fisheye view of the Space Shuttle Atlantis leaving launch pad 39B at the Kennedy Space Center on 9 July 2006.

Graz, Austria, where he had a teaching position. Fortunately Tycho ➤ *Brahe*, by that time Imperial Mathematician in Prague, invited Kepler to join him in 1600 to work on 20 years of observations of Mars. It took him eight years to find the reason for the puzzling motion of Mars: that its orbit is an ellipse. Brahe's precise data enabled Kepler to become the first person to explain planetary motion correctly. On Tycho's death in 1601, Kepler took over his position but in 1611, the Emperor fell from power in a civil war and Kepler moved to Linz. He stayed in Linz for 14 years before taking up his final appointment in Silesia. There his patron provided him with the facilities he needed to write and study in return for producing horoscopes.

Johannes Kepler.

In optics, Kepler was the first to explain the formation of images and the principles of how a telescope works. He was also the first to explain how the Moon causes tides. He invented the word satellite to describe a small body orbiting a much larger one.

Kepler's laws Three fundamental rules about planetary motion, discovered by Johannes ➤ *Kepler* from his study of detailed observations of the planets made by Tycho ➤ *Brahe*:

1. The orbit of each planet is an ellipse with the Sun at one of the foci.
2. Each planet travels around the Sun such that the line connecting the planet to the Sun sweeps out equal areas in equal times.
3. The squares of the sidereal periods of the planets are proportional to the cubes of their mean distances from the Sun.

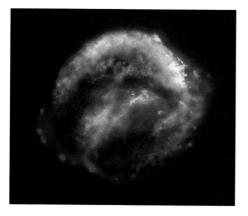

A composite image of the remnant of Kepler's Supernova constructed with data from the Hubble Space Telescope, the Spitzer Space Telescope, and the Chandra X-ray Observatory.

The first two were published in 1609 and the third in 1619. The physical basis for the laws was not understood until ➤ *Isaac Newton* formulated his law of gravity.

Kepler's Supernova (Kepler's Star) A ➤ *supernova* that appeared in the constellation Ophiuchus in October 1604. ➤ *Johannes Kepler* observed it and determined its position. At its brightest it reached a magnitude of about −2.5 and the light curve shows that it was a Type I supernova. The remnant is a source of radio emission, and there is also a faint visible nebula.

Keyhole Nebula (NGC 3372) A dark dust nebula near the center of the ➤ *Carina Nebula*. The circular part of the keyhole shape is a bubble expanding at 40 km/s.

Keystone The four stars Epsilon, Zeta, Eta and Pi in the constellation ➤ *Hercules*.

Kids A group of three stars, Epsilon, Zeta and Eta, in the constellation Auriga. Their name comes from the fact that Alpha Aurigae is named Capella, meaning "the little she-goat."

kiloparsec (symbol kpc) A unit of distance equal to one thousand ➤ *parsecs*, which is 3261.61 light years.

Kirkwood gaps Gaps in the ➤ *asteroid belt* due to the influence of Jupiter. There are notable gaps where the orbital periods of the asteroids would be in the proportion to that of Jupiter by factors of 4:1, 3:1, 5:2, 7:3 and 2:1. Asteroids cannot remain in such orbits because of regular gravitational interactions with Jupiter. The explanation was first given by Daniel Kirkwood in 1857.

Kitt Peak A mountain-top observatory site near Tucson, Arizona, which is home to one of the largest collection of astronomical research instruments in the world. These include the ➤ *Kitt Peak National Observatory*, facilities of the US ➤ *National Solar Observatory*, and the ➤ *WIYN Telescope* operated by the US ➤ *National Optical Astronomy Observatories*. A number of universities and other research organizations also lease space on Kitt Peak for astronomical work.

An infrared image of the Keyhole Nebula.

Telescopes on Kitt Peak, Arizona.

Kitt Peak National Observatory (KPNO) A facility of the US ➤ *National Optical Astronomy Observatories* located at ➤ *Kitt Peak* in Arizona. The largest telescope is the 4-m (160-inch) ➤ *Mayall Telescope*.

Kiviuk A small outer moon of Saturn in a very elliptical orbit. It was discovered in 2000 and is about 14 km (9 miles) across.

Kleinmann–Low Nebula An extended source of infrared radiation in the ➤ *Orion Nebula*. It is a region of star formation lying behind the luminous gas.

216 Kleopatra An asteroid discovered by Johann Palisa in 1880. Radar observations reported in 2000 showed Kleopatra's shape to be very unusual. It resembles a distorted dumbbell or "dog-bone" measuring $217 \times 94 \times 81$ km ($135 \times 58 \times 50$ miles) and its composition is metallic.

Kohoutek, Comet A comet discovered in 1973 when it was still near the orbit of Jupiter. Predictions that it would prove spectacular when it got nearer to Earth turned out to be incorrect. However, it was the subject to an extensive, coordinated observing program by professional astronomers and much new information about comets was obtained as a result.

Koronis family One of the ➤ *Hirayama families* of asteroids, at a mean distance of 2.88 AU from the Sun. The members have very similar compositions so probably come from the break-up of a single parent body. The largest member is 208 Lacrimosa, which is about 45 km (28 miles) in diameter. The family is named after 158 Koronis, which has a diameter of 35 km (22 miles) and was discovered in 1876.

Kreutz group ➤ *sungrazer*.

Krüger 60 A faint binary star in the constellation Cepheus. The two members of the system, magnitudes 10 and 11, orbit each other in a period of 44 years. Since their orbits are face-on to us, and the two stars are easily resolved, their relative motion can easily be followed over a period of decades. Both stars are dwarf ➤ *M stars* and the fainter component is a ➤ *flare star*. At a distance of 13 light years, it is one of the nearest stars to the solar system.

KSC Abbreviation for ➤ *Kennedy Space Center*.

K star A star of ➤ *spectral type* K. K stars have surface temperatures in the range 3500–4900 K and are orange in color. Absorption lines of neutral and ionized calcium are strong in their spectra. There are also numerous absorptions due to neutral metals and molecules, particularly in the spectra of cooler K stars. ➤ *Arcturus* and ➤ *Aldebaran* are examples of K stars.

Kuiper Airborne Observatory (KAO) A 0.915- m (36-inch) Cassegrain reflecting telescope mounted in a Lockheed C141 Starlifter jet transport aircraft, which was operated between 1975 and 1996 by NASA as a national facility in the USA. Important discoveries made with the observatory include the ring system around the planet Uranus.

Kuiper, Gerard Peter (1905–1973) Kuiper was a pioneering Dutch–American planetary scientist. Born and educated in the Netherlands, he emigrated to the United States, where we worked for the rest of his life. He held appointments at the Lick Observatory, Harvard University, the Yerkes Observatory and the McDonald Observatory. From 1960 until his death he was the head of the Lunar and Planetary Laboratory of the University of Arizona.

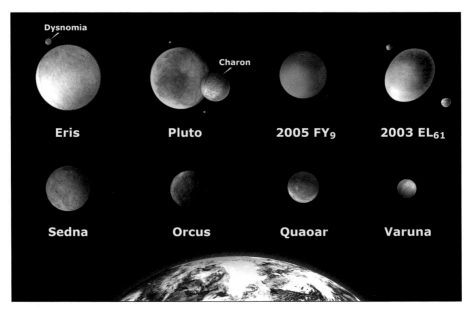

The Largest Kuiper Belt objects known in 2006, with part of Earth as a size comparison.

Kuiper discovered Uranus's moon ➤ *Miranda* in 1948 and ➤ *Nereid* a new moon of Neptune, in 1949. He also discovered that Saturn's moon ➤ *Titan* has an atmosphere containing methane, and identified carbon dioxide in the martian atmosphere. In 1951 he predicted the existence of the belt of small, icy planetary bodies beyond Neptune, which is known as the ➤ *Kuiper Belt*.

Kuiper Belt A population of small, icy, solar-system bodies beyond Neptune, most of which are similar in size to ➤ *asteroids*, though some are larger than Pluto. They occupy a ring-shaped region in the plane of the solar system extending from the orbit of Neptune (30 AU from the Sun) out to 100 or 150 AU. This population is believed to be the source of ➤ *short-period comets*. The name of Gerard ➤ *Kuiper*, a distinguished Dutch–American planetary scientist, is attached to the belt because he predicted its existence in 1951. However, an Irish writer and theorist, Kenneth E. Edgeworth, had published a similar idea in 1943 and 1949. In recognition of his contribution, some astronomers call it the Edgeworth–Kuiper Belt.

The first observational evidence for the existence of the Kuiper Belt was the discovery in 1992 of a faint object known as 1992 QB1 in a near circular orbit about 50 AU from the Sun. Hundreds more members have been found since. ➤*Pluto* is one of the largest known members of the Kuiper Belt. The discovery of ➤ *Eris*, which is slightly larger than Pluto, was announced in 2005.

Kuiper Belt object (KBO) A body belonging to the ➤ *Kuiper Belt*.

L

Lacerta (The Lizard) A small, inconspicuous constellation between Cygnus and Andromeda. It was introduced by Johannes Hevelius in the late seventeenth century and contains only one star brighter than fourth magnitude.

Lagoon Nebula (M8; NGC 6523) A luminous nebula in the constellation Sagittarius. It is a complex region of ➤ *ionized hydrogen*, gas and dust with hot, recently formed stars. A star cluster, NGC 6530, lies near the center of the nebula. The light from two naked-eye stars in the cluster, 7 and 9 Sagittarii, is responsible for ionizing the gas. The nebula is estimated to lie at a distance of 4500 light years.

Lagrangian points Equilibrium points in the orbital plane of two large bodies circling about each other, where a much smaller object could remain. There are five Lagrangian points for two bodies in circular orbits around each other, called L_1 to L_5. They are named for the great French mathematician, Joseph Louis Lagrange (1736–1813), who discovered that they exist.

L_4 and L_5, in the orbit of the less massive of the two large bodies and 60° either side of it, are stable as long as the ratio of the masses of the two large bodies exceeds 24.96. The ➤ *Trojan asteroids*, which share the orbit of Jupiter, are examples of objects trapped at L_4 and L_5 in the Sun–Jupiter system.

Objects at the other three Lagrangian points are not stable and are easily displaced by small perturbations. Nevertheless, L_1 and L_2 in the Sun–Earth system are important locations for stationing spacecraft, because they remain at fixed distances from Earth as it orbits the Sun. Regular course corrections are needed to keep spacecraft at these locations. In practice, a spacecraft

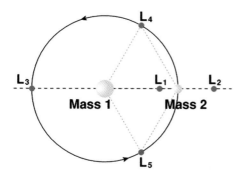

The Lagrangian points.

Pierre-Simon Laplace.

normally follows a circular "halo orbit" around the Lagrangian point.
➤ *Roche lobe*.

Laplace, Marquis Pierre Simon de (1749–1827) Laplace was a French applied mathematician who extended the work of Isaac ➤ *Newton* on motion in the solar system. His greatest work was his *Celestial Mechanics*, published in five volumes between 1799 and 1825. In it he proved that the solar system is dynamically stable, which Newton had not been able to do. Laplace also argued in favor of the so-called "nebular hypothesis" for the origin of the solar system, according to which the planets formed from the same rotating mass of gas as the Sun.

Large Binocular Telescope (LBT) A telescope consisting of two 8.4-m mirrors on a single mount, located at the ➤ *Mount Graham International Observatory* in Arizona. The project is a collaboration between the Italian Istituto Nazionale de Astrofisica, a consortium of US universities, and German astronomical research institutes. The binocular arrangement gives the telescope a light-gathering power equivalent to a single 11.8-m mirror, and a resolution corresponding to a 23-m telescope. The telescope began operating in 2006.

Large Magellanic Cloud (LMC) ➤ *Magellanic Clouds*.

Larissa A small satellite of Neptune discovered during the flyby of ➤ *Voyager 2* in August 1989. It measures about 208×178 km (129×111 miles).

Las Campanas Observatory An observatory in the Sierra del Condor, Chile, at a height of 2300 m (7500 feet). It is operated by the Carnegie Institution of Washington and is the site of the ➤ *Magellan Telescopes*. The other main instruments are the Irénée du Pont 2.5-m (100-inch) and Henrietta Swope 1-m (40-inch) reflectors.

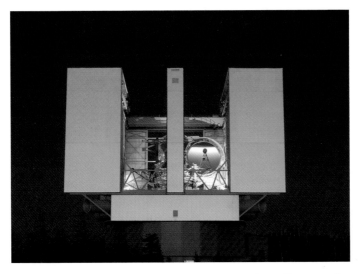

The Large Binocular Telescope in 2005 when the first of its two mirrors came into operation.

Laser Interferometer Gravitational-Wave Observatory (LIGO) Two ➤ *gravitational wave* detectors constructed by a consortium led by the California Institute of Technology and the Massachusetts Institute of Technology. The two widely separated detectors are located in Washington state and Louisiana. Scientific operation began in 2001.

La Silla Observatory One of the two sites of the ➤ *European Southern Observatory*. It is located in the southern part of the Atacama desert, about 600 km (370 miles) north of Santiago de Chile, at an altitude of 2400 m (7900 feet). The instruments include a 3.6-m (142-inch) telescope, the 3.5-m (138-inch) ➤ *New Technology Telescope* and the 15-m (50-foot) Swedish/ESO Submillimeter Telescope.

last quarter The ➤ *phase* of the Moon when it is waning and half of its disk is bright. Last quarter is formally defined as the time when the Moon's celestial ➤ *longitude* is 270° greater than the Sun's. It occurs about seven days after full Moon.

latitude Angular distance measured north or south of an equator. In celestial ➤ *equatorial coordinates*, "latitude" is known as ➤ *declination*.

leap second ➤ *Universal Time*.

Leavitt, Henrietta Swan (1868–1921) Leavitt was an American astronomer best remembered for discovering that there is a relationship between the luminosity of ➤ *Cepheid variable* stars and the periods with which they vary. She graduated from Radcliffe College in 1892. Three years later she became an unpaid assistant at Harvard College Observatory, receiving a salary only from 1902. She devoted herself to the Observatory's program of measuring the

La Silla Observatory.

magnitudes of stars from photographs. She discovered about 2400 variable stars. By 1912 she had shown that dimmer Cepheid variables have longer periods of variability. This relationship proved to be of immense importance for determining the distances to galaxies.

Leda A small moon of Jupiter discovered by Charles Kowal in 1974. It is about 20 km (12 miles) across. With Elara, Himalia and Lysithea it belongs to a family of four moons with closely spaced orbits. Their average distances from Jupiter are between 11.1 and 11.7 million km (6.9 and 7.3 million miles).

lenticular galaxy A galaxy intermediate in shape between an elliptical and a spiral. Lenticular galaxies are so-called because their shape resembles a convex lens. In the ➤ *Hubble classification* of galaxies they are type S0.

Leo (The Lion) One of the 12 constellations of the traditional ➤ *zodiac*. The pattern made by the brightest stars of this large and conspicuous constellation resemble the shape of a lion in profile. The asterism outlining the head is known as "the Sickle." There are 10 stars brighter than fourth magnitude, the brightest being ➤ *Regulus* and ➤ *Denebola*. Leo also contains numerous galaxies.

Leo Minor (The Little Lion) A small and very inconspicuous constellation between Leo and Ursa Major. It was introduced by Johannes ➤ *Hevelius* in the late

The lenticular galaxy NGC 5866, which is in the constellation Draco, and lies at a distance of 44 million light years.

seventeenth century and contains only one star brighter than fourth magnitude.

Leonids An annual ➤ *meteor shower* that radiates from "the Sickle" in the constellation ➤ *Leo*. It lasts about four days, peaking on November 17. Though a small number of meteors are detected each year, spectacular displays are seen occasionally. For example, in 1966, observers in the USA saw up to 40 meteors a second. The Leonids are associated with Comet 55P/Tempel–Tuttle, which was first recorded in 1865 and has an orbital period of 33 years. The dust stream giving rise to the meteor shower is concentrated near the comet rather than being evenly spread around the comet's orbit. Because of this, good displays are possible only every 33 years, though they are not necessarily seen even then if Earth does not happen to pass through the dust.

Lepus (The Hare) One of the 48 ancient constellations listed by ➤ *Ptolemy*. Lying just south of Orion, it is small but distinctive and contains seven stars brighter than fourth magnitude.

Leverrier, Urbain Jean Joseph (1811–1877) The French astronomer Leverrier made the calculations that led to the discovery of the planet Neptune on

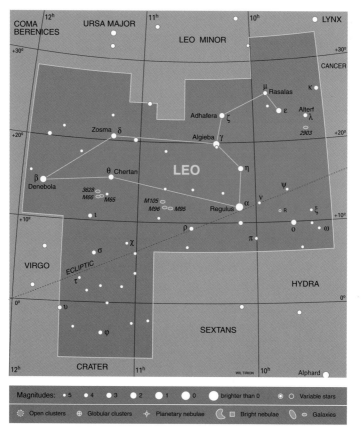

A map of the constellation Leo.

September 23, 1846, based on observations that Uranus was deviating from its expected course. He further developed the work in celestial mechanics done by Pierre Simon ➤ *Laplace*. In 1859 he announced his discovery that the ➤ *perihelion* point of the orbit of Mercury does not remain fixed but gradually moves around in space. Leverrier wrongly believed this was due to the gravity of an undiscovered planet between Mercury and the Sun, which he called Vulcan. Sixty years later this phenomenon would be a key test of ➤ *general relativity*, which provided the correct explanation. In 1854, Leverrier became director of the Paris Observatory.

Lexell's Comet A comet discovered by Charles ➤ *Messier* in 1770, but named after Anders Lexell (1740–84) who investigated its orbit. He showed that a close approach to Jupiter in 1767 caused a large change in the comet's orbit. After this, it came close enough to Earth to be visible. The comet passed within 1.2 million km (0.75 million miles) of Earth, the nearest a comet has ever been

Urbain-Jean-Joseph Leverrier.

known to come. However, another close approach to Jupiter in 1779 perturbed its orbit so drastically that it was never seen again.

Libra (The Scales) One of the twelve constellations of the ➤ *zodiac*, in ancient times known as the claws of the Scorpion. Libra is one of the least conspicuous constellations in the zodiac, with just five stars brighter than fourth magnitude.

libration An effect that slightly alters which part of the Moon's surface is visible from Earth. Though the Moon very nearly keeps the same face towards the Earth all the time, a total of 59 percent of the Moon's surface can be viewed from Earth at some time as a result of libration. There are several reasons for it. Physical libration is a real irregularity in the Moon's rotation; a greater effect is geometrical libration, affecting both latitude and longitude. Libration in latitude results from the Moon's orbit being inclined to the ecliptic by an angle of $5°\ 9'$. The elliptical shape of the Moon's orbit means that its orbital velocity is not constant, and this produces libration in longitude of $7°\ 45'$. Diurnal libration is a small additional effect that results from observing the Moon at different times of day.

Lick Observatory An observatory belonging to the University of California. The observatory site is on Mount Hamilton in the Californian Diablo Range at a height of 1300 m (4200 feet). The funds for the observatory were donated by a

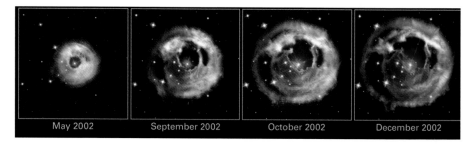

| May 2002 | September 2002 | October 2002 | December 2002 |

The development of the light echo around a variable star, V838 Monocerotis, following an outburst in 2002.

millionaire businessman, James Lick (1796–1876). The building and a 92-cm (36-inch) refracting telescope were completed in 1888, twelve years after Lick's death. He is buried at the base of the telescope. The main research telescope is the Shane 3-m (120-inch) reflector, in operation since 1959.

light curve A graph on which the brightness of a ➤ *variable star* (or other varying astronomical object) is plotted against time.

light echo A reflection by neighboring interstellar clouds of a sudden burst of light from a ➤ *supernova*, ➤ *nova* or other stellar outburst. The echo is a ring of light surrounding the star, which expands over time.

light pollution The scattering of artificial light into the night sky. Light pollution increases the background brightness of the sky above its natural level and interferes with astronomical observations. It is worst close to major centers of habitation. Legislation has been enacted in parts of the USA to protect important observatory sites from the damaging effects of unnecessary artificial lighting in nearby cities. However, the problem is a growing one and a matter of worldwide concern for both amateur and professional astronomers.

light year (l.y.) The distance traveled through a vacuum by light (or any other form of electromagnetic radiation) in one year. A light year is equivalent to 9.4607×10^{12} km, 63 240 astronomical units or 0.306 60 parsecs.

LIGO Abbreviation for ➤ *Laser Interferometer Gravitational-Wave Observatory*.

limb The extreme edge of the visible disk of a body such as the Sun, Moon or a planet.

LINEAR Abbreviation for Lincoln Near Earth Asteroid Research, a project of the Massachusetts Institute of Technology's Lincoln Laboratory, funded by the United States Air Force. The project makes use of a telescope on the White Sands Missile Range in Socorro, New Mexico and began operations in 1998. It employs technology originally developed for the surveillance of Earth satellites, which has been adapted for detecting ➤ *near-Earth objects*.

Linné A small lunar crater, 2.4 km (1.5 miles) in diameter, situated in the Mare Serenitatis. A claim made in the mid-nineteenth century that a neighboring crater called Linné B disappeared was never confirmed. Though small, Linné is

The Little Dumbbell planetary nebula.

relatively conspicuous because it is surrounded by a bright area, probably ejecta from the crater.

Little Dumbbell A popular name for M76 (NGC 650), a ➤ *planetary nebula* in Perseus. It is the faintest object in the ➤ *Messier Catalogue*.

Little Dipper A popular North American name for the constellation ➤ *Ursa Minor*, describing the figure formed by its seven brightest stars.

Local Group The group of galaxies to which our own Milky Way ➤ *Galaxy* belongs. It is dominated by the ➤ *Andromeda Galaxy* (M31), which is the largest and most massive member, and our own Galaxy. Next in size are the spiral galaxy in Triangulum, M33, which is a near companion of M31, and the Large ➤ *Magellanic Cloud*, near our Galaxy. The other members of the Local Group are small elliptical and irregular galaxies plus a number of faint, dwarf spheroidal galaxies, resembling isolated globular clusters. There is no central concentration of galaxies, but two subgroups centered around the two most massive galaxies. Four small elliptical galaxies (NGC 221, 205, 185 and 147) are satellites of M31; the Magellanic Clouds and various dwarf galaxies are satellites of our own Galaxy.

The Local Group occupies a volume of space with a radius of about 3 million light years (1 megaparsec). The next nearest galaxies are two or three time this distance away.

Local Hot Bubble An expanding bubble of low-density, hot gas in which the Sun is located. Its boundary is nearest to the solar system about 90 light years away in a direction roughly towards the galactic center. The furthest extent of the bubble is about 360 light years away. Its shape is like that of an hourglass or peanut, because of pressure from another adjacent bubble. Both these bubbles were blown out by shock waves from ➤ *supernova* explosions that took place in our part of the Galaxy in the remote past and radiate low-energy X-rays.

Local Interstellar Cloud (LIC) A small, diffuse cloud of interstellar material in which the Sun lies. It is about 20–30 light years across and the Sun is currently located towards one edge. The cloud is moving relative to the Sun and will pass the Sun completely in the next few thousand years. The Local Interstellar Cloud is within the ➤ *Local Hot Bubble*.

Local Supercluster A ➤ *supercluster* of galaxies, centered on the ➤ *Virgo Cluster*. It is more than a hundred million light years across and the ➤ *Local Group* is on its periphery.

Lockyer, Sir Joseph Norman (1836–1920) The British astronomer Lockyer is particularly noted for his discovery and naming of the chemical element helium, which he found in the atmosphere of the Sun through his study of the Sun's spectrum. His astronomical work was mainly concerned with the Sun and spectroscopy and between 1870 and 1905 Lockyer organized eight eclipse expeditions. He also founded the scientific journal *Nature* and made significant contributions to ➤ *archeoastronomy*.

Loki One of the most active volcanoes on Jupiter's moon ➤ *Io*, and the most powerful volcano in the solar system.

long-baseline interferometry A technique in radio astronomy in which two or more radio telescopes separated by up to 1000 km or so are linked in real time by signals transmitted via microwaves or cable in order to form a ➤ *radio interferometer*. ➤➤ *very-long-baseline interferometry*.

The plume of the volcano Loki on an image of Io taken by Voyager 1.

longitude Angular distance around an equator (or a circle parallel to the equator) measured from an arbitrary place. In the celestial ➤ *equatorial coordinate* system, the counterpart of longitude on Earth is ➤ *right ascension*.

long-period comet A ➤ *comet* with a period of revolution round the Sun greater than 200 years. Some have periods of millions of years. ➤➤ *short-period comet*.

long-period variable A variable star with a period between about 100 and 1 000 days. The periods of long-period variables, and their change in brightness (typically several magnitudes), both vary considerably from one cycle to another. These variables are ➤ *red giant* stars. ➤ *Mira* is one of the best-known examples.

Lost City Meteorite A meteorite that fell in Oklahoma in 1970. The ➤ *fireball* observed as it passed through the atmosphere was photographed and the images were used to locate the fallen meteorite, which was recovered a few days later.

Lovell Telescope ➤ *Jodrell Bank Observatory*.

Lowell, Percival (1855–1916) Lowell was an American astronomer who devoted much of his personal fortune to searching for evidence of life on Mars. He founded the ➤ *Lowell Observatory* in Flagstaff, Arizona, for that purpose. He began work there in 1894 and spent 15 years observing Mars with an excellent 24-inch refractor. His interest had been stimulated by reports of channels or "canali" reportedly seen by the Italian astronomer ➤ *Giovanni Schiaparelli*. Lowell claimed to see canals, oases, and seasonal vegetation, and became convinced that Mars was inhabited. However, his claims were discredited within decades. Lowell also became interested in searching for a planet beyond

Percival Lowell.

Neptune. Though he was unsuccessful in his lifetime, the research program he began led to the discovery of Pluto at the Lowell Observatory in 1930, some 14 years after his death.

Lowell Observatory A private observatory in Flagstaff, Arizona. It was founded in 1894 by Percival ➤ *Lowell*, who was particularly interested in the possibility of intelligent life on Mars. ➤ *Pluto* was discovered there by Clyde ➤ *Tombaugh* in 1930. There are five modest telescopes at the Flagstaff site together with the 33-cm (13-inch) telescope used by Tombaugh to search for Pluto, restored and returned to its original dome in 1996. In addition, the observatory operates three other telescopes at Anderson Mesa, 24 km (15 miles) south-east of Flagstaff, including the 1.8-m (72-inch) Perkins Telescope. The construction of the 4.2-m Discovery Channel Telescope began in 2005 at a site 65 km (40 miles) south-east of Flagstaff.

luminous blue variable (LBV) A rare kind of star that is extremely massive and luminous. The most massive stars known are of this type. Examples include ➤ *Eta Carinae* and ➤ *P Cygni*. LBVs undergo irregular eruptions during which they lose material 10 or 100 times faster than they do in their quiescent state. Though their visual magnitude changes, their total luminosity does not. Rather, the change is in the wavelength of the energy they emit. Many are surrounded by nebulosity created by gas they have ejected.

Luna A series of Soviet space missions to the Moon launched between 1963 and 1976. The first three were named ➤ *Lunik*. Luna 9 achieved the first soft landing on the Moon, in the Oceanus Procellarum in January 1966. In March 1966, Luna 10 became the first lunar orbiting satellite. Luna 16 in September 1970, Luna 20 in 1972 and the last of the series, Luna 24 in August 1976, returned soil samples. Lunas 17 and 21 landed the ➤ *Lunokhod* roving vehicles on the Moon. Successes were also achieved with Lunas 11, 12 and 13 (1966), 14 (1968), 19 (1971) and 22 (1974).

lunar Pertaining to the Moon.

lunar eclipse ➤ *eclipse*.

Lunar Orbiter A series of American spacecraft launched in 1966 and 1967 with the primary objective of mapping the Moon and locating suitable landing sites for the manned ➤ *Apollo program*. They undertook the first systematic exploration of the Moon's surface and all five craft in the series were very successful.

Lunar Prospector A NASA mission to orbit the Moon launched in November 1997. It spent 18 months acquiring the most detailed and complete maps ever obtained of the chemical composition of the lunar surface and the Moon's magnetic and gravity fields. In July 1999 it was deliberately crashed near to the Moon's south pole in the hope that the signature of water might be detected in material vaporized by the impact. In the event, nothing was detected.

Astronaut Dave Scott on the Apollo 15 Lunar Roving Vehicle.

Lunar Reconnaissance Orbiter A NASA mission to the Moon scheduled for launch in October 2008. Its main purpose is to collect information needed in planning for astronauts to travel to the Moon in the future. A second payload will share the launch. The Lunar Crater Observation and Sensing Satellite will travel independently of the orbiter and crash into the lunar surface in an experiment to search for water ice.

Lunar Roving Vehicles (LRV) Battery-powered vehicles for traveling on the Moon's surface taken with the last three ➤ *Apollo program* missions (15, 16 and 17). The journeys made by the astronauts of Apollos 15, 16 and 17 were 28, 27 and 35 km (17.5, 16.5 and 22 miles), respectively.

lunar transient phenomenon (LTP) A suspected temporary appearance of colored patches or obscuration on the surface of the Moon. (The term *transient lunar phenomenon*, or TLP, is also used.) Reported observations are associated particularly with the craters Aristarchus, Gassendi and Alphonsus. It remains unclear whether or not real phenomena have been observed.

lunation A complete cycle of the phases of the Moon. One lunation lasts 29.530 59 days, which is also called a synodic month.

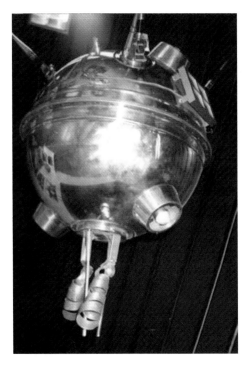

Lunik (alternatively Luna) 1 in 1959 became the first spacecraft to reach the Moon.

Lunik The name of the first three Moon missions launched by the Soviet Union in January, September and October 1959. Lunik 1 missed the Moon by 5000 km (3000 miles). Lunik 2 crashed near the crater Archimedes, but Lunik 3 returned the first pictures of the lunar farside. Subsequent spacecraft in the series were named ➤ *Luna*, starting with Luna 4.

Lunokhod An automated roving vehicle landed on the Moon during two unmanned Soviet missions, Luna 17 and Luna 21. Lunokhod 1 was landed at a site in the western part of Mare Imbrium by Luna 17 on November 17, 1970 and operated for 10 months. Lunokhod 2 was delivered on January 16, 1973, by Luna 21 to the eastern part of Mare Serenitatis, where it worked for four months. The total distances traveled were 10.5 km and 37 km (6.5 and 23 miles), respectively. Each eight-wheeled vehicle carried cameras, a communication system, a laser reflector, a magnetometer, solar panels and a cosmic ray detector.

Lupus (The Wolf) An ancient southern constellations between Scorpius and Centaurus. It contains eight stars brighter than fourth magnitude.

l.y. Abbreviation for ➤ *light year*.

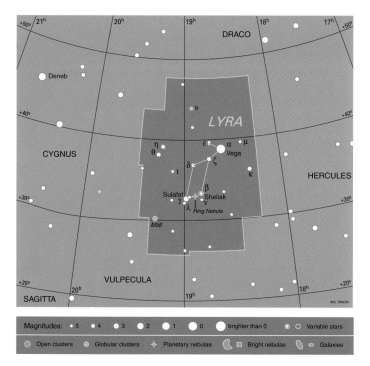

A map of the constellation Lyra.

Lynx An obscure northern constellation introduced in the late seventeenth century by Johannes ➤ *Hevelius* to fill a gap between Auriga and Ursa Major. It contains only two stars brighter than fourth magnitude.

Lyra (The Lyre) A small but prominent constellation in the northern hemisphere. Its brightest star, ➤ *Vega*, is zero magnitude and the fifth-brightest star in the sky. There are three other stars brighter than fourth magnitude. Epsilon Lyrae is a "double double," consisting of a widely spaced pair of close double stars. Lyra also has one of the best-known of all ➤ *planetary nebulae*, the ➤ *Ring Nebula*.

Lyrids An annual ➤ *meteor shower*, sometimes called the April Lyrids. Its radiant lies on the border between the constellations Lyra and Hercules. The shower peaks around April 22, and its normal limits are April 19–25. The meteor stream responsible is associated with Comet Thatcher. Though usually sparse, this shower has occasionally been good in the past. Records of it date back 2500 years.

Lysithea A small satellite of Jupiter, discovered by Seth B. Nicholson in 1938. It is 36 km (22 miles) across. With Elara, Leda and Himalia it belongs to a family of four moons with closely spaced orbits. Their average distances from Jupiter are between 11.1 and 11.7 million km (6.9 and 7.3 million miles).

M

M Abbreviation for the ➤ *Messier catalog* of galaxies, nebulae and star clusters.

Mab A small inner satellite of Uranus, discovered in 2003. It is about 25 km (16 miles) across. Mab orbits Uranus at the same distance as one of Uranus's rings and may be the source of the material for that ring.

Magellan A US spacecraft placed in orbit around ➤ *Venus* to map the surface by means of ➤ *synthetic aperture radar*. It was launched from the Space Shuttle Atlantis on May 4, 1989. The use of radar was essential because Venus is perpetually covered by opaque cloud. Magellan arrived at Venus on August 10, 1990, and completed its first phase of operations in May 1991, having mapped 84 percent of the surface. The next phase of observation involved filling in gaps and making more detailed observations. Magellan burned up in the venusian atmosphere in 1994.

Magellanic Clouds Two small, irregular galaxies, which are satellites of our own ➤ *Galaxy*. They are visible as hazy patches in the southern sky. The Large Magellanic Cloud (LMC) is in the constellation Dorado and is about 170 000 light years away. The Small Magellanic Cloud (SMC), in Tucana, is about 210 000 light years distant.

Magellanic Stream A long streamer of neutral hydrogen gas apparently spanning the 200 000 light years between the ➤ *Magellanic Clouds* and our own ➤ *Galaxy*. It forms an arc 150° long in the southern sky.

Magellan Telescopes Two 6.5-m telescopes at the ➤ *Las Campanas Observatory* in Chile built by a consortium led by the Carnegie Institution of Washington. The first instrument, called the Walter Baade Telescope, was completed in 2000 and the second, the Landon Clay Telescope, was finished in 2002. They are designed to observe large areas of sky simultaneously.

magnetar A ➤ *neutron star* that has a much stronger magnetic field than normal. Magnetars possess magnetic fields far more intense than any other known type of object. The powerful field results in X-ray emission and periodic "starquakes" that crack the surface crust of the star and release bursts of energy in the form of gamma-ray flashes. Occasionally, extremely intense gamma-ray flares are observed, thought to be when the entire crust of the star is shattered. ➤ *Soft gamma repeaters* are a type of magnetar.

magnetic star A star with an exceptionally strong magnetic field. Magnetic fields more than a thousand times stronger than the Sun's general field have been measured for a group of ➤ *A stars*, which also have peculiar spectra.

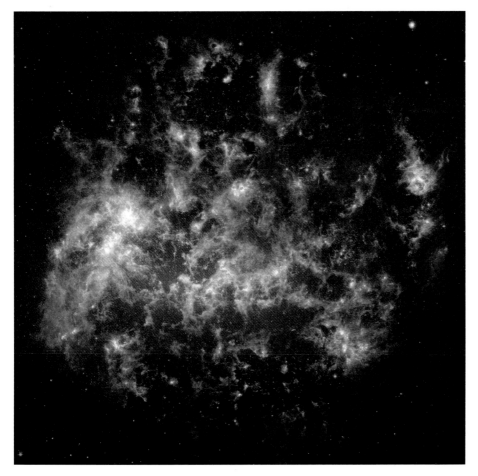

An infrared mosaic of the Large Magellanic Cloud from the Spitzer Space Telescope.

magnetic storm A major world-wide disturbance in the ➤ *geomagnetic field*, caused by an enhancement of the ➤ *solar wind*, such as a ➤ *coronal mass ejection*. The onset of a storm is marked by a sudden increase in Earth's magnetic field when it is compressed by the oncoming solar wind. The magnetic field varies rapidly during a storm, which normally lasts for about a day. Major magnetic storms can induce electric currents in power lines on Earth, sometimes disrupting the supply when safety cut-outs are triggered. ➤➤ *magnetic substorm*.

magnetic substorm A disturbance lasting about half an hour in the ➤ *geomagnetic field* in the region of Earth's geomagnetic poles. Substorms may take place during a ➤ *magnetic storm*, or when there is no other magnetic disturbance. An effect of a magnetic substorm is to propel charged particles in the ➤ *magnetotail*

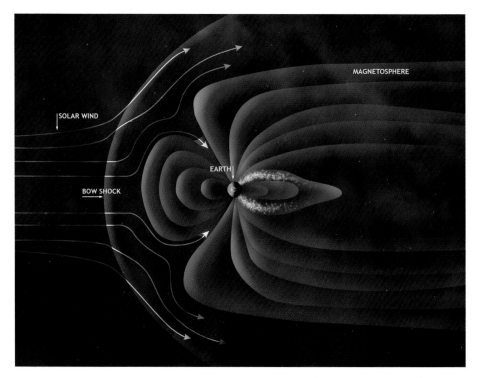

Earth's magnetosphere.

both away from and towards Earth. The particles directed towards Earth produce ➤ *auroras*.

magnetograph An instrument used in solar astronomy for mapping the strength, direction and distribution of magnetic field across the surface of the Sun.

magnetopause A region between 100 and 200 km (60 and 120 miles) thick at the boundary between the ➤ *magnetosphere* and the ➤ *solar wind*.

magnetosphere The region around the Earth, or any other planet, within which its natural magnetic field is constrained by the ➤ *solar wind*.

magnetotail The part of the ➤ *magnetosphere* of Earth, or of any other planet, extending like a tail in the direction opposite the Sun on the planet's nightside. Earth's magnetotail begins about 8–10 Earth radii away on the nightside and stretches for between 80 and 1000 Earth radii according to the state of the solar wind.

magnitude A measure of the brightness of a star or other celestial object. On the magnitude scale, the lowest numbers mean greatest brightness. The brightness of stars as observed from the Earth – their ➤ *apparent magnitude* – depends on their luminosity and their distance. ➤ *Absolute magnitude* is a measure of luminosity on the magnitude scale.

The magnitude system grew from early attempts to classify the apparent brightness of stars. The Greek astronomer ➤ *Hipparchus* (*c.* 120 BC) ranked stars on a scale of magnitude from first, for the brightest stars, to sixth, for those just detectable in a dark sky by naked eye. This qualitative approach was standardized in the mid-nineteenth century so that a difference in magnitude of 5 corresponds to a brightness ratio of 100:1. If two stars differ by one magnitude, their brightnesses differ by a factor equal to the fifth root of 100, which is 2.512. The zero point of the scale was set by assigning standard magnitudes to a small group of stars near the north celestial pole.

The magnitude of an object varies with the wavelength of radiation observed. Visual magnitude corresponds to the normal sensitivity of the human eye. Bolometric magnitudes take account of all radiation, both visible and outside the visible range. Magnitudes measured over a defined wavelength range are often described as "colors." The accurate determination of magnitudes is called ➤ *photometry.*

main sequence ➤ *Hertzsprung–Russell diagram.*

major planet Any of the eight planets Mercury, Venus, Earth, Mars, Jupiter, Saturn, Uranus or Neptune. Before 2006, Pluto was included with the major planets but since then has been considered a ➤ *dwarf planet* according to criteria adopted by the ➤ *International Astronomical Union.*

mantle The layer inside a planet or satellite lying below the crust but above the core. Earth's mantle contains 65 percent of the mass of our planet.

mare (pl. maria) The Latin for "sea," which is applied to the extensive dark areas on the Moon. Its use for features on the Moon dates from a time when it was believed that the darker areas were liquid water.

The lunar maria are actually "seas" of solidified lava. They were created more than 4000 million years ago when the Moon was volcanically active. The molten lava flowed into huge basins that had been excavated by the impacts of large meteorites. By this stage in the Moon's history, the frequency of meteoritic impacts had fallen. There are fewer craters on the lunar maria than on the brighter "highland" areas.

Margaret A small moon of Uranus discovered in 2003. It is about 11 km (7 miles) across.

Maria Mitchell Observatory An observatory in Nantucket, Massachusetts, that was founded in 1908 as a memorial to Maria ➤ *Mitchell* (1818–88), a pioneering scientist and teacher in an era when very few women took on such academic work. She won international fame after discovering a comet in 1847. The observatory houses 18-cm (7-inch) and 20-cm (8-inch) telescopes used primarily for education.

Mariner A series of spacecraft launched by the USA during the 1960s and 1970s in a program to explore the planets Mercury, Venus and Mars.

Part of a lunar mare region, the Mare Tranquilitatis (Sea of Tranquility), photographed from orbit by the crew of Apollo 15.

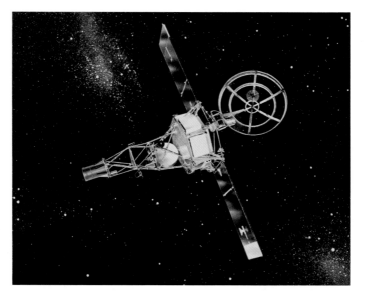

Mariner 2, the first successful interplanetary spacecraft.

In 1962, Mariner 2 achieved the first successful flyby of Venus, to be followed by Mariner 5 in 1967. Mariner 4, launched in 1964, was the first successful mission to Mars and revealed the existence of martian craters. Mariners 6 and 7 followed in 1969. Mariner 9 was put into orbit around Mars in 1971 and returned over 7000 images. Mariner 10, launched in 1974, was the first two-planet mission. It made three separate encounters with Mercury, providing 10 000 images, and a flyby of Venus. There were a total of seven successful missions in the Mariner series. Numbers 11 and 12 were renamed ➤ *Voyagers 1 and 2*.

Mars The fourth major planet from the Sun, often known as the Red Planet because of its distinctive color, noticeable even to the naked eye. It takes 687 days to complete one orbit at an average distance of 1.52 AU from the Sun.

The Hubble Space Telescope captured this view of Mars on Jun 26, 2001 when Mars was 68 million km (43 million miles) away. Several dust storms are active, one large one over the north polar cap and another over the Hellas impact basin in the southern hemisphere (to the lower right).

Mars is one of the terrestrial planets with a diameter of 6786 km (4217 miles), which is just over half the size of Earth. It has long been regarded as the planet (other than Earth) most likely to have life, a view encouraged by the presence of polar ice caps and observations of seasonal changes. Nineteenth century observers, notably Percival ➤ *Lowell*, convinced themselves that they could make out systems of straight channels and speculated that they might be artificial constructions. Exploration of the planet by spacecraft has so far failed to produce any evidence that life exists currently on Mars. However, studies of meteorites believed to be of martian origin have fuelled speculation that microscopic life at least may have existed on Mars in the remote past when the climate was wetter and warmer, and the possibility of microscopic life in subsurface rocks cannot be excluded.

Successful missions to Mars include ➤ *Mariner* 4 in 1965, Mariners 6 and 7 in 1969, Mariner 9 in 1971, ➤ *Vikings* 1 and 2 in 1976, ➤ *Mars Pathfinder*, which landed in July 1997, ➤ *Mars Global Surveyor*, ➤ *Mars Odyssey*, ➤ *Mars Express* and ➤ *Mars Exploration Rovers*.

The relatively low density of Mars (3.95 times that of water) suggests that 25 percent of its mass is contained in an iron core. There is a weak magnetic field, about 2 percent the strength of the Earth's. The crust is rich in olivine and ferrous oxide, which gives the rusty color.

The tenuous martian atmosphere is composed of 95.3 percent carbon dioxide, 2.7 percent molecular nitrogen and 1.6 percent argon, with oxygen as a major trace constituent. The atmospheric pressure at the surface is only 0.7 percent that at the surface of the Earth. However, strong winds in the atmosphere cause extensive dust storms, which occasionally engulf the entire planet.

Clouds and mist sometimes appear. Early morning fog forms in valleys and clouds occur over the high mountains of the Tharsis region. In winter, the north polar cap is swathed in a veil of icy mist and dust, known as the polar hood. A similar phenomenon is seen to a lesser extent in the south.

The polar regions are covered with a thin layer of ice, which is a mixture of water ice and solid carbon dioxide. High-resolution images show a spiral formation and strata of wind-borne material. The north polar region is surrounded by stretches of dunes. The polar ice caps grow and recede with the seasons, which Mars experiences because its rotation axis is tilted by 25° to its orbital plane.

The martian year is about twice as long as Earth's year, so the seasons are also longer. However, the relatively high eccentricity of Mars's orbit makes them of unequal duration: southern summers, which occur when Mars is near perihelion, are shorter and hotter than those in the north. Seasonal changes in the appearance of features as observed from Earth are explained as physical and chemical changes.

There is a marked difference in the nature of the terrain in the two halves of Mars roughly either side of a great circle tilted at 35° to the equator. The more southerly part consists largely of ancient, heavily cratered terrain. Major impact basins – the Hellas, Argyre and Isidis planitiae – are located in this hemisphere. The north is dominated by younger, more sparsely cratered terrain, lying 2–3 km lower. The highest areas are the large volcanic domes of the Tharsis and Elysium planitiae. Both areas are dominated by several huge extinct volcanoes, the largest of which is ➤ *Olympus Mons*.

These volcanic areas are located at the east and west ends of an immense system of canyons, the Valles Marineris, which stretches for more than 5000 km (3000 miles) around the equatorial region and has an average depth of 6 km. It is believed to have been caused by faulting associated with the upthrust of the Tharsis dome.

There is evidence, in the form of flow channels, that liquid water once existed on the surface of Mars. Channels from the Valles Marineris appear to have been created in some kind of sudden flood. There are also sinuous, dried-up river beds with many tributaries, found only in the heavily cratered terrain.

Mars has two small moons, ➤ *Phobos* and ➤ *Deimos*, which are in near-circular orbits in the equatorial plane, close to the planet. They are very difficult to see from Earth. They are so different from Mars that it seems likely they are captured asteroids.

Mars Exploration Rovers Two robotic rovers that were landed on Mars by NASA in 2004. Each weighs 185 kg (408 lb) and carries instruments to investigate the

Mars Exploration Rovers. An artist's concept of one of the rovers on the surface of Mars.

geology of Mars and return images. The rovers travel slowly on six wheels, controlled by radio signals from Earth. The rover called "Spirit" was launched on June 10, 2003 and landed in Gusev Crater on January 3, 2004. Its twin, named "Opportunity," was launched on July 7, 2003 and landed on January 25, 2004 in Meridiani Planum, on the side of Mars opposite Gusev Crater. They bounced down protected by airbags. The performance of the rovers far exceeded initial expectations that they would operate for about three months. Both were still working in mid-2007.

Mars Express An ESA mission to Mars, launched on June 2, 2003. The spacecraft carried seven instruments to study Mars over at least one martian year. It was successfully put in orbit on December 25, 2003 and later maneuvered into its operational near-polar orbit. It carried a lander, Beagle II, equipped to search for signs of past or present life, but the attempted landing on Mars failed.

Mars Global Surveyor A NASA mission to Mars, launched on November 7, 1996 for arrival in September 1997. It was successfully put into a high elliptical orbit around Mars on September 11, 1997. In the following months, ➤ *aerobraking* was used to maneuver it gradually into an almost circular near-polar orbit from which to carry out systematic mapping beginning in March 1999. Many thousands of high-resolution images of the martian surface were returned.

2001 Mars Odyssey A Mars orbiter launched by NASA on April 7, 2001. It arrived at Mars on October 24, 2001, and its main program of scientific work was carried out between February 2002 and August 2004 though its mission was extended beyond that date. The spacecraft mapped the distribution of minerals and chemical elements over the martian surface and discovered that there are large amounts of water just below the surface in the polar regions. Another of its tasks was to study the radiation environment near Mars to see what risks there may be for any future astronauts. 2001 Mars Odyssey also acts as a communications relay station for lander missions, including the Mars Exploration Rovers.

Mars Pathfinder A NASA Mars mission, which was launched on December 4, 1996 and arrived at Chryse Planitia on July 4, 1997. The main objective was to test a low-cost means of sending a spacecraft and a surface roving vehicle to land on the martian surface. It carried a 10-kg (22-lb) miniature rover, named "Sojourner," equipped to measure the chemical composition of the surface rocks and soil and to take images around the landing site in Ares Vallis.

The lander's impact was cushioned by airbags, which bounced several times before coming to rest. The performance of both the lander and the rover exceeded expectations and they were able to continue operating for nearly three months, until September 27. Panoramic views of the landscape were returned and Sojourner successfully traveled on expeditions covering about 80 m. Instruments on the lander monitored atmospheric conditions at the

An artist's concept of the Mars Reconnaissance Orbiter.

surface. Measurements of the martian atmosphere also were made during the parachute descent.

Mars Reconnaissance Orbiter A NASA mission to Mars, launched on August 12, 2005. It arrived in orbit in March 2006. Science operations began in November 2006, after the spacecraft's orbit had been gradually adjusted by ➤ *aerobraking*. The orbiter is equipped with more advanced instruments than any previous mission and is expected to return several times more data than all previous Mars missions combined. A particular objective of the mission is to investigate the history and distribution of water on Mars.

Mars Science Laboratory A robotic rover NASA plans to send to Mars in December 2009 to arrive in October 2010. It will be twice as long and three times as heavy as the ➤ *Mars Exploration Rovers* and will carry more ambitious equipment. It will be able to collect samples of martian soil and rock and analyze them.

mascon An area of anomalously strong gravity on the Moon. The word is a contraction of "mass concentration." Mascons are presumed to indicate the presence of rocks denser than average, though there is no consensus about exactly how they formed. The areas are roughly circular and are associated with ➤ *mare* areas.

mass transfer The flow of material from one star to another in a close binary system. The material transferred may stream directly onto the star's surface or form an ➤ *accretion disk.*

253 Mathilde An asteroid that was imaged by the ➤ *Near-Earth Asteroid Rendezvous* (NEAR) mission in a close flyby on June 27, 1997. It is a uniformly dark ➤ *C-type*

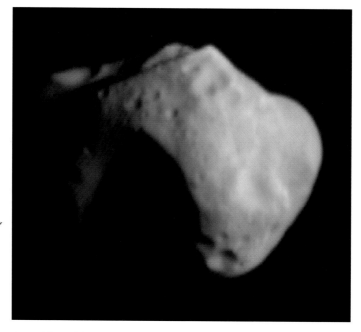

Asteroid Mathilde imaged by the NEAR spacecraft.

asteroid. NEAR found its mean diameter to be 52 km (33 miles). Five craters with diameters in excess of 20 km (12 miles) were identified on the side of the asteroid in sunlight at the time of NEAR's encounter. Mathilde's rotation period was also measured and found to be unexpectedly long at 17.4 days.

Mauna Kea Observatories Observatories on a mountain-top on the island of Hawaii at a height of 4200 m (13 800 feet). The site is one of the best in the world for optical, infrared and submillimeter-wave astronomy and there have been observatories there since 1970. The first large telescope to be installed was a 2.24-m (88-inch) reflector belonging to the University of Hawaii. In 1979, three more major telescopes began operation: the ➤ *United Kingdom Infrared Telescope* (UKIRT), NASA's ➤ *Infrared Telescope Facility* (IRTF) and the ➤ *Canada–France–Hawaii Telescope*. The ➤ *James Clerk Maxwell Telescope* working in the millimeter wave region, and the Caltech 10.4-m (34-foot) Submillimeter Telescope were opened in 1987. The first of the two telescopes of the ➤ *Keck Observatory* was completed in 1992, and the second in 1996. Two 8-m instruments, the Japanese ➤ *Subaru Telescope* and one of the international ➤ *Gemini Telescopes*, were completed in 1999. The ➤*Submillimeter Array* became fully operational in 2003.

Maunder diagram ➤ *butterfly diagram*.

An aerial view at sunrise of Mauna Kea Observatories with Gemini North (the large silver dome) on the foreground ridge.

Maunder minimum A period of about 70 years, starting around 1645, during which ➤ *solar activity* was consistently at a low level and sunspots were rare. For 37 years no aurora was recorded.

Maxwell Montes The highest mountain peaks on Venus, located on Ishtar Terra. They are up to 11.5 km (38 000 feet) high.

Mayall Telescope A 4-m (158-inch) optical reflecting telescope at ➤ *Kitt Peak National Observatory*, belonging to the US ➤ *National Optical Astronomy Observatories*. It has been in operation since 1973.

McDonald Observatory An observatory belonging to the University of Texas located on Mount Locke in the Davis Mountains near Fort Davis. It was established in 1932, financed by a bequest from a wealthy Texan banker and amateur astronomer, William J. McDonald. The original instrument, completed in 1938 and still in use, is a 2.08-m (82-inch) reflector, known as the Otto Struve Telescope after the observatory's first director. In 1969, a 2.72-m (107-inch) reflector came into use. The latest addition is the ➤ *Hobby–Eberly Telescope*, completed in 1996. There are also several smaller telescopes and a 5-m (16-foot) millimeter-wave dish.

McMath–Pierce Solar Telescope Facility A large solar observatory located at ➤ *Kitt Peak* and belonging to the US ➤ *National Solar Observatory*. The main telescope was completed in 1962 and consists of a 1.6-m (60-inch) mirror, mounted on a tower, which directs the sunlight down a long shaft inclined at 32° to the horizontal. Much of it is below ground. The telescope produces a

The McMath–Pierce Solar Telescope Facility.

high-resolution image of the Sun 75 cm (30 inches) in diameter. The entire building is encased in copper, and coolants are piped through the outer skin to maintain a uniform temperature inside.

Megaclite A small outer moon of Jupiter discovered in 2000. It is about 5 km (4 miles) across.

megaparsec (symbol Mpc) A unit of distance equal to one million ➤ *parsecs*.

Mensa (The Table or Table Mountain) A faint southern constellation introduced in the mid-eighteenth century by Nicolas L. de Lacaille with the longer name Mons Mensae, the Table Mountain. It contains no stars brighter than fifth magnitude but part of the Large ➤ *Magellanic Cloud* lies within its boundaries.

Merak (Beta Ursae Majoris) One of the two stars of the ➤ *Big Dipper* in Ursa Major, called the ➤ *Pointers*. The stars in the Plough were designated by position rather than in brightness order. It is therefore actually the fifth-brightest star in the constellation, with a magnitude of 2.4. Merak is an ➤ *A star* 79 light years away and its name, derived from Arabic, means "the loin."

Mercury The nearest major planet to the Sun and the smallest of the terrestrial planets.

Observation of Mercury from Earth is very difficult, partly because it is a small planet and partly because it is never more than 28° from the Sun on the sky. Because it lies closer to the Sun than Earth, Mercury (like Venus) goes through a cycle of phases, similar to those of the Moon. Hardly any surface detail can be discerned and very little was known about the planet until the flybys of ➤ *Mariner* 10 in 1974 and 1975. Mariner 10 followed an orbit around the Sun that made it pass close to Mercury three times before it ran out of attitude-control gas. The images it returned cover about 35 percent of the surface of Mercury.

A mosaic of Mariner 10 images of Mercury.

Ancient, heavily cratered terrain accounts for 70 percent of the area
surveyed by Mariner 10. The most significant single feature is the Caloris Basin,
a huge impact crater 1300 km (800 miles) across – a quarter the diameter of the
planet. The basin has been filled in and is now a relatively smooth plain. The
impact took place 3800 million years ago and temporarily revived volcanic
activity, which had mostly ceased 100 million years earlier. This volcanism
created the smoother areas inside and around the basin. At the point on Mercury
diametrically opposite the impact site, there is curious chaotic terrain that must
have been created by the shock wave. Cliffs between a few hundred and 3000 m
high are characteristic feature on Mercury. It is believed that they formed when
the planet's crust shrank as it cooled. In places they cut across craters.

The length of Mercury's orbital and rotation periods means that a "day" on
Mercury lasts two mercurian "years." This leads to immense temperature
contrasts: at perihelion, it reaches 430 °C where the Sun is overhead; the
nighttime temperature plunges to −170 °C. The high daytime temperatures
and Mercury's small mass make it impossible for an atmosphere to be retained.
Small amounts of helium that are detected may be the product of radioactive
decay of surface rocks or captured from the ➤ *solar wind*.

The average density of Mercury is only slightly less than that of Earth. Taking
account of its smaller size and lower interior pressure leads to the conclusion
that Mercury has a substantial iron core accounting for 70 percent of its mass and
75 percent of its total diameter. There is also a magnetic field of about 1 percent
the strength of the Earth's field, providing further evidence for a metallic core.

A NASA spacecraft to study Mercury, called ➤ *Messenger*, was launched in
August 2004 and will enter orbit around Mercury in 2011.

meridian (1) The great circle on the celestial sphere passing through the celestial
poles and the zenith.

meridian (2) A line of longitude on the Earth, or on another astronomical body.
On Earth, the meridian through Greenwich marks the zero of longitude and is
sometimes called the prime meridian.

MERLIN Acronym for Multi-Element Radio Linked Interferometer Network, a
network of radio telescopes at various locations in the UK, operated by the
University of Manchester from ➤ *Jodrell Bank Observatory*. MERLIN is used for
➤ *long-baseline interferometry* and can be linked to other telescopes and networks
worldwide to carry out ➤ *very-long-baseline interferometry*.

Merope One of the brighter stars in the ➤ *Pleiades*.

mesosphere A region of Earth's atmosphere above the stratosphere, between
heights of about 50 and 85 km (30 and 50 miles), in which the temperature
decreases with height to −90 °C at its upper boundary, the mesopause.

Messenger A NASA mission to Mercury launched on August 3, 2004. The flight
plan involves flybys of Venus in October 2006 and June 2007, and three flybys

of Mercury in January and October 2008 and September 2009. It will go into orbit around Mercury in March 2011. Messenger carries eight instruments to investigate Mercury's structure, composition, ➤ *magnetosphere* and ➤ *exosphere* and is expected to operate for a year.

Messier catalog A catalog of about a hundred of the brightest galaxies, star clusters and nebulae, compiled by the French astronomer Charles Messier (1730–1817). His initial list, published in 1774, contained 45 objects but it was supplemented later with additional discoveries and contributions from Messier's colleague, Pierre Méchain. Objects in the catalog, which is still widely used, are identified by the prefix "M" and their catalog number.

The list was not compiled systematically. Messier's main interest was searching for comets, and he noted hazy objects he spotted during comet searches. Some were first recorded by Messier, but others were already known.

There are some errors and discrepancies in the list as published. M40 is a double star and M73 a group of four stars, but not a true cluster. The identification of M91 is uncertain from the original source, and M102 was a duplication of M101. Messier's own list stopped at number 103 but a further seven objects were added in the twentieth century.

meteor The brief luminous trail observed as a particle of dust or a piece of rock from space enters Earth's upper atmosphere. The popular name for a meteor is "shooting star" or "falling star."

Earth is constantly bombarded with material from space. The individual objects range in size from rocks of several kilograms down to microscopic particles weighing less than one millionth of a gram. It is estimated that more than 200 million kilograms (200 000 tons) of meteoric material is swept up by the Earth in the course of a year. One-tenth of it reaches the ground, in the form of ➤ *meteorites* and ➤ *micrometeorites*. The remainder burns up in the atmosphere, becoming visible as *meteor trails*.

A meteor trail recorded during the Leonid meteor shower in 1998.

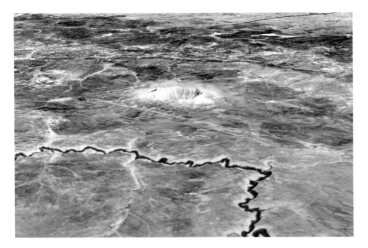

Meteor Crater imaged by a remote sensing instrument aboard the Terra
Earth observation satellite.

Meteoric material typically enters the atmosphere at speeds of around
15 km/s. Heating due to friction causes medium-sized particles to vaporize.
They give off light, leaving a temporary trail of ionized gas. This trail is capable
of reflecting radar signals. Radar has been used to detect meteors during the
daytime when they are too faint to be seen by eye.

Much of the meteoric material in the solar system orbits the Sun in distinct
streams. Many meteor streams have the same orbits as known comets. The
particles may be strung out all along the orbit or concentrated in a swarm. When
Earth happens to cut through a stream, a ➤ *meteor shower* is observed. In addition
to the dozens of regular meteor showers, a background of *sporadic meteors* is
observed throughout the year. They may come from any direction. ➤ *fireball*.

Meteor Crater (also known as Barringer Crater; Canyon Diablo Crater; Coon Butte)
The best-preserved and most famous ➤ *meteorite* crater on the Earth, formed
about 50 000 years ago. Located between Flagstaff and Winslow in Arizona,
USA, it was discovered in 1891. It is a bowl-shaped depression in the ground,
1200 m (4000 feet) across and 183 m (600 feet) deep, surrounded by a wall
30–45 m (100–150 feet) high. The original meteorite is thought to have been of
the iron type and to have weighed more than 10 000 tonnes but most of it was
destroyed on impact. The many scattered fragments of what is called the
Canyon Diablo meteorite total only about 18 tonnes.

meteorite A piece of a ➤ *meteoroid* that has survived its passage through Earth's
atmosphere and has landed on the ground. Individual meteorites are normally
named after the place where they fell. Studies of the paths of a small number
of meteorites observed as ➤ *fireballs*, and subsequently recovered, show they
were in orbits originating in the ➤ *asteroid belt*. The chemical composition of

A piece of the stony-iron Huckitta meteorite from the Northern Territory of Australia. This sliced and polished piece is about 2.5 cm (1 inch) across.

meteorites and the minerals they contain provide clues to the origin and evolution of the solar system.

There are three main classes of meteorite: irons, stony-irons and stones. Stony meteorites are further divided into two important categories: *achondrites* and *chondrites*. Chondrites contain chondrules, small spherical inclusions, made of metal or of silicate or sulfide materials. Chondrules are not present in achondrites.

The chemical composition of chondrites is very similar to that of the Sun and they are thought to represent primitive solar system material that has not been altered significantly by heating. Carbonaceous chondrites have the highest proportion of volatile elements and their composition is closest to the Sun's. Among the achondrites, there are many subtypes with differing chemical and mineralogical composition.

Stony-iron meteorites contain free metal and stony material in roughly equal proportions. Iron meteorites consist almost entirely of iron and nickel. Over 40 different minerals have been identified in them, though the basic constituents are two forms of iron–nickel alloy, kamacite and taenite.

meteoroid A small piece of rock or dust in space, especially one with the potential to become a ➤ *meteor* or ➤ *meteorite*. The word "meteoroid" is normally used to mean an object about 100 m (300 feet) across, or smaller.

meteor shower ➤ *Meteors* appearing to come from one point in the sky over a period of several hours or days. Meteor showers result when Earth cuts through a stream of meteoric material in space. Many showers recur around the same time each year when Earth reaches the same place in its orbit. Dozens of such annual showers are known, though only a handful give good regular displays. Very occasionally, a particularly dense swarm of particles produces tens or hundreds of meteors every minute. A more typical, average rate for a good regular shower would be between 20 and 50 meteors per hour.

The trails of shower members, if traced back on a sky map, appear to intersect at a point, called the radiant. However, this is only an effect of perspective. The meteors are actually traveling along parallel tracks when they enter the atmosphere. Meteor showers are normally named after the constellation in which their radiant lies. For example, the radiant of the Perseids is in Perseus.

meteor stream An extended swarm of meteoric material in orbit around the Sun. Many meteor streams are known to be spread out around the orbits of particular comets. The particles tend to become distributed more or less evenly along the orbit over time. A meteor stream that has formed relatively recently may still be concentrated near its parent comet. ➤ *meteor, meteor shower.*

Methone A tiny moon of Saturn, about 3 km (2 miles) across, orbiting at a distance of 194 000 km (120 546 miles) between Mimas and Enceladus. It was discovered by the ➤ *Cassini* spacecraft in 2004, but may have been imaged once before, in 1981, by ➤ *Voyager.*

9 Metis An asteroid discovered in 1848 by Andrew Graham from Ireland. Its diameter is 190 km (118 miles).

Metis Jupiter's innermost known moon, discovered by Stephen P. Synott in 1979. It is about 40 km (25 miles) in diameter, irregularly shaped and reddish in color.

Metonic cycle A period of 19 tropical ➤ *years*, after which the phases of the Moon recur on the same days of the year. This happens because 19 tropical years equals 6 939.60 days, which is almost exactly 235 synodic ➤ *months* (6939.69 days). The discovery of the cycle is attributed to the Greek astronomer Meton, who worked in the fifth century BC.

Mice A popular name for the pair of interacting galaxies, NGC 4676 A and B. Long, tail-like streamers of material extend from the galaxies, giving them shapes reminiscent of a pair of mice.

microlensing The result of an invisible stellar-sized object acting as a ➤ *gravitational lens* and amplifying the light of a more distant star directly behind it. The star brightens temporarily in a characteristic way. Surveys for microlensing events have been used to search for invisible dim stars or ➤ *brown dwarfs*, which may account for as much as 90 percent of the total mass of the Galaxy.

micrometeorite A particle of meteoritic material so small that it does not burn up in the atmosphere. Micrometeorites fall to Earth as a rain of minute dust particles. It is estimated that four million kilograms of them reach the ground each year. The size of the particles is typically less than 120 μm. ➤ *meteor, meteorite.*

The interacting galaxies nicknamed "the Mice."

micrometer (1) In general, any instrument for measuring small distances accurately. In astronomy micrometers attached to telescopes are used by visual observers to measure the angular separations of pairs of objects, such as ➤ *binary stars*.

micrometer (2) (micron, symbol μm) A unit of measurement equal to one millionth of a meter.

Microscopium (The Microscope) A small, insignificant, southern constellation introduced in the mid-eighteenth century by Nicolas L. de Lacaille. Its brightest star is magnitude 4.7.

microquasar A stellar-mass ➤ *black hole*, which is physically similar to a ➤ *quasar*, but on a much smaller scale. Microquasars are characterized by powerful jets traveling almost at the speed of light.

microwave astronomy The study of radio waves from astronomical sources across a wide band of the electromagnetic spectrum from the far infrared at 1 mm wavelength to short-wave radio at about 6 cm. At the shorter wavelength end, these waves are absorbed by Earth's atmosphere. ➤ *Cosmic Background Explorer, millimeter-wave astronomy, radio astronomy, submillimeter-wave astronomy.*

microwave background radiation ➤ *cosmic background radiation.*

Milankovic cycles Small variations in the tilt of Earth's rotation axis and the eccentricity of its orbit around the Sun. They have been linked with long-term climate variations and the incidence of ice ages.

Milk Dipper An ➤ *asterism* formed by the stars Zeta, Tau, Sigma, Psi and Lambda in the constellation Sagittarius. It is presumably named because of its ladle shape and the fact that the most intense part of the Milky Way lies in Sagittarius.

Milky Way A band of hazy light circling the sky. It is the combined light of vast numbers of stars in our own ➤ *Galaxy*. The term Milky Way is also used to mean the Galaxy itself.

The band of light we see around the sky is the disk of the Galaxy viewed from within. The Sun is situated two-thirds of the way out towards the edge of the galactic disk, and the Milky Way appears brightest in the direction of the bulge at the galactic center, which lies in the constellation Sagittarius. Clouds of obscuring dust, such as the ➤ *Coalsack* near the Southern Cross, gives the Milky Way a patchy appearance in places.

The main constellations through which the Milky Way passes are Perseus, Cassiopeia, Cygnus, Aquila, Sagittarius, Scorpius, Centaurus, Vela, Puppis, Monoceros, Orion, Taurus and Auriga.

millimeter-wave astronomy Astronomical observations of radio waves with wavelengths between about 1 and 10 mm. This part of the radio spectrum contains many features caused by complex molecules and it is specially important in the study of ➤ *molecular clouds*, regions of star formation, ➤ *circumstellar disks* and ➤ *comets*. ➤➤ *Atacama Large Millimeter Array*, *submillimeter-wave astronomy*.

Mimas A moon of Saturn, discovered by William Herschel in 1789. It orbits 185 520 km (115 277 miles) from Saturn and measures 418×392×383 km (260×244×238 miles). Its surface is heavily cratered. The largest crater, Herschel, is 130 km in diameter, one-third the size of Mimas, and has a central peak. The impact that created Herschel must very nearly have shattered Mimas.

Mimosa (Beta Crucis) The second-brightest star in the constellation Crux. It is a giant ➤ *B star* of magnitude 1.3 and is 350 light years away. Mimosa is slightly variable, changing by 0.1 magnitude in about 6 hours.

Minkowski's Footprint A double-lobed nebula surrounding a star in Cygnus, regarded as a ➤ *planetary nebula* in the process of formation. It was discovered by Rudolph Minkowski in 1946.

Mimas imaged by the Cassini spacecraft.

The Mir space station photographed by the crew of the Space Shuttle Endeavor in 1998.

minor planet An alternative name for an ➤ *asteroid*.

Mintaka (Delta Orionis) One of the three stars forming the belt of Orion. At magnitude 2.2, it is the seventh-brightest star in the constellation. It is actually an ➤ *eclipsing binary* and varies in brightness by 0.1 magnitude in a period of 5.7 days. In addition, it has a seventh-magnitude visual companion. The primary star is a ➤ *supergiant* ➤ *O star* lying 915 light years away. Mintaka is a name of Arabic origin meaning "the belt."

minute of arc ➤ *arc minute*.

Mir A Russian space station, launched into Earth orbit in 1986. It was occupied by astronauts regularly, sometimes for many months at a time, before finally being abandoned. In a controlled maneuver, it was allowed to re-enter the atmosphere and crash into the Pacific Ocean in March 2001.

Mira (Mira Ceti; Omicron Ceti) The prototype of a class of long-period variable stars. The name is Latin for "wonderful." Mira was the first variable star to be discovered. The Dutch astronomer David Fabricius noted it at third magnitude in 1596, but found it to be invisible to the naked eye a few months later. He noted it again at third magnitude in 1609. Mira is a giant ➤ *M star* that varies between about second and tenth magnitudes in a period of roughly 332 days. It is shedding large amounts of gas and dust, which form a strong ➤ *stellar wind*. Mira also has a ➤ *white dwarf* companion and material is being transferred onto it.

Mirach (Beta Andromedae) The second-brightest star in the constellation Andromeda. It is a giant ➤ *M star* of magnitude 2.1 and is 200 light years away. Derived from Arabic, Mirach means "girdle."

229

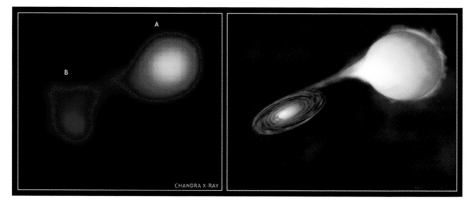

A Chandra X-ray image and an artist's impression of Mira. The Chandra image shows Mira A (right), a highly evolved red giant star, and Mira B (left), a white dwarf. To the right of the image is an artist's concept of the Mira star system.

A Voyager 2 spacecraft image of Miranda taken in 1986

Miranda A small moon of Uranus. It was discovered by Gerard Kuiper in 1948. The ➤ *Voyager 2* spacecraft passed Miranda at a distance of only 3000 km (1800 miles) in 1986, returning very detailed images of its surface.

Though only 472 km (293 miles) in diameter, Miranda has several contrasting types of terrain. Alongside cratered areas, typical of planets and satellites, there are large tracts of grooves and ridges. It seems unlikely that such variety could have been caused by geological activity in such a small satellite. One theory suggests that the satellite was once shattered by a massive impact into several parts that subsequently coalesced again.

Maria Mitchell.

Mirfak (Alpha Persei) The brightest star in the constellation Perseus. It is a yellow supergiant ➤ *F star* of magnitude 1.8 lying 200 light years away. The name comes from Arabic and means "the elbow."

Mirzam (Beta Canis Majoris) The second-brightest star in the constellation Canis Major. It is a giant ➤ *B star* of magnitude 2.0 and the prototype of a class of slightly variable stars. It changes in brightness by just a few hundredths of a magnitude every six hours. Mirzam is 500 light years away.

Mitchell, Maria (1818–1889) Mitchell was the first woman in the USA to be appointed to a professional position in astronomy. She was born on the island of Nantucket, then the world center for the whaling industry, into a Quaker society that expected capable women to work independently. Her father taught her astronomy. With his telescope she discovered a comet in 1847, for which the King of Denmark awarded her a gold medal, and this led to worldwide fame. In 1849 she became the first woman to work full time for the US Nautical Almanac Office, where she computed the position of Venus. She was the very first person to be appointed to the academic staff of Vassar College, Poughkeepsie, New York.

Mizar (Zeta Ursae Majoris) The fourth-brightest star in the constellation Ursa Major. It is an ➤ *A star* of magnitude 2.3 and is about 78 light years away. The Arabic name means "girdle." In the sky, it happens to lie very close to the fourth magnitude star ➤ *Alcor*. However, it is not known whether the two stars are linked. Mizar does have a known fourth-magnitude companion. Both it and Mizar are ➤ *spectroscopic binaries*.

MMT Observatory ➤ *Multiple Mirror Telescope*.

Mneme A small outer moon of Jupiter discovered in 2003. Its diameter is about 2 km (1 mile).

mock Sun (parhelion; sundog) A circular patch of light in the sky, 22° away from the real Sun. Mock Suns usually appear in pairs, one either side the real Sun, on a circular halo of light, though one may be brighter than the other. The effect is caused by ice crystals in Earth's atmosphere.

molecular cloud A cloud of interstellar matter consisting mainly of molecules of gas. There are two distinct types, both found close to the plane of the Galaxy, within the Milky Way. These are small molecular clouds and giant molecular clouds (GMCs).

The small clouds are typically a few light years across, with 1000 to 10 000 molecules per cubic centimeter and temperatures of around 10–20 K. They may contain even colder condensed "cores," with densities ten or a hundred times greater. These small clouds contain mostly molecular hydrogen (H_2). They are very cold because there is no radiation from stars within to heat them.

Giant molecular clouds are made up primarily of molecular hydrogen and carbon monoxide (CO), but they also contain many other ➤ *interstellar molecules*. They are the most massive entities within our Galaxy, containing up to ten million solar masses, and are typically 150–250 light years across. Their density can be as high as ten million molecules per cubic centimeter. Infrared emission from these clouds is evidence that they are regions of star formation. GMCs are nearly always associated with clusters of hot, massive, young stars. Luminous clouds of ➤ *ionized hydrogen* (H II regions) are created by young stars near the edges of a GMC. The ➤ *Orion Nebula* is one example. A GMC lies behind the glowing nebula. Up to 4 000 GMCs are thought to exist in the Galaxy.

Molonglo Observatory An Australian radio astronomy observatory, located near Canberra, belonging to the University of Sydney.

Monoceros (The Unicorn) A faint constellation, but one rich in stars and nebulae by virtue of its location in the ➤ *Milky Way*, straddling the celestial equator next to Orion. It is not one of the ancient constellations but seems to have gained general acceptance in the mid-seventeenth century. Its brightest two stars are third magnitude; it contains the ➤ *Rosette Nebula*, the ➤ *Cone Nebula* and ➤ *Hubble's Variable Nebula*.

month The time the Moon takes to complete one orbit of the Earth, or to complete one cycle of phases. There are various kinds of month, their length depending on what is used as the start and end point for the measurement.

Type of month	How measured	Length in days
Anomalistic	Perigee to perigee	27.554 55
Draconic	Node to node	27.212 22
Sidereal	Relative to fixed stars	27.321 66
Synodic	Cycle of phases	29.530 59
Tropical	Equinox to equinox	27.321 58

The Moon photographed from space by the Apollo 11 astronauts.

moon A natural ➤ *satellite* of a planet.

Moon The Earth's only natural satellite. It is a barren, heavily cratered world, without liquid water or an atmosphere. It was explored by American astronauts who traveled there during the ➤ *Apollo program* landings of 1969–72 and has been extensively mapped both from orbiting craft and from Earth.

Tidal forces are responsible for the fact that the same side of the Moon now always faces Earth, apart from the minor effects of ➤ *libration*. As the Moon travels around Earth over the course of a month, it goes through its familiar cycle of ➤ *phases*. The Moon shines only by reflected sunlight. The Moon's phase changes according to the proportion of the sunlit side visible from Earth, and it depends on the relative positions of the Sun, Earth and Moon

The terrain on the nearside consists of two basic types: the heavily cratered, light-colored highlands, and the darker, more sparsely cratered "maria" (seas). The maria have roughly circular outlines, a relic of their formation early in the history of the Moon by the impact of large meteorites. A further type of surface terrain is formed by ➤ *ejecta*. Significant areas are marked by material ejected from the large Imbrium and Orientale basins.

The way the Moon formed is uncertain, but it has existed as a separate body for around 4500 million years. The most favored explanation is that it was created in a giant impact event when an asteroid collided with the recently formed Earth. Evidence from the ➤ *Lunar Prospector* mission suggests that the Moon has a very small core about 700 km (450 miles) in diameter, consistent with the impact theory of its origin.

Early in its life the Moon became hot and molten. As it cooled, the crust formed but it was heavily cratered by the impact of large numbers of meteorites, the largest of which created the mare basins. These subsequently filled with dark basaltic lavas. Significant volcanic activity then ceased, at least 2000 million years ago. The farside of the Moon differs from the nearside in that it lacks any large lava-flooded mare areas.

Morehouse, Comet A comet discovered from the USA in 1908. It was the first comet to be studied extensively by photography. Remarkable changes were observed in the structure of its tail. Throughout September 30, 1908 the tail changed continuously. On October 1, it broke off and a tail could not be seen visually, though a photograph of October 2 showed three tails. The breaking off and subsequent growth of tails occurred repeatedly. ➤ *disconnection event.*

morning star Venus or Mercury when it appears in the eastern sky in the early morning before sunrise.

Mount Graham International Observatory An observatory site located on Mount Graham near Safford in south-eastern Arizona. There are three telescopes there: the 1.8-m Vatican Advanced Technology Telescope (VATT), the ➤ *Heinrich Hertz Submillimeter Telescope* and the ➤ *Large Binocular Telescope*.

Mount Stromlo and Siding Spring Observatories The astronomy research facilities of the Australian National University (ANU), operated by the university's Institute of Advanced Studies. All the telescopes at the Mount Stromlo site, near Canberra, were destroyed in a bush fire in January 2003. At the Siding Spring site, which was established in 1962, the Observatories have four telescopes, including a 2.3-m (90-inch) reflector. Siding Spring Mountain is at an altitude of 1000 m (3200 feet) in the Warrumbungle range, most of which forms a national park. This observatory site now also houses a number of telescopes owned by organizations other than the ANU, including the ➤ *United Kingdom Schmidt Telescope*, and the ➤ *Anglo-Australian Telescope*.

Mount Wilson Observatory An observatory near Pasadena, California, located on Mount Wilson at an altitude of 1750 m (5700 feet). The first instrument located there was a horizontal solar telescope built in 1904. Two solar tower telescopes were added in the next few years. A 1.5-m (60-inch) reflecting telescope was begun in 1904 and brought into service in 1908. It was the largest telescope in the world until the opening of the 100-inch (2.5-m) ➤ *Hooker Telescope* at the site in 1917. The observatory is owned by the Carnegie Institution of Washington.

The Mount Wilson site is now also home to optical and infrared ➤ *interferometers*. The largest, completed in 2000, is Georgia State University's CHARA array. (CHARA stands for Center for High Angular Resolution Astronomy.) It consists of six 1-m telescopes arranged in a Y-shape on a 400-m-diameter circle. The University of California's Infrared Spatial Interferometer (ISI) consists of three 1.65-m (65-inch) telescopes mounted on trailers which can be positioned up to 85 m (280 feet) apart.

M star A star of ➤ *spectral type* M. M stars have surface temperatures in the range 2400–3480 K and are red in color. Molecular bands are prominent in their spectra, particularly due to titanium oxide (TiO). Examples of M-type stars include the nearest star, ➤ *Proxima Centauri*, which is a dwarf, and the supergiant ➤ *Antares*.

The coolest M dwarfs of spectral types M7 to M9.5 and temperatures of 2400 K and below, include both stars and substellar objects with masses below the lower limit for a true stars (0.08 solar masses). ➤➤ *brown dwarf*.

Mullard Radio Astronomy Observatory (MRAO) The radio astronomy observatory of the University of Cambridge. It was founded in 1957 and Sir Martin ➤ *Ryle*, who had begun research in radio astronomy at Cambridge in 1946, was its director until 1982. The major telescopes to have been built at the site, just outside Cambridge, include the Ryle Telescope, consisting of eight 13-m (43-foot) dishes on an east–west baseline 5 km (3 miles) long.

Historically, the MRAO specialized in cataloging radio sources, producing the Third, Fourth, Fifth, Sixth and Seventh Cambridge Catalogues (abbreviated to 3C, 4C, etc.) at different frequencies. These led to the discovery of many ➤ *quasars* and ➤ *radio galaxies*. The first ➤ *pulsars* were detected at the MRAO in 1967.

Since the 1990s, the focus of research has shifted, with the development of an optical ➤*interferometer* (COAST) and instruments for studying the ➤ *cosmic background radiation*.

Multiple Mirror Telescope (MMT) A telescope of unique design that operated at the Fred Lawrence Whipple Observatory on Mount Hopkins in Arizona between 1977 and 1997 as a joint venture of the Smithsonian Institution Astrophysical Observatory and the University of Arizona. It combined six individual 1.8-m (72-inch) mirrors in a circular arrangement on an altazimuth mount. Together they had the light-gathering power of a single mirror 4.5 m (176 inches) in diameter. In 1989, in view of developments in mirror technology, the decision was taken to substitute a single light-weight mirror of 6.5 m diameter in order to double the light-gathering power and substantially increase the field of view. The conversion was completed in 1999 and the facility renamed the MMT Observatory.

Mundilfari A small outer moon of Saturn in a very elliptical orbit. It was discovered in 2000 and is about 6 km (4 miles) across.

Murzim Alternative spelling of the star name ➤ *Mirzam*.

Musca (The Fly) A small southern constellation containing one second-magnitude star and three of third magnitude. Its origin is obscure, but it has been attributed to Johann Bayer.

N

nadir The point on the celestial sphere diametrically opposite the ➤ *zenith*.

Naiad A small moon of Neptune discovered during the flyby of ➤ *Voyager 2* in August 1989. Its diameter is about 340 km (211 miles).

nanometer (symbol nm) One thousand-millionth (10^{-9}) of a meter.

Narvi A small outer moon of Saturn, discovered in 2004. It is about 7 km (4 miles) across.

NASA Abbreviation for ➤ *National Aeronautics and Space Administration*.

Nasmyth focus A point where the image made by an altazimuth-mounted reflecting telescope can be located on one side of the telescope tube. Light is brought to a focus there by an extra mirror in the optical system, which directs the beam along the altitude axis, and through a hole in the supporting trunnions. It was first used by the inventor James Nasmyth in the nineteenth century. It has the advantage of remaining at a fixed position relative to the telescope wherever it is pointed. This means that bulky or heavy instruments can be mounted there on a permanent platform, which rotates only in ➤ *azimuth*. In practice, there are two possible Nasmyth foci, one either side of the telescope tube. ➤➤ *altazimuth mount*.

National Aeronautics and Space Administration (NASA) The US government agency responsible for civilian manned and robotic activities in space, including launch vehicle development and operations, scientific satellites (such as orbiting observatories) and spacecraft, and advanced satellite technology development.

   NASA was created on July 29, 1958, when US President Dwight D. Eisenhower signed the National Aeronautics and Space Act of 1958. This legislation was widely acknowledged to have been passed by the US Congress in response to the successful and unexpected launch of the first artificial Earth satellite (Sputnik 1) by the Soviet Union. The headquarters of NASA are in Washington, DC, and it operates field centers and other facilities at various locations in the USA as well as several tracking stations around the world.

National Optical Astronomy Observatories (NOAA) An organization formed in the USA in 1984 to bring under one administration the national facilities for optical astronomy at ➤ *Kitt Peak National Observatory*, ➤ *Cerro Tololo Inter-American Observatory* and the ➤ *National Solar Observatory*. NOAA is also responsible for the operation of the ➤*WIYN Telescope* at Kitt Peak, and the US contribution to the international ➤ *Gemini 8-meter Telescopes* project.

National Radio Astronomy Observatory (NRAO) Combined radio astronomy facilities operated in the USA by a private consortium of universities. It obtains its funding under a cooperative agreement with the National Science Foundation. The NRAO runs the ➤ *Very Large Array* (VLA) in New Mexico, and the telescopes at ➤ *Green Bank*, West Virginia.

National Solar Observatory (NSO) The solar observation facilities of the US ➤ *National Optical Astronomy Observatories*. They consist of the ➤ *McMath–Pierce Solar Telescope Facility* at ➤ *Kitt Peak*, Arizona, and the Sacramento Peak Observatory in New Mexico.

NEAR Abbreviation for the ➤ *Near-Earth Asteroid Rendezvous mission*. ➤➤ *NEAR Shoemaker*.

Near-Earth Asteroid Rendezvous mission ➤ *NEAR Shoemaker*.

near-Earth asteroid ➤ *near-Earth object*.

Near-Earth Asteroid Tracking system (NEAT) A NASA survey for ➤ *asteroids* approaching relatively close to Earth. It started in 1995, using a CCD camera installed on a 1-m (39-inch) telescope operated on Mount Haleakala in Maui, Hawaii, by the US Air Force. From 2001 it has also used the Schmidt telescope at ➤ *Palomar Observatory*.

near-Earth object An ➤ *asteroid* or ➤ *comet* in an orbit that can bring it exceptionally close to Earth. They are believed to be asteroids from the main ➤ *asteroid belt* that have been perturbed by the gravitational attraction of the major planets, together with inactive nuclei of comets.

NEAR Shoemaker A NASA mission, launched on February 17, 1996, to rendezvous with the asteroid ➤ *Eros*. The original plan was for the craft to enter orbit around Eros in January 1999. The misfiring of a rocket made this impossible. NEAR flew past Eros on December 23, 1998 and, after completing a further orbit of the Sun, entered orbit around Eros in February 2000, enabling it to study the asteroid for a period of about one year, from distances as close as 24 km (15 miles). At the end of the mission, it was allowed to descend to the surface of Eros. Though it was never designed for a landing, the touch down was soft enough for some instruments to continue to operate. On the way to Eros, the spacecraft flew past asteroid 253 ➤ *Mathilde* in June 1997.

The initials NEAR stand for "Near Earth Asteroid Rendezvous." The craft was named in honor of the planetary scientist Eugene M. Shoemaker (1928–1997) in March 2000, after it had successfully entered orbit around Eros.

nebula (pl. nebulae or nebulas) A cloud of interstellar gas and dust. Nebula was at one time also used for objects we now know to be galaxies, such as the Andromeda Galaxy, which used to be called "the great nebula in Andromeda." An ➤ *emission nebula* glows because of the effects of ultraviolet radiation on the gas; a ➤ *reflection nebula* shines by reflecting starlight. An ➤ *absorption nebula* is dark and is usually visible only in silhouette against the background of a luminous nebula or starfield.

Other objects consisting of luminous gas are also known as nebulae, in particular ➤ *planetary nebulae* and ➤ *supernova remnants*.

Neptune A major planet of the solar system, eighth in order from the Sun. Its average distance from the Sun is 30.06 AU. It is one of the four giant planets, having a diameter of 49 528 km (30 777 miles), almost four times Earth's. It has a small rocky core surrounded by an icy mantle of frozen water, methane and ammonia. The outer atmosphere is mainly molecular hydrogen with 15–20 percent helium (by mass) and some methane.

Neptune was discovered by Johann G. Galle of the Berlin Observatory on September 23, 1846 following predictions made independently by John Couch Adams in England and Urbain J. J. Leverrier in France. Their calculations were based on discrepancies between the observed and predicted orbits of Uranus since its discovery in 1781, which were attributed to gravitational perturbations by an unknown planet.

Viewed from Earth, Neptune is seventh or eighth magnitude and so not visible to the naked eye. With high magnification and larger telescopes, it is seen as a faintly bluish disk, the color being due to the presence of methane in the upper atmosphere.

Close-up images were obtained by ➤ *Voyager 2* during its flyby of Neptune in August 1989. Observations with the Hubble Space Telescope (HST), capable of resolving atmospheric detail, began in 1994. Neptune has distinctive and varying cloud features in a highly dynamic atmosphere. The most prominent feature found by Voyager 2 was termed the Great Dark Spot. Located about 20° south of the equator, it rotated anticlockwise in a period of about 16 days. Bright cirrus-like clouds had formed over this and other small dark spots. However, it had completely disappeared when observations were made with the HST in 1994. Meanwhile, another dark spot, not seen by Voyager, had formed in the northern hemisphere. It too was accompanied by bright clouds. Subsequent observations with the HST have revealed that the pattern of clouds is constantly changing, though the underlying banded structure of the atmosphere remains stable. The atmospheric features rotate at different rates, and also drift in latitude. Wind speeds up to 2200 km/hr (1400 mph) have been measured. Neptune's average temperature is 59 K. It is not understood why Neptune radiates 2.7 times more energy than it receives from the Sun.

There are two main cloud layers in Neptune's upper atmosphere. The highest consists of crystals of methane ice, and this lies over a lower opaque blanket of cloud that may contain frozen ammonia or hydrogen sulfide. There is also a high-altitude haze of hydrocarbons produced by the action of sunlight on methane.

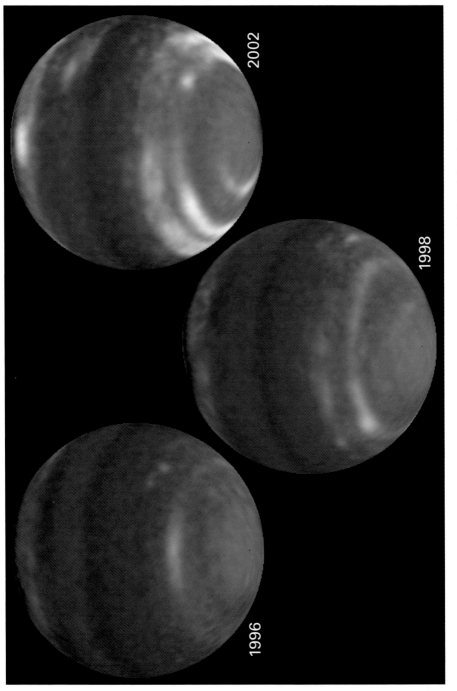

An increasing amount of cloud on Neptune between 1996 and 2002 revealed by visible light and infrared data from the Hubble Space Telescope, which were combined to make these false color images.

Regular radio bursts detected by Voyager 2 revealed that Neptune has a magnetic field and is surrounded by a magnetosphere. The bursts occurred at intervals of 16.11 hours, apparently the rotation period of the planetary core. The magnetic axis is tilted at 47° to the rotation axis and it is thought that the asymmetric magnetic field may originate in the planet's mantle rather than in its core.

Observations made from the ground during occultations by Neptune had suggested the presence of incomplete ring "arcs." Voyager 2 detected four tenuous rings, one of which is "clumpy" in a way that can account for the occultation observations. The mission also discovered six new moons around Neptune, bringing the total number known, with ➤ *Triton* and ➤ *Nereid*, to eight. Five more were found subsequently, in 2002 and 2003.

Nereid The small outermost moon of Neptune, discovered by Gerard Kuiper in 1949. The best ➤ *Voyager 2* image obtained was from a distance of 4.7 million km (2.9 million miles), not sufficiently close to reveal surface detail but good enough to determine its diameter as 170 km (105 miles).

neutrino astronomy The collection and analysis of neutrinos from cosmic sources, especially the Sun. Neutrinos are elementary particles with no electric charge and almost no mass, which interact only very weakly with other matter. They travel essentially at the velocity of light and are produced in vast quantities by the nuclear reactions that take place in the centers of stars and in ➤ *supernova* explosions.

Because they hardly interact with matter at all, neutrinos are very difficult to detect. The first long-term experiment to search for solar neutrinos, at Homestake Mine, South Dakota, took advantage of the fact that neutrinos sometimes interact with a chlorine atom, converting it to a radioactive isotope of the gas argon. The detector consisted of a tank containing 400 000 liters of the cleaning fluid, carbon tetrachloride. It was located in a mine because neutrino detectors need to be deep underground to protect them from ➤ *cosmic rays*.

In another form of neutrino detector, sensors in a large tank of water pick up radiation generated by the interaction of electrons with solar neutrinos. Detectors of this type, the Kamiokande experiment in Japan and a similar detector in Ohio, made the first observation of neutrinos from a supernova – those from ➤ *SN1987A*. In 1996 Kamiokande was superseded by a larger version, ➤ *Super-Kamiokande*. In 1999, the ➤ *Sudbury Neutrino Observatory* (SNO) opened in Canada, and work began on ➤ *AMANDA*, a cosmic neutrino "telescope" in the Antarctic ice at the South Pole. Following the success of AMANDA, a much larger neutrino experiment, called ➤ *IceCube*, was constructed in the Antarctic ice around AMANDA. A European collaboration (GALLEX) and a Russian experiment have made use of the interaction of neutrinos with gallium.

The early neutrino experiments detected fewer neutrinos from the Sun than they expected if standard particle physics theory was correct. The discrepancy was called the "neutrino problem." It is now known that neutrinos are capable of changing from one form to two others, and the early experiments picked up only one kind. The occurrence of this phenomenon implies that neutrinos do have some mass, albeit very tiny indeed.

neutron star A star with a mass between about 1.5 and 3 solar masses that has collapsed under gravity to the point where it consists almost entirely of neutrons. Neutron stars are formed in ➤ *supernova* explosions. They are typically about 10 km across and have a density of 10^{17} kg/m^3. The greater a neutron star's mass, the smaller its diameter. Neutron stars are detected by the radiation they emit, as ➤ *pulsars* or ➤ *magnetars*.

Once a star has exhausted the nuclear fuel available in its core, the core starts to cool and the internal pressure falls. This leads to the core contracting. It is a sudden and catastrophic event for stellar cores of more than 1.4 solar masses, which implode until the pressure between neutrons balances the inward pull of gravity. In the resulting supernova, much of the original star is blown off into space leaving behind a dense core that forms a neutron star or – if it contains more than three solar masses – a ➤ *black hole*. The process leaves the neutron star spinning rapidly and with a strong magnetic field.

A neutron star is believed to consist of a solid crust of heavy atomic nuclei and electrons, about a kilometer thick, overlying a superconducting liquid of neutrons, with a few protons and electrons. There may be a small solid core.

Newcomb, Simon (1835–1909) Newcomb became a mathematician in the US navy and in 1877 was made the superintendent of the American Nautical Almanac. He is best remembered for his work on new tables for calculating the positions of the planets and the Moon and for introducing a unified, more accurate system of standard astronomical data. His tables were the most accurate ever made without the aid of a digital computer and remained in use by professional astronomers for a century.

New General Catalogue of Nebulae and Star Clusters (NGC) A catalog of non-stellar objects compiled by John L. E. Dreyer of Armagh Observatory and published in 1888. It listed 7840 objects. A further 1529 were listed in a supplement that appeared seven years later, called the *Index Catalogue* (IC). The *Second Index Catalogue* of 1908 extended the supplementary list to 5386 objects. The original *General Catalogue of Nebulae* was compiled by John ➤ *Herschel* and published in 1864.

New Horizons A NASA flyby mission to ➤ *Pluto*, launched on January 19, 2006. After a close flyby of Jupiter in February 2007 designed to increase its speed, the spacecraft is expected to fly past Pluto in 2015 and may then go on to one or more other bodies in the ➤ *Kuiper Belt*.

Artist's impression of the New Horizons spacecraft at its encounter with Pluto and Charon

new Moon The Moon's phase when it is at the same celestial ➤ *longitude* as the Sun. At new Moon, all of the illuminated half of the Moon faces away from Earth and the Moon is not visible from Earth.

New Technology Telescope (NTT) A 3.5-m (138-inch) reflecting telescope of the ➤ *European Southern Observatory*, located at the ➤ *La Silla Observatory* in Chile. Regular observations with the telescope started in 1990. The name reflects a number of features incorporated in the design that were innovative at the time. The relatively thin mirror is kept in the correct shape by means of an ➤ *active optics* system, which analyzes a stellar image and controls the mirror supports about once per second. The altazimuth mount and special enclosure are designed for maximum stability and pointing accuracy, and minimum disturbance of images from air turbulence. The telescope can be operated remotely from ESO's headquarters in Germany via satellite link.

Newton, Sir Isaac (1642–1727) Newton was one of the greatest scientists of all time, remembered for his work on gravitation and mechanics. Born prematurely after the death of his father, he had an unsettled childhood. In 1661 he went to Trinity College Cambridge, where he studied mathematics, graduating in 1665. His Cambridge career was interrupted by the plague. He spent 18 months at his mother's house, developing the ideas to do with gravity, optics, and calculus for which he is most famous.

On his return to Cambridge in 1667, he was elected a Fellow of Trinity and two years later was elected Lucasian Professor of Mathematics. He invented the methods of calculus but published very little until 1693. In optics Newton

Sir Isaac Newton.

investigated the refraction of light, breaking white light into its constituent colours using a prism. He concluded that an astronomical ➤ *telescope* made with mirrors rather than lenses would give superior images. As a demonstration, he presented a 6-inch reflector to the Royal Society in 1672.

His first research on gravity, in 1665, concerned the motion of the Moon, which he described as being in free fall around the Earth. He returned to this in 1684, when he began to write his greatest work, now known as the *Principia Mathematica*, which was published in 1687. He set out three laws of motion as well as the law of universal gravitation. Importantly, Newton showed that the laws of physics governing the universe are the same as those applying on Earth.

Newtonian telescope A simple type of reflecting telescope, designed by Isaac ➤ *Newton* (1642–1727), who demonstrated it to the Royal Society in London in 1671. The main mirror is a paraboloid (or spherical if very small). The secondary mirror is flat and is positioned in the reflected beam at an angle of 45° to form an image just outside the main tube. The design is suitable for small amateur instruments but not for large telescopes.

NGC Abbreviation for ➤ *New General Catalogue*.

Nix A small moon of Pluto discovered in 2005. Its diameter is estimated at 45 km (30 miles).

nm Abbreviation for ➤ *nanometer*.

noctilucent clouds (NLC) Luminous bluish clouds high in the atmosphere that are sometimes seen in the summer twilight sky. They form at a height of about 80 km (50 miles). Noctilucent clouds are very thin and cannot be seen from the ground during daytime or in bright twilight. They occur in summer so are difficult to observe from the highest latitudes, where the sky never gets dark enough at that time of year. However, they are also a high-latitude phenomenon, so the range of latitudes from which they can be seen – between 50° and 65° – is very limited. The number of sightings decreases when ➤ *solar activity* increases.

nocturnal A simple instrument for telling the time at night by observing the position of the two stars in the constellation Ursa Major known as the ➤ *Pointers*. The continuation of the line joining these stars passes very close to the north celestial pole, and it acts like a giant clock-hand in the sky that sweeps around daily as Earth rotates.

A nocturnal consists of two concentric disks and a sighting arm, fixed at the center by an eyelet through which the Pole Star can be sighted. The lower disk is graduated with the days of the year and the upper one with the 24 hours of the day. The sighting arm is aligned with the Pointers and then the time can be read off.

node A point on the celestial sphere where the plane of an orbit intersects a reference plane.

Norma (The Rule) A small and insignificant southern constellation introduced by Nicolas L. de Lacaille in the mid-eighteenth century. It contains no star brighter than fourth magnitude.

normal galaxy Any spiral or elliptical ➤ *galaxy* that does not have unusual structure, a disturbed or active nucleus or non-thermal radio emission.

North America Nebula (NGC 7000) A complex region of nebulosity in the constellation Cygnus. It includes ➤ *emission nebulae*, ➤ *reflection nebulae* and ➤ *absorption nebulae*. It was discovered by William ➤ *Herschel* in 1786 and was first photographed in 1890 by Max Wolf. The photographs showed its shape to be suggestive of the North American continent. It is about 1° across and is just detectable with the naked eye under ideal conditions. Its estimated distance is 2300 light years.

Northern Cross A name sometimes given to the constellation ➤ *Cygnus* and, in particular, the cross formed by the stars Alpha, Beta, Gamma, Delta, Epsilon, and Eta.

northern lights A popular name for an ➤ *aurora* when observed from northern latitudes.

North Star A popular name for the star ➤ *Polaris*.

nova (pl. novae) A star that suddenly increases in brightness by about 10 magnitudes, then declines gradually over a period of months. The word nova is

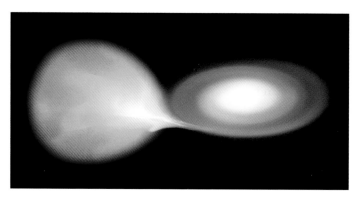

An artist's impression of how a nova occurs as a result of material streaming onto a white dwarf in a double star system

a shortening of the Latin "nova stella" – a new star. In any one galaxy, typically a few tens of novae occur in a year.

Novae are close binary systems in which one star is a ➤ *white dwarf*. When the companion star evolves and expands to fill its ➤ *Roche lobe*, material streams towards the white dwarf, forming an ➤ *accretion disk* around it. Material accumulates in a layer on the surface of the white dwarf until the temperature and pressure at the base of the layer become high enough for nuclear reactions to be sparked. The energy produced is unable to escape as more material is deposited in the overlying layers. The temperature may rise to 100 million degrees and, at some point, explosive nuclear reactions are triggered, producing the nova outburst.

Some novae are surrounded by an expanding envelope of gas. At speeds of up to 1500 km/s, the envelope soon disperses into space. It is estimated that the mass of material lost is about one ten-thousandth (10^{-4}) the mass of the Sun, and the energy released is only a millionth of that released in a ➤ *supernova*.

"Classical" novae are observed to erupt only once, though outbursts may recur every 10 000 to 100 000 years. Recurrent novae, such as ➤ *P Cygni*, repeat their outbursts on timescales of 10–100 years. Recurrent novae differ from classical novae in that their brightness increase of six to eight magnitudes is smaller, and their spectra during outbursts are different. At other times their spectra seem to indicate the presence of a ➤ *red giant* in the binary system. ➤➤ *dwarf nova*.

nucleosynthesis The making of chemical elements in naturally occurring nuclear reactions. Nucleosynthesis takes place in the interiors of stars, in ➤ *supernova* explosions and other astrophysical situations where high-energy collisions can take place between atomic nuclei and elementary particles.

Nunki (Sigma Sagittarii) The second-brightest star in the constellation Sagittarius. It is a ➤ *B star* of magnitude 2.0 and is 225 light years away.

nutation A relatively short-period oscillation superimposed on the ➤ *precession* of the rotation axis of a spinning body. The nutation of Earth's axis amounts to a maximum of 15 arc seconds over a period of about 18.6 years and is caused by changes in the Moon's orbit over this time.

44 Nysa An asteroid discovered in 1857 by Herman Goldschmidt. Its diameter is 68 km (42 miles) and it is notable for its high ➤ *albedo*, which is nearly 40 percent. It is one of two large members of the Nysa ➤ *Hirayama family*, the other being 135 Hertha.

O

Oberon The second largest moon of Uranus, diameter 1523 km (946 miles) which was discovered by William ➤ *Herschel* in 1787. Its surface is covered by numerous impact craters and seems not to have been altered very much.

objective The main light-collecting lens at the front of a refracting telescope.

obliquity of the ecliptic The angle between the plane of Earth's equator and the ➤ *ecliptic* or the "tilt" of Earth's axis. Its present value is approximately 23° 26'. The effects of ➤ *precession* and ➤ *nutation* cause it to change between extreme values of 21° 55' and 24° 18'.

Observatorio del Roque de los Muchachos An observatory on the island of La Palma in the Canary Islands Group. The observatory site, regarded as one of the best in the world, is operated by the Instituto de Astrofisica de Canarias, which is host to a number of different organizations and countries that have telescopes there. It is the site of the 10-m ➤ *Gran Telescopio Canarias*. Other telescopes there include the ➤ *William Herschel Telescope* and the ➤ *Telescopio Nationale Galileo*. The observatory occupies an area of nearly two square kilometers at an altitude of 2400 m (7900 feet).

observatory A place where astronomical observations are (or were formerly) made, or an administrative center for astronomical research.

occultation The passage of one astronomical object directly in front of another. ➤ *grazing occultation, eclipse.*

Octans (The Octant) A faint and obscure constellation containing the south celestial pole. It was introduced in the mid-eighteenth century by Nicolas L. de Lacaille and contains only one star brighter than fourth magnitude.

A Voyager 2 image of Oberon taken in 1986.

Olbers' Paradox The question "Why is the sky dark at night?" In 1826, Heinrich W. M. Olbers (1758–1840) drew attention to the fact that the sky should be a continuous blaze of light if the universe is infinitely old and filled more or less uniformly with stars because every line of sight from an observer would ultimately encounter a star. In fact of course, the sky is dark at night. This turns out to be an important cosmological observation. It supports the idea that the universe is not infinitely old but began with the event called the ➤ *Big Bang*. The finite age of the galaxies means that there has been insufficient time to fill the universe with light. In addition, the expansion of the universe means that the observed sky brightness due to remote objects is reduced.

2201 Oljato A small asteroid, diameter 2.8 km (1.7 miles), discovered in 1947 by Henry Giclas then lost until recovered in 1979. It is in a highly elliptical Earth-crossing orbit and has a unique spectrum that does not resemble that of any other known asteroid, meteorite or comet. Its nature is not known, but it could be the "dead" nucleus of a comet that has ceased to be active.

Olympus Mons The highest mountain on Mars, and the largest volcano in the solar system. It rises to a height of 27 km (17 miles). This gigantic ➤ *shield volcano* is 700 km (435 miles) across and is similar in nature to volcanoes on Earth but its volume is at least 50 times greater than its nearest terrestrial equivalent. The caldera at the summit is 90 km (60 miles) across and a cliff at

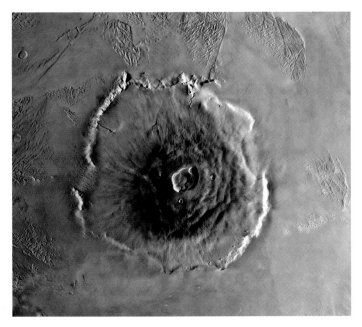

The huge martian volcano Olympus Mons.

An infrared image of the globular star cluster Omega Centauri.

least 4 km high rings the mountain. Older volcanic rocks, fractured and eroded by the wind, surround the main peak.

Omega Centauri (NGC 5139) A particularly bright ➤ *globular cluster* of stars in the southern constellation Centaurus. With a diameter of 620 light years it is the largest globular cluster known.

Omega Centauri lies at a distance of 16 500 light years. It spreads over about 1° of the sky and its total magnitude is 3.6, making it easily visible to the naked eye. Its shape is distinctly elliptical with axes in the ratio 5:4.

The name Omega Centauri is of a type normally given to single stars. The cluster was mistaken for a single star by early observers at Mediterranean latitudes, who could never see it more than about 10° above the horizon.

Omega Nebula (M17; NGC 6618) A luminous nebula in the constellation Sagittarius, also known as the Horseshoe Nebula and the Swan Nebula. It is 4800 light years away and 27 light years in diameter. It is a region of ➤ *ionized hydrogen*, stimulated to glow by a group of at least five hot stars. A dark dust cloud lies on the western edge of the luminous region.

Oort, Jan Hendrik (1900–1992) The Dutch astronomer Oort proved that galaxies rotate, and he was a pioneer in using radio astronomy to trace the spiral structure of our ➤ *Galaxy* and probe its center. In 1950, he proposed that the solar system is surrounded by a huge, distant swarm of comets, which came to be called the ➤ *Oort Cloud*. He became a professor at the University of Leiden in the Netherlands in 1935 and was director of the Leiden Observatory from 1945 to 1970.

Oort Cloud (Oort–Öpik Cloud) A spherical shell of billions of ➤ *comets* surrounding the solar system at a distance of about 1 light year (50 000 AU). The total mass of the objects in the cloud is thought to be about that of the Earth.

An infrared
mosaic of the
Omega Nebula.

Though it is impossible to observe directly with present technology, there are strong theoretical and practical reasons to believe it exists. The idea was first put forward on theoretical grounds by Ernst Öpik in 1932, and developed by Jan ➤ *Oort* in the 1950s. It is likely that the comets were formed near the present location of the outer planets and subsequently ejected to their much greater distance by gravitational interactions.

open cluster A type of star cluster containing several hundred to several thousand stars distributed in a region a few light years across. The member stars are much more spaced out than in ➤ *globular clusters*.

Open clusters are relatively young, typically containing many hot, highly luminous stars. Open clusters within our own ➤ *Galaxy* are located within the Galaxy's disk, so appear to lie within the ➤ *Milky Way* on the sky. Well-known open clusters include the ➤ *Pleiades*, the ➤ *Hyades* and the ➤ *Jewel Box*.

Ophelia A small moon of Uranus, 43 km (27 miles) in diameter, discovered by the ➤ *Voyager 2* spacecraft in 1986. Together with the moon Cordelia, Ophelia acts as a "shepherd" of Uranus's Epsilon ring.

Ophiuchus (The Serpent Bearer) A large constellation straddling the celestial equator. The mythological figure of the serpent holder is sometimes identified with the healer Aesculapius. Though Ophiuchus is not traditionally a zodiacal constellation, the ➤ *ecliptic* passes through its southern part. It contains five stars of second magnitude and seven of third magnitude.

The open cluster of stars, M39, in the constellation Cygnus.

opposition The position of one of the planets orbiting farther from the Sun than Earth when it is opposite the Sun in the sky, that is, when its ➤ *elongation* is 180°. When a planet is at opposition, its disk is fully illuminated and it reaches its highest point in the sky at midnight. It also comes closest to the Earth around opposition. As the orbits of the planets are elliptical rather than perfectly circular, some oppositions bring planets closer to Earth than others. This is particularly the case with Mars.

optical double star A pair of stars that lie close to each other in the sky by chance, but are not physically associated with each other as is the case for a true ➤ *binary star*.

orbit The path followed by a body moving in a gravitational field.

90482 Orcus A large ➤ *Kuiper Belt object* discovered in 2004. It has an orbit similar to Pluto's and is approximately 1600 km (1000 miles) across.

Orientale Basin A huge impact feature on the Moon's extreme western limb as viewed from Earth. It is visible only at times of favorable ➤ *libration*. Photographs taken from lunar orbit by spacecraft show a structure of at least three concentric rings. Unlike many other impact basins on the Moon, it is not extensively filled by dark ➤ *mare* material.

A radar image of the Orientale Basin on the Moon.

Orion (The Hunter) A brilliant constellation straddling the celestial equator, widely considered to be the most magnificent and interesting in the sky. Its pattern is interpreted as the hunter brandishing a raised club and a shield. Three bright stars mark his belt, and several fainter ones a sword hanging from it. Orion contains five stars of the first magnitude or brighter and a further ten brighter than fourth magnitude. The most spectacular diffuse nebula in the sky, the ➤ *Orion Nebula*, is faintly visible to the naked eye in the "sword."

Orion Arm The spiral arm of the ➤ *Galaxy* in which the Sun is located.

Orionids An annual ➤ *meteor shower*, with multiple radiants, which lie on the border of Orion and Gemini near the star Gamma Geminorum. The peak of the shower occurs around October 22, and its normal limits are October 16–27. The shower is produced by meteoroids that have come from ➤ *Halley's Comet*.

Orion Molecular Cloud ➤ *Orion Nebula*.

Orion Nebula (M42 and M43; NGC 1976 and NGC 1982) A bright emission nebula surrounding the multiple star Theta[1] Orionis in the "sword" of Orion.

   The luminous nebula is just part of a complex region of interstellar matter at a distance of 1300 light years occupying much of the constellation of Orion. The Orion cloud is the largest dark cloud of its kind known in the Galaxy. Millimeter-wave observations of the emission from the molecules CO (carbon monoxide), HCHO (formaldehyde), and many others, reveal the presence behind the visible part of a large ➤ *molecular cloud*, known as the Orion Molecular Cloud (OMC-1). This is an important region of star formation, and the four young hot stars that make up Theta[1] Orionis, also known as "the

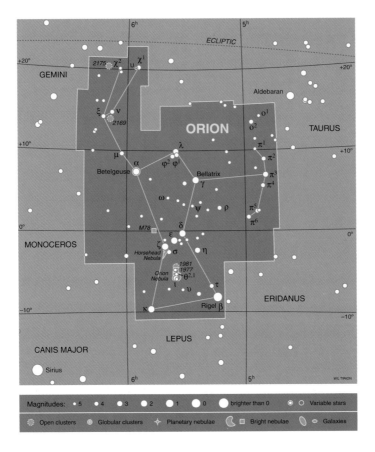

A map of the constellation Orion.

Trapezium," are believed to be less than 100 000 years old. The ➤ *Becklin-Neugebauer object* and the ➤ *Kleinmann–Low Nebula*, both detected through their infrared emission, are sites of current star formation in the complex.

The Trapezium stars are creating an expanding spherical cavity near the edge of the dark cloud. Their ultraviolet radiation is ionizing the gas and blowing away the dust. Relatively recently, in astronomical terms, the bubble broke through on our side of the dark cloud revealing the stars and ionized hydrogen within. The sharper edges of the nebula are produced by remnants of dust. M43 (NGC 1982) is a northern section of the nebula separated from the larger part (M42; NGC 1976) by a lane of dust.

In color images, the dominant color of the luminosity is red from the ➤ *hydrogen alpha* light. Observed visually, the nebula appears greenish because of the eye's low sensitivity to red light. The green emission is due to oxygen. The nebula occupies an area of sky about one degree across and is faintly visible to the naked eye.

The Orion Nebula. This mosaic was created from images taken by the Hubble Space Telescope.

orrery A working model of the solar system showing the planets, possibly with some of their moons, in their orbits around the Sun. The term "orrery" comes from a model made in 1713 for an Irish nobleman, the Fourth Earl of Cork and Orrery.

Orthose A small outer moon of Jupiter discovered in 2001. Its diameter is about 2 km (1 mile).

Oschin Telescope The 1.2-m (48-inch) ➤ *Schmidt camera* at ➤ *Palomar Observatory*. It has been in operation since 1948.

O star A star of ➤ *spectral type* O. O stars have surface temperatures in the range 28 000–50 000 K and are bluish white in color. Their spectra are characterized by lines of both neutral and ionized helium; emission lines are also commonly present. The four brightest O stars in the sky are Delta and Zeta Orionis, the easternmost stars in Orion's belt, and the southern stars Zeta Puppis and Gamma² Velorum.

outer planets The major planets beyond the ➤ *asteroid belt*, namely Jupiter, Saturn, Uranus and Neptune.

Owens Valley Radio Observatory The radio astronomy observatory of the California Institute of Technology (Caltech), located 400 km (250 miles) north of Los Angeles at an altitude of 1200 m (4000 feet). The instruments in use are a 40-m (130-foot) dish built in 1965 and an array of dishes used to observe the Sun. The Millimeter Wavelength Array, consisting of six 10.4-m (34-foot) dishes, was moved in 2005 from the original site to a new, nearby location at a higher altitude to form part of CARMA, the Combined Array for Research in Millimeter-Wave Astronomy. The interferometer is used particularly for solar observation and the 40-m dish is used for ➤ *very-long-baseline interferometry*.

Owl Nebula (M97; NGC 3587) A ➤ *planetary nebula* in the constellation Ursa Major. It is one of the largest planetary nebulae known, with a diameter of 1.5 light years, and lies at a distance of 1600 light years.

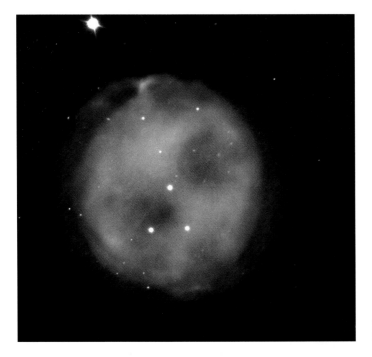

The Owl Nebula.

P

PA (p.a.) Abbreviation for ➤ *position angle*.

Paaliaq A small outer moon of Saturn, in a very elliptical orbit. It was discovered in 2000 and is about 19 km (12 miles) across.

2 Pallas A large asteroid discovered by Heinrich W. M. Olbers in 1802. With a diameter of 533 km (331 miles), it is the second largest asteroid. It is of the carbonaceous type, similar to the largest asteroid, ➤ *Ceres*. Its orbit is at the unusually steep inclination of 35° to the plane of the solar system.

Pallene A small moon of Saturn discovered in 2004 by the ➤ *Cassini* team. It orbits Saturn at a distance of 211 000 km (131 000 miles) between Mimas and Enceladus.

Palomar Observatory The observatory on Palomar Mountain in California where the 5-m (200-inch) ➤ *Hale Telescope* is sited. It is owned and operated by the California Institute of Technology (Caltech). The other principal instruments at the observatory are the 1.2-m (48-inch) Oschin Telescope (a ➤ *Schmidt camera*), a 46-cm (18-inch) Schmidt camera and the 1.5-m (60-inch) reflector owned jointly by Caltech and the Carnegie Institution of Washington.

Pan A small moon of Saturn orbiting in the ➤ *Encke Division* in the planet's ring system at a distance of 133 583 km (83 005 miles) from Saturn. It was found by Mark R. Showalter in 1990 from studies of images taken by the spacecraft ➤ *Voyager 1* and ➤ *Voyager 2*. Its existence had been predicted as an explanation for observed structure in the rings around the Encke Division. Pan is about 20 km (12 miles) in diameter.

Pandora A small moon of Saturn, measuring about 110×88×62 km (68×55×39 miles). It was discovered by ➤ *Voyager 2* in 1980. With ➤ *Prometheus*, it acts as a

This close-up view of Pandora from the Cassini spacecraft shows that the surface is coated in fine dust-sized icy material.

"shepherd" to keep the F-ring in place, orbiting 141 520 km (88 050 miles) from Saturn.

parallax The apparent change in the relative positions of objects at different distances when they are viewed from different places. In astronomy, there is a formal definition of the annual parallax of a star, based in the change in its apparent position over a period of six months, when Earth moves from one side of its orbit to the other. Only the parallaxes of relatively nearby stars are measurable with any accuracy. However, the determination of parallaxes is important since it is the most direct method of finding a star's distance.

Paranal Observatory The site in Chile of the ➤ *European Southern Observatory*'s ➤ *Very Large Telescope* (VLT). It is situated in the Atacama desert, 120 km (75 miles) south of Antofagasta, at an altitude of 2632 m (8500 feet).

Paris Observatory The French national astronomical research institute, based at the original site in Paris where it was founded in 1667. This is the oldest astronomical observatory still in use for research. There is an astrophysics section, located at the Observatoire de Meudon, just outside Paris, and a radio astronomy station at Nançay.

At the Paris site, there are some historical nineteenth century instruments, including the telescope built for the ➤ *Carte du Ciel* project. The Observatoire de Meudon was founded in 1876. It became the Astrophysics Section of the Observatoire de Paris in 1926 when the two institutions were merged. The Nançay radio astronomy station, established in 1953, is a large site with many instruments.

The Parkes Observatory radio dish.

Parkes Observatory An Australian radio astronomy observatory located 20 km (12 miles) north of Parkes, New South Wales. The main telescope there is an altazimuth mounted 64-m (210-foot) single dish, originally commissioned in 1961. In 1988, it became a unit of the ➤ *Australia Telescope National Facility*. It can serve as a stand-alone telescope or as a member of a long-baseline ➤ *array*.

    The Parkes dish was the first radio telescope to be built in the southern hemisphere. It identified the first quasar in 1963, and has discovered many ➤ *interstellar molecules* and ➤ *pulsars*. It has also been used as part of the ➤ *Deep Space Network* for tracking spacecraft.

parsec (symbol pc) A unit of distance used in professional astronomy. It is defined as the distance at which an object would have an annual ➤ *parallax* of one arc second. It is equivalent to 3.0857×10^{13} km, 3.2616 light years or 206 265 astronomical units.

Pasiphae One of the small outer moons of Jupiter, with a diameter of 60 km (37 miles). It was discovered in 1908 by Philibert J. Melotte.

Pasithee A small outer moon of Jupiter discovered in 2001. Its diameter is about 2 km (1 miles).

451 Patientia An asteroid 230 km (143 miles) across, discovered by Auguste Charlois in 1899.

Pavo (The Peacock) A southern constellation introduced in the 1603 star atlas of Johann Bayer. It contains one first-magnitude star, sometimes itself called "Peacock."

Pavonis Mons One of the three giant ➤ *shield volcanoes* on the ➤ *Tharsis Ridge* on Mars. It is about 400 km (250 miles) in diameter and rises to a height of 27 km (17 miles), 17 km above the level of the surrounding ridge.

Payne-Gaposchkin, Cecilia Helena (1900–1979) In 1925, Payne-Gaposchkin became the first person to receive a doctorate in astronomy from Harvard University. In her thesis she established for the first time that hydrogen is by far the most abundant chemical element in the Sun's atmosphere, with helium the second most abundant.

    Payne-Gaposchkin was born in England but moved to Harvard University after graduating from the University of Cambridge. She spent all her professional life at Harvard. For many years she had no formal position but in 1956 she was appointed to a professorship. She also became the first female Chair of the astronomy department at Harvard.

pc Abbreviation for ➤ *parsec*.

P Cygni An unusual variable star, which is a ➤ *recurrent nova*. It was recorded as third magnitude in August 1600 and stayed at this brightness for six years before fading slowly. A second outburst occurred in about 1655, which was again followed by slow fading. It subsequently fluctuated in brightness around

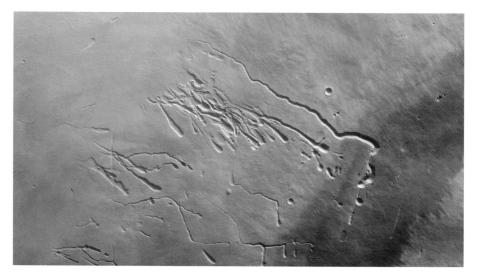

Collapsed lava tubes on the side of the Pavonis Mons volcano imaged by the Mars Express spacecraft.

Cecilia Payne-Gaposchkin.

sixth magnitude and has been about fifth magnitude, with only small variations, since 1715.

The lines in the spectrum of P Cygni are all double, consisting of a broad ➤ *emission line* with a narrower ➤ *absorption line* on the blue side. The absorption comes from starlight passing through surrounding shells of material, while the emission comes from the portions of the shells either side of the central star as

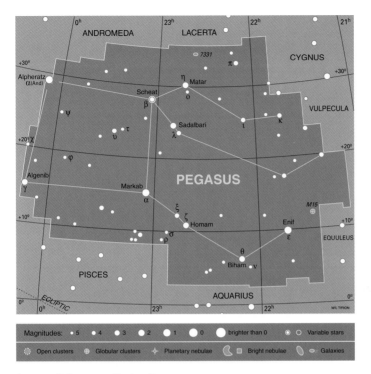

A map of the constellation Pegasus.

viewed from the Earth. The emission and absorption lines are displaced from each as a result of the ➤ *Doppler effect* because the shells are expanding. There are three distinct shells. The outermost one is pulsating with a 114-day period.

Peacock The brightest star in the constellation ➤ *Pavo*.

peculiar galaxy Any ➤ *galaxy* that does not obviously fit into the ➤ *Hubble classification* of galaxies, shows signs of unusual energetic activity or is interacting with neighboring galaxies.

Pegasus (The Winged Horse) A large northern constellation. The prominent Square of Pegasus is formed by its three brightest stars and Alpha Andromedae (Alpheratz), all of which are second magnitude. Alpha Andromedae used to be considered as belonging to Pegasus and was known as Delta Pegasi.

Pele One of the most active volcanoes on Jupiter's moon ➤ *Io*.

Pelican Nebula Popular name for the diffuse nebulae IC 5067 and 5070 in the constellation Cygnus. They form part of the ➤ *North America Nebula* (NGC 7000) complex.

penumbra (1) A region of partial shadow. During a solar ➤ *eclipse*, when the Moon's shadow sweeps across the surface of the Earth, observers in the penumbral zone see only a partial eclipse.

261

The volcano Pele deposited sulfur to create the large red ring seen in this image of Io taken by the Galileo spacecraft.

Part of the Pelican Nebula.

penumbra (2) The outer zone of a ➤ *sunspot*, surrounding the darker ➤ *umbra*. In the penumbra the magnetic field is horizontal and spreads out radially from the sunspot.

Perdita A small inner satellite of Uranus, which is about 27 km (17 miles) across. It was first imaged by ➤ *Voyager 2* in 1986 and seen again in 1999 but the discovery was not confirmed until further observations in 2003 made with the ➤ *Hubble Space Telescope*.

periastron In the orbital motion of a ➤ *binary star system*, the point where the two stars are closest.

perigee In orbital motion, the point where the Moon or an artificial satellite is closest to Earth.

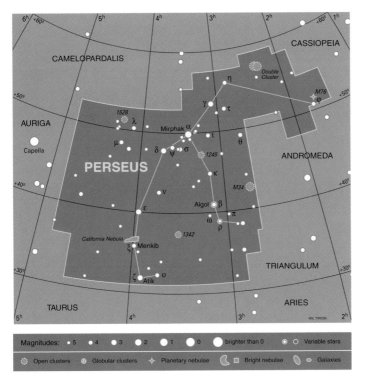

A map of the constellation Perseus.

perihelion (pl. perihelia) In orbital motion in the solar system, the point where a body is closest to the Sun.

period The time after which a cyclical phenomenon repeats itself.

periodic comet A ➤ *comet* in a closed, elliptical orbit within the solar system. Periodic comets are observed at their regular returns to Earth's vicinity, if they come near enough to be seen. This term is usually applied to comets with periods of less than 200 years, more strictly called ➤ *short-period comets*.

Perseids A major annual ➤ *meteor shower*. Its radiant lies near the star Eta Persei and it peaks on August 12. The normal limits are July 23–August 20. The meteor stream responsible is associated with Comet 109P/Swift–Tuttle. The Perseids are one of the best, most reliable annual showers, with peak rates typically between 50 and 100 meteors per hour. Records of it date back around 2000 years.

Perseus A large constellation of the northern hemisphere lying in a rich part of the ➤ *Milky Way*. It contains 10 stars brighter than fourth magnitude, including the variable star ➤ *Algol*. Perseus also includes a magnificent pair of ➤ *open clusters* visible to the naked eye, known as the ➤ *Double Cluster in Perseus*.

263

Perseus Arm One of the spiral arms of the Milky Way ➤ *Galaxy*. It winds around from the far side of the galactic center to the region of the Galaxy beyond the Sun.

Petavius A large lunar crater, 176 km (110 miles) in diameter, near the south-east limb of the Moon. A prominent rille runs across the crater floor between the multiple central peak and the terraced walls.

3200 Phaethon An asteroid, diameter 6 km (4 miles), discovered in 1983 by the ➤ *Infrared Astronomical Satellite* (IRAS). It is in a highly eccentric, Earth-crossing orbit and appears to be the parent body of the ➤ *Geminid* meteor shower. It is probably the inactive nucleus of a former comet.

phase The ratio between the illuminated area of the disk of a celestial body as it presents to an observer on Earth and the area of its entire disk, taken as a circle. The phases of the Moon are the repeating cycle of illuminated shapes presented by the Moon.

   The Moon and planets show phases because they emit no light of their own and shine only by reflected sunlight. The hemisphere of a moon or planet facing the Sun is bright, its other hemisphere dark. The phase as seen from Earth depends on the relative positions of Earth, the Sun and the body in question since this determines what proportion of the body's illuminated half is visible from Earth.

Phekda (Phecda; Gamma Ursae Majoris) The third-brightest star in the constellation Ursa Major. It is an ➤ *A star* of magnitude 2.4 located 84 light years away.

Phobos The nearer to Mars of its two small moons, which were discovered by Asaph Hall in 1877. Images from the ➤ *Viking* spacecraft of 1977 showed Phobos to be ellipsoidal in shape (28×20 km, 17×12 miles) and covered with craters. The largest crater, Stickney, is 10 km (6 miles) in diameter, one third of the moon's largest dimension. A series of striations emanating from Stickney appear to be fractures caused by the impact that created the crater. Data collected by ➤ *Mars Global Surveyor* suggests the surface may be covered by fine dust up to a meter (3 feet) thick.

Phocaea group A group of asteroids at a distance of 2.36 AU from the Sun, with orbits inclined at 24° to the plane of the solar system. The group is separated from the ➤ *asteroid belt* by one of the ➤ *Kirkwood gaps* and is not a true family with a common origin. The group is named after 25 Phocaea, which has a diameter of about 70 km (45 miles).

Phoebe The largest of Saturn's outer moons, discovered by William Pickering in 1898. Phoebe is 220 km (137 miles) in diameter and roughly spherical. Its surface is largely covered by very dark material that reflects only 6 percent of incident light, though there are also bright spots. Phoebe's orbit is very elliptical, tilted, and ➤ *retrograde*, which suggests it was captured by Saturn.

Phobos imaged by the Mars Express spacecraft.

Phoenix A southern constellation introduced in the 1603 star atlas of Johann Bayer. Though not particularly conspicuous, it does contain seven stars brighter than fourth magnitude.

Phoenix Mars Scout A NASA spacecraft launched on August 4, 2007. It will land in Mars's north polar region in May 2008. It is expected to operate for three months, and will be equipped to dig into the ground. It will particularly be looking for organic material and sub-surface water. Its name includes "Phoenix" because the spacecraft uses instruments constructed for planned Mars missions that have been cancelled.

5145 Pholus An asteroid with a diameter of 190 km (118 miles), discovered in 1991 by David L. Rabinowitz. It follows a highly unusual, remote orbit, on which it ranges between 8.7 and 32 AU from the Sun. With ➤ *Chiron*, and five other asteroids in orbits with similar characteristics, it forms the group termed ➤ *Centaurs*. Pholus has a low ➤ *albedo* of 4.4 percent, and is very much redder in color than typical asteroids.

photometry The accurate determination of the ➤ *magnitudes* of stars, or other astronomical objects, over specified wavelength bands in their spectra. Photometric measurements are used to find out broad physical characteristics

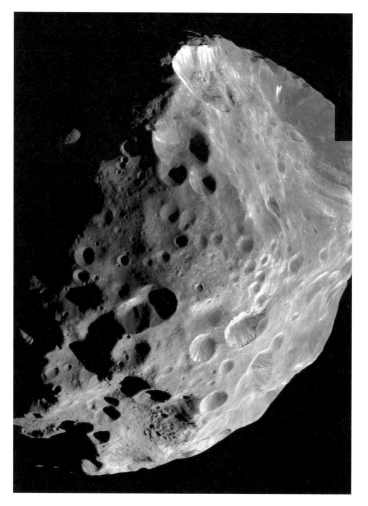

A high-resolution mosaic of Phoebe made from images taken by the Cassini spacecraft.

of stars, without the need to study their spectra in detail. Several systems are used for this employing standard filters. Photometric measurements are also important for determining the ➤ *light curves* of variable stars.

photosphere The visible "surface" of the Sun or a star. About 500 km (300 miles) thick, the photosphere is a zone where the character of the gaseous layers changes from being completely opaque to being transparent. It is the layer that emits the light we actually see. The temperature of the Sun's photosphere is about 6000 K on average, falling to about 4000 K at the base of the ➤ *chromosphere*.

Pic du Midi Observatory An observatory in the French Pyrenees at an altitude of 2877 m (9350 feet). The largest telescope is a 2-m (79-inch) reflector, which started operation in 1980.

The Pinwheel
Galaxy.

Pico An isolated mountain peak in the Mare Imbrium on the Moon.

Pictor (The Painter's Easel or Painter) An inconspicuous southern constellation introduced in the mid-eighteenth century by Nicolas L. de Lacaille. Its original name was Equuleus Pictoris, the painter's easel, but this was subsequently shortened to Pictor. Its two brightest stars are third magnitude.

Pierre Auger Observatory A pair of installations for detecting ➤ *cosmic rays*, run by an international consortium. The first is in western Argentina. A second is expected to be constructed later in Colorado. At each site there will be 1600 particle detectors in an area of about 4800 km^2 (1860 square miles). The detectors are 3000-gallon water tanks equipped with instruments. By 2006, the installation in Argentina was very nearly complete.

Pinwheel Galaxy (M101; NGC 5457) The popular name for a large spiral galaxy in Ursa Major, which we see face on. It is estimated to be at a distance of 15 million light years.

Pioneer A series of 11 American spacecraft, launched between 1958 and 1973. Pioneers 1 to 4 were directed towards the Moon and all failed. Pioneers 5 to 9 were put in orbits around the Sun and used to study the Sun and conditions in interplanetary space. Pioneers 10 and 11 were highly successful flyby missions, Pioneer 10 to Jupiter and Pioneer 11 to both Jupiter and Saturn.

Pioneer 10 was launched on March 3, 1972 and passed Jupiter on December 4, 1973 at a distance of 132 000 km (82 000 miles), returning the best

Pioneer 11 returned this image of Jupiter in December 1974.

pictures of the planet up to that time. It continued to transmit information continuously while heading out of the solar system. On March 30, 1997, when it was more than 66 AU from the Sun, its scientific program was ended. The last signals from it were received in February 2003. Pioneer 11 was launched on April 6, 1973 to encounter Jupiter on December 3, 1974 at a distance of 42 800 km (26 600 miles) and Saturn on September 1, 1979 at 20 800 km (13 000 miles). The last signals from it were received on September, 30, 1995.

Pioneer Venus Two American spacecraft sent to Venus in 1978. Pioneer Venus 1 was an orbiter, launched on May 20, 1978, which obtained radar maps of the surface and returned visual images and other data. Pioneer Venus 2, launched on August 8, 1978, was an atmospheric probe, with five small landers not intended to transmit data after impact.

Pisces (The Fishes) A large but faint constellation in the ➤ *zodiac*. Its three brightest stars are only fourth magnitude.

Piscis Austrinus (or Piscis Australis; The Southern Fish) A small southern constellation. It contains the first-magnitude star ➤ *Fomalhaut*, but no others brighter than fourth magnitude.

Pistol Star A ➤ *luminous blue variable star* near the galactic center surrounded by nebulosity where it has ejected large amounts of material. Its name describes the shape of the nebula. It is one of the most luminous stars known, thought to have had an initial mass around 200 times the Sun's.

Piton An isolated mountain peak in the Mare Imbrium on the Moon.

plage A bright emission region in the solar ➤ *chromosphere*. Plages coincide with ➤ *faculae* in the photosphere beneath them and enhancements of the ➤ *corona*

The Pistol Star.

above, from which there is increased X-ray, extreme ultraviolet and radio emission. They occur in ➤ *active regions* on the Sun.

Planck A European Space Agency spacecraft to image anisotropies in the ➤ *cosmic background radiation* over the whole sky with unprecedented sensitivity and resolution. Launch is scheduled for 2008, jointly with the ➤ *Herschel* telescope. It is named in honor of the German physicist Max Planck (1858–1947). Planck will operate in an orbit around the ➤ *Lagrangian point* L$_2$ between Earth and the Sun.

planet An astronomical body, with not enough mass to become a star or a ➤ *brown dwarf*. The upper mass limit for a planet is about 0.013 solar masses (equivalent to about 13 Jupiter masses). Though planets have traditionally been considered as objects in orbit around parent stars, isolated bodies of very low mass discovered in regions of star formation have also been described as "free-floating planets." To qualify as a planet in the solar system, a body must be in orbit around the Sun, and massive enough both to take on a shape close to spherical and to have swept away most smaller objects from the vicinity of its orbit. Under this definition, there are eight planets in the solar system.

Planets may be basically rocky objects, such as the inner planets – Mercury, Venus, Earth and Mars, or primarily liquid and gas with a small solid core like the outer planets – Jupiter, Saturn, Uranus and Neptune. These eight are regarded as the major planets of the solar system. Historically, Pluto was also considered to be a major planet, but that categorization was called into question by the discovery of other ➤ *transneptunian objects* similar in size to Pluto, or even larger. In 2006, the ➤ *International Astronomical Union* adopted the term ➤ *dwarf planet* to describe Pluto, the largest asteroid Ceres, and other similarly sized bodies orbiting the Sun. Within the solar system, there are large numbers of minor planets, or ➤ *asteroids*, and a population of small icy bodies in the region beyond Neptune, known as the ➤ *Kuiper Belt*. Pluto is one of the larger known members of this population.

So far, the presence of planets around stars other than the Sun has largely been inferred, either from the measurement of small cyclical changes in ➤ *radial velocity*, as revealed through the ➤ *Doppler effect*, or by the small dip in light observed when the planet crosses in front of its parent star. Observations of disks around newly forming stars, which could provide the material to form planetary systems, strengthen the argument that planetary systems probably accompany at least some stars comparable with the Sun. In addition, there is strong evidence (from small variations in pulse frequency) that at least two ➤ *pulsars* have planetary-sized companions. ➤➤ *extrasolar planet*.

planetarium A dome-shaped building housing a special projector that is used to simulate the appearance of the night sky. Planetaria are widely used for both educational purposes and entertainment. An ➤ *orrery* is sometimes called a planetarium.

planetary nebula An expanding shell of gas surrounding a star in a late stage of ➤ *stellar evolution*. It was ➤ *William Herschel* who coined the term. He thought that some circular nebulae he saw looked rather like the disks of the planets as seen through a small telescope. There is no other connection between planets and planetary nebulae.

Planetary nebulae are formed when ➤ *red giant* stars lose mass in the process of becoming ➤ *white dwarfs* and form a gaseous shell. Typically, a planetary nebula contains a few tenths of a solar mass of material and is

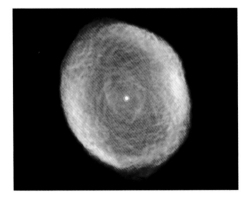

A planetary nebula nicknamed the "Spirograph Nebula". This false color image from the Hubble Space Telescope is assembled from views through three different color filters.

moving outwards with a velocity of 20 km/s. It lasts for perhaps 35 000 years before the shell becomes too tenuous to be visible. The central star is essentially a burnt-out core in the process of becoming a white dwarf.

Planetary nebulae are seen in a variety of forms, often with complex structure resulting from the outflow of stellar winds at different rates and different episodes of mass loss. Some planetary nebulae have been found to be dumbbell-shaped rather than spherical, probably due to the central star being a binary, or the existence of a ring around the star, constraining the direction of outflow. Well-known planetary nebulae include the ➤ *Ring Nebula*, the ➤ *Helix Nebula* and the ➤ *Dumbbell Nebula*.

planetesimal A body smaller than about 10 km (6 miles) across made of rock and/or ice, that formed in the primordial ➤ *solar nebula*. Planetesimals coalesced to form larger planetary bodies.

planisphere A round star map, which is a projection of part of the celestial sphere onto a plane. Planispheres usually have a rotating overlay with an aperture in it that reveals the portion of the sky visible at a given date and time. A planisphere is only useful over the latitude range for which it is designed.

plasma An ionized gas, consisting of a mixture of electrons and atomic nuclei. All of the matter in the interior of stars is in the form of plasma, as is ➤ *ionized hydrogen*.

plasmasphere A region of ionized gas surrounding Earth above the ➤ *ionosphere*, at altitudes greater than about 1000 km (600 miles). It extends out to between 25 000 and 40 000 km (about four and six Earth radii). The upper boundary is marked by a sudden drop in plasma density, known as the plasmapause. The particles in the plasmasphere are almost all protons and electrons.

plasma tail ➤ *ion tail*.

Plato (*c.* 428 BC–347 BC) The Greek philosopher Plato established the Academy in Athens, which survived for over 900 years as the greatest institution of learning in the ancient classical world. Plato's contribution to science was to establish that natural phenomena, such as the apparent motion of the planets, could be analyzed mathematically, particularly by using geometry.

Platonic year The period of 25 800 years it takes the Earth's rotation axis to sweep out a complete cone in space as a result of ➤ *precession*.

Pleiades (M45; NGC 1432) An ➤ *open cluster* of stars in the constellation Taurus, clearly visible to the naked eye. It is thought to contain about 1000 stars within a sphere 30 light years across, and is 440 light years away. The stars are embedded in a ➤ *reflection nebula* of cold gas and dust that appears blue in color photographs. The cluster is young by astronomical standards – only about 50 million years old – and contains some very massive bright stars.

In Greek mythology, the Pleiades were the seven daughters of Atlas and Pleione, who were called Alcyone, Asterope, Celaeno, Electra, Merope, Maia

A color-composite image of the Pleiades star cluster taken by the Palomar 48-inch Schmidt telescope.

and Taygete. These names, along with Atlas and Pleione, have been given to brighter stars in the cluster. Though the popular name for the Pleiades is the Seven Sisters, most people are able to distinguish only six stars with the naked eye.

Pleione One of the brighter stars in the ➤ *Pleiades*. It is slightly variable in brightness and was observed in 1938 and 1970 to throw off shells of gas.

plerion A ➤ *supernova remnant* with no clear shell structure. The ➤ *Crab Nebula* is the chief example. About 10 percent of supernova remnants are like this.

Plough (Dipper or Big Dipper) A popular English name for the ➤ *asterism* formed by the stars Alpha, Beta, Gamma, Delta, Epsilon, Zeta and Eta in the constellation ➤ *Ursa Major*.

plume eruption A type of volcanic activity observed on ➤ *Io*. Plumes arise from rifts or vents on the surface. The eruptive center is surrounded by a deposit of white or dark red material. The eruption may be violent and short-lived or a longer-lasting eruption of white material, rather like a geyser, consisting of liquid sulfur or sulfur dioxide. Hot liquid from underground changes to a gas as it rushes up through the eruptive center then condenses again as it falls. There is evidence for similar eruptions on ➤ *Triton* in images returned by ➤ *Voyager 2*.

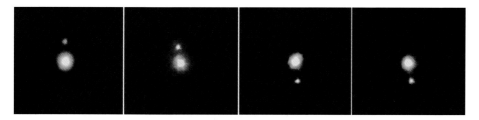

Pluto and its moon Charon, in a time sequence of images.

Plutino A ➤ *transneptunian object* which, like Pluto, follows an orbit in a 2:3 ➤ *resonance* with Neptune so that it completes two orbits of the Sun in the time it takes Neptune to make three orbits (248.5 years).

Pluto A ➤ *dwarf planet* in the solar system, discovered as a fifteenth-magnitude object on February 18, 1930, from the ➤ *Lowell Observatory* by Clyde ➤ *Tombaugh*. Searches for a planet beyond Neptune had started in 1905, stimulated by apparent discrepancies between the calculated and observed orbits of Uranus and Neptune. However, it is now known that the mass of Pluto is less than one-fifth that of the Moon, insufficient to have any gravitational effect on Uranus and Neptune.

From its discovery until 2006, Pluto was categorized as the ninth major planet of the solar system. However, discoveries of other objects beyond Neptune made from the early 1990s onwards, revealed that Pluto is only one of the larger known members of the ➤ *Kuiper Belt*. It was known from early on that Pluto's orbit is more highly inclined to the ecliptic and more eccentric than that of any major planet. Its distance from the Sun ranges between 30 and 50 AU. Its most recent perihelion passage was in 1989 and between 1979 and 1999, Pluto was nearer to the Sun than Neptune.

The discovery of the moon ➤ *Charon*, in 1978, made it possible to obtain improved values for Pluto's diameter and mass. Its diameter is 2300 km (1429 miles) and its mass is 1.27×10^{22} kg (0.0021 of Earth's mass). Pluto's overall density is approximately twice that of water and it is thought likely to consist of a thick layer of water ice overlying a core of partially hydrated rock. Charon and Pluto are locked in synchronous rotation with a period of 6.39 days. Charon keeps the same face towards Pluto and its orbital period is the same as Pluto's rotation period, so it is always over the same part of Pluto. Pluto's rotation axis is inclined at 122° to the plane of the ecliptic. As is the case for Uranus, Pluto's rotation is retrograde, so it seems to orbit while "lying on its side."

A rare series of mutual occultations and transits of Pluto and Charon took place between 1985 and 1990. They happen only twice in Pluto's 248-year orbital period. They made it possible to distinguish the individual spectra of Pluto and Charon and to construct the first rough maps of Pluto's surface. These confirmed previous suspicions of a highly non-uniform and variegated surface. In contrast with Charon, which is gray, Pluto's surface is reddish in color. Methane ice was detected on Pluto in 1976. The occultation of a star by Pluto in 1988 revealed the presence of an extended tenuous atmosphere. Nitrogen and carbon monoxide ices were discovered on the surface in 1992. The surface temperature is about 40 K. In 1996, observations with the Hubble Space Telescope resolved broad light and dark features on Pluto's surface for the first time.

In 2005, observations made with the Hubble Space Telescope in preparation for the launch of NASA's ➤ *New Horizons* mission to Pluto revealed two previously unknown moons orbiting Pluto between two and three times farther away than Charon. They were estimated to be around 50 km (30 miles) in size and were named Hydra and Nix.

Pointers The stars Alpha and Beta in the constellation Ursa Major, so called because the line joining them points almost directly to the Pole Star.

polar A type of short-period, variable ➤ *binary star* that emits X-rays. The light from polars is strongly polarized and the polarization varies over the orbital period, which is between one and four hours. These close systems appear to consist of a normal star and a strongly magnetic ➤ *white dwarf*. Matter is transferred from the normal star to the white dwarf but, because of the strong magnetic field, an ➤ *accretion disk* cannot form. Instead, the material is channeled along the magnetic field lines and is deposited at the poles. Polars are also known as AM Herculis stars after their prototype.

Intermediate polars are similar, but they have longer orbital periods lasting several hours. A pulse of radiation comes from them each time the white dwarf spins, which is typically in less than one hour. Their white dwarfs are thought to have weaker magnetic fields, making it possible for an outer accretion disk to form, though material close to the white dwarf is channeled onto the magnetic poles. The pulsed emission is a searchlight effect seen as the accreting pole of the white dwarf sweeps across the line of sight. Intermediate polars are also known as DQ Herculis stars after their prototype.

polar axis One of the two rotation axes about which a telescope on an ➤ *equatorial mount* can turn. The polar axis must be accurately oriented parallel to Earth's rotation axis, which means that it has to be set at an angle to the horizontal equal to the latitude of the place where it is located, and also in the north–south plane. Rotation about the polar axis results in a change in the right ascension of the direction in which the telescope is pointing, but not in declination.

polar cap A roughly circular area of limited extent around a pole of rotation of a planet. In the case of Earth and Mars, the term is applied to the areas covered by ice or frost in the two polar regions.

Polaris (Alpha Ursae Minoris) The brightest star in the constellation Ursa Minor. It lies within one degree of the north celestial pole. Polaris is a ➤ *Cepheid variable*

The north polar cap of Mars.

and its magnitude changes between about 1.95 and 2.05 over a period of four days. Its distance is 430 light years.

polarization (of light) The non-random distribution of electric field direction in a beam of ➤ *electromagnetic radiation.*

polar motion A small "wobble" in the position of Earth's geographic poles relative to its surface (not relative to the stars). It arises because the axis around which Earth rotates does not quite coincide with Earth's axis of symmetry. The wobble typically amounts to about 0.3 arc seconds, and there are regular variations over periods of 433 days and one year. Much smaller variations also take place over short timescales, ranging between two weeks and three months, due to surface air pressure changes. ➤ *Chandler wobble.*

polar plume A bright stream of gas flowing away from the Sun. Plumes usually come from the Sun's poles but they can appear wherever there is a ➤ *coronal hole.* They follow the direction of the magnetic field out of the hole.

polar-ring galaxy An ➤ *elliptical galaxy* or ➤ *lenticular galaxy* that has dust, gas and stars orbiting in a ring around it, more or less at right angles to the main plane of the galaxy. Polar rings can form when two galaxies merge or pass close to each other.

Pole Star Popular name for the star ➤ *Polaris.*

Pollux (Beta Geminorum) The brightest star in the constellation Gemini. It is an orange-colored, giant ➤ *K star.* Castor and Pollux were the twin sons of Leda in classical mythology.

Polydeuces A small moon of Saturn discovered in 2004 by Carl Murray of the ➤ *Cassini* team. With ➤ *Helene,* it shares the orbit of the larger moon ➤ *Dione,* 377 400 km (234 500 miles) from Saturn. It measures 13 km (8 miles).

NGC 4450A is a rare example of a polar-ring galaxy. It lies 130 million light years away. This image was taken by the Hubble Space Telescope.

Population I Stars and star clusters in our ➤ *Galaxy* that are relatively young and are in the ➤ *galactic plane*, especially in the spiral arms. The terms Population I and ➤ *Population II* (for older stars) were introduced by Walter ➤ *Baade* in 1944. Hotter stars, ➤ *open clusters*, and stellar ➤ *associations* are all typical Population I objects. Interstellar material is also associated with Population I. Population I stars are relatively rich in heavier elements because the material from which they formed includes elements formed by ➤ *nucleosynthesis* in earlier generations of stars. ➤➤ *astration*.

Population II Old stars and star clusters in our ➤ *Galaxy*. They are found all around the Galaxy throughout a spherical halo, rather than just in the ➤ *galactic plane* like ➤ *Population I* stars. Population II stars contain less of the elements heavier than helium than Population I. They move at high speeds around the Galaxy on very elliptical orbits inclined at steep angles to the plane of the Galaxy. ➤ *Globular clusters* belong to Population II. Population II stars formed when the Galaxy was spherical before its disk formed, and before the gas and dust could be enriched with heavier elements from previous generations of stars.

pore A small ➤ *sunspot* without a ➤ *penumbra* that lasts for about a day.

Porrima (Gamma Virginis) The second-brightest star in the constellation Virgo, which is in reality a visual ➤ *binary star* made up of two almost identical ➤ *A stars*. Their combined magnitude is 2.8; individually they are each magnitude 3.6. They orbit around each other in 170 years. At a distance of 38 light years, Porrima is relatively close to the solar system. Porrima was a Roman goddess of prophecy.

Portia A small moon of Uranus, discovered in 1986 by the ➤ *Voyager 2* spacecraft. Its diameter is 135 km (84 miles).

position angle (PA, p.a.) An angle specifying the position of one astronomical object or feature relative to another. The position angle of an object B relative to another one, A, is the angle between the line from A pointing due north and the line from A to B. It is measured in the sense north–east–south–west, on a scale from 0° to 360°. For double stars, the position angle is given for the fainter component relative to the brighter one.

potentially hazardous asteroid (PHA) An ➤ *asteroid* that can pass within 0.05 of an astronomical unit (5 million miles) of Earth. The first to be discovered was 1862 Apollo. In practice, hardly any pose a threat to Earth over timescales of many thousands of years.

Praesepe (M44; NGC 2632) An ➤ *open cluster* of at least 200 stars in the constellation Cancer lying at a distance of 500 light years. Its brightest stars are about sixth magnitude. In many ways it is similar to the ➤ *Hyades*, even sharing the same direction and speed of motion through space; this suggests that the two clusters originated in the same interstellar cloud.

Praxidike A small outer moon of Jupiter discovered in 2000. Its diameter is about 7 km (4 miles).

precession The uniform motion of the rotation axis of a freely rotating body when it is subject to a turning force (torque).

Precession causes Earth's rotation axis to sweep out a cone in space over a period of 25 800 years. Earth's axis is always tilted at an angle of about 23.5° to the plane of Earth's orbit (the ➤ *ecliptic*), but the direction it points in slowly changes. The main torques acting on Earth are the gravitational pulls of the Sun and Moon on Earth's equatorial bulge. The Moon's contribution is about twice as large as the Sun's, because it is so close. Precession would not occur if Earth were a perfect sphere. However, the Earth's equatorial radius is 0.3 percent greater than its polar radius because of its rotation.

Because of precession, the celestial poles trace out circles in the sky over 25 800 years. So, for example, about 13 000 years from the present, the nearest bright star to the north celestial pole will be Vega rather than Polaris.

The zero point of ➤ *right ascension*, one of the ➤ *equatorial coordinates*, normally used to give the positions of celestial objects, is one of the two points where the celestial equator and the ecliptic cross. For centuries this point has been called "The First Point of Aries." However, the equator is "sliding around" the ecliptic because of precession so the intersection points are constantly moving. As a result, the First Point of Aries is no longer in the constellation Aries, but has moved into Pisces and will soon be in Aquarius. This phenomenon is known as the precession of the equinoxes. Its effect on the right ascension and declination of an object is noticeable from year to year. Tabulated values of right ascension and declination are given for a particular date or ➤ *epoch*, at which they were precisely correct. ➤➤ *equinox, nutation.*

primary mirror The main light-collecting mirror in a reflecting telescope.

prime focus The point at which the primary mirror in a reflecting telescope brings light to a focus if there is no secondary mirror.

prime meridian The great circle on the surface of a planetary body adopted as the zero of longitude measurement. On the Earth, the prime meridian is the circle of longitude passing through Greenwich, London – the Greenwich Meridian.

Procyon (Alpha Canis Minoris) The brightest star in the constellation Canis Minor. At magnitude 0.38, it is the fifth-brightest in the sky, and it lies 11.25 light years away. Procyon was discovered to be a binary system by John M. Schaeberle in 1896. The primary is a normal ➤ *F star* and its faint companion an eleventh-magnitude ➤ *white dwarf*. Their orbital period is 41 years. The name Procyon comes from Greek and means "before the dog." It refers to the fact that Procyon rises before the "Dog Star," ➤ *Sirius.*

Project Ozma The first serious scientific attempt to contact ➤ *extraterrestrial intelligence* by radio waves. The experiment was conducted in 1960 at ➤ *Green Bank* and unsuccessfully looked for unnatural radio signals from the nearby stars Tau Ceti and Epsilon Eridani. ➤➤ *SETI.*

Prometheus (1) A small moon of Saturn, discovered in 1980 by the ➤ *Voyager 1* team. Its diameter is about 148 × 100 × 68 km (92 × 62 × 42 miles). It is one of the "shepherd" moons that keep the F-ring in place, orbiting 139 350 km from Saturn.

Prometheus (2) One of the most active volcanoes on Jupiter's moon ➤ *Io.*

prominence A term used for a variety of flame-like streams of gas in the ➤ *chromosphere* and ➤ *corona* of the Sun, which have a higher density and lower temperature than their surroundings. When seen at the edge of the Sun they look bright but when seen against the Sun's disk they appear as dark ➤ *filaments.*

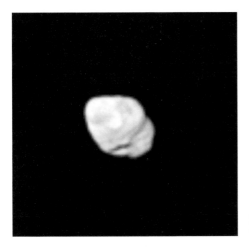

Saturn's moon Prometheus as seen by the Cassini spacecraft.

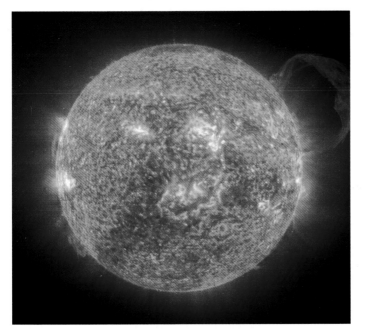

A solar prominence observed by the SOHO spacecraft on April 11, 2003. It extends for over 30 times the size of the Earth.

Quiescent prominences occur away from active regions and are stable over many months. They may extend upwards for tens of thousands of kilometers. Active prominences are associated with ➤ *sunspots* and ➤ *flares*. They appear as surges, sprays and loops, have violent motions, and last up to a few hours.

Proplyds in the Orion Nebula, imaged by the Hubble Space Telescope.

proper motion The apparent annual motion of a star across the celestial sphere due to its real motion through space relative to the solar system.

proplyd A recently formed star, surrounded by a cloud of gas and dust, which may ultimately produce a planetary system. The word "proplyd" is a contraction of "protoplanetary disk."

Prospero A small outer moon of Uranus discovered in 1999. Its diameter is about 30 km (19 miles).

Proteus A moon of Neptune discovered during the flyby of ➤ *Voyager 2* in August 1989. It measures about 436×416×402 km (271×259×250 miles).

protogalaxy The earliest stage in the formation of a galaxy from primordial gas and dust.

proton–proton chain A series of nuclear reactions that convert hydrogen to helium inside stars and are a major source of the energy stars generate.

protoplanet A body in the early stage of the process of becoming a planet.

protostar A star in the earliest observable stage of formation.

Proxima Centauri The nearest star to the solar system lying at a distance of 4.26 light years. It is an eleventh-magnitude dwarf red ➤ *M star* in the constellation Centaurus. It appears to be associated physically with the bright binary star Alpha Centauri, which is two degrees away in the sky, and about 0.11 light years more distant from the solar system. It is estimated that Proxima may take a million years to orbit its companions.

Psamathe A small moon of Neptune discovered by David Jewitt in 2003. Its diameter is about 24 km (15 miles).

16 Psyche An asteroid with a diameter of 248 km (154 miles), discovered by Annibale de Gasparis in 1852. It is of the metallic type and its surface appears to be an almost pure alloy of iron and nickel.

Ptolemaeus A large shallow crater, flooded with dark lava in the Moon's southern upland area. Its diameter is 153 km (95 miles) and there are many smaller craters within its walls.

The Ptolemaic system, showing the Sun, Moon and planets in orbit around the Earth.

Ptolemaic system A concept of the solar system with Earth at the center, described by the Greek astronomer ➤ *Ptolemy* (*c.* AD 100–170) in his book, the ➤ *Almagest*. It was the model generally accepted in the Arab and Western worlds for more then 1300 years, until superseded by the ➤ *heliocentric model*.

Ptolemy, Claudius (*c.* AD 100–170) Little is known about the life of the Egyptian astronomer Ptolemy who worked in the city of Alexandria between about AD 127 and 151 but his great book, the *Almagest*, remained the most influential work on astronomy for 1300 years until his assumption that the Earth is the center of the solar system was overtaken by the idea put forward by ➤ *Copernicus* of a Sun-centered solar system. The *Almagest* included a star catalog and rules for calculating the future positions of the planets based on complicated geometry. It was not all original work by Ptolemy but a synthesis of Greek astronomy from over 500 years.

Puck A moon of Uranus, discovered by the ➤ *Voyager 2* spacecraft in 1985. Its diameter is 162 km (101 miles), and a Voyager image shows several relatively large craters on its surface.

Pulkovo Observatory An observatory near St Petersburg in Russia, originally established in 1718. The present site dates from 1835; the buildings were destroyed during World War II and subsequently rebuilt in their old style. The observatory is associated particularly with the ➤ *Struve* family, six members of which became well-known astronomers. F. G. W. Struve was director from 1839 to 1862 and his son, Otto, director from 1862 to 1889.

pulsar A star from which we detect regular bursts of radio waves in rapid sequence. The time between successive pulses ranges from milliseconds for pulsars belonging to binary systems up to 4 seconds for the slowest. Some pulsars emit radiation in other bands of the electromagnetic spectrum, including X-rays, gamma rays and visible light, as well as radio waves.

A pulsar is a rotating, magnetic ➤ *neutron star*, with a mass similar to the Sun's but a diameter of only about 10 km (6 miles). The pulses occur because the neutron star is rotating very rapidly and a beam of radio emission sweeps by once per rotation. The pulses are very regular, apart from the occasional ➤ *glitch*, and all single pulsars are slowing down as they gradually lose energy.

Some X-ray pulsars are in binary systems where complex dynamical effects cause their spin rate to speed up, and these millisecond pulsars are the fastest known. Millisecond pulsars not currently in binary systems are thought to have once belonged to pairs that have been split apart. Most have been discovered in ➤ *globular clusters*, where stars are densely packed and gravitational interactions can easily occur. At least one pulsar appears to have another neutron star as a companion, and another has two or three planet-sized companions.

Pulsars are formed in ➤ *supernova* explosions, though only two – the ➤ *Crab Pulsar* and the ➤ *Vela Pulsar* – are within currently observable supernova remnants. ➤➤ *magnetar*.

pulsating star A ➤ *variable star* with an unstable internal structure that causes it to pulsate in a regular way. ➤➤ *Cepheid variable*, *RR Lyrae star*.

Puppis (The Poop or Stern) A large southern constellation lying in rich starfields of the ➤ *Milky Way*. It is the largest of the three sections into which the ancient constellation Argo Navis was divided by Nicolas L. de Lacaille in the mid-eighteenth century. It contains ten stars of the second and third magnitudes.

Pythagoras A large lunar crater near the north-west limb of the Moon. It is 129 km (80 miles) in diameter and has high walls and a central peak.

Pyxis (The Compass) A small and insignificant southern constellation introduced by Nicolas L. de Lacaille in the mid-eighteenth century. It contains only one star that is brighter than fourth magnitude.

Q

QSO Abbreviation for quasi-stellar object. ➤ *quasar*.

quadrant An instrument for measuring the altitudes of stars and the angular separation between celestial objects. It consists of a quarter-circle marked in degrees and a movable sighting arm. Prior to the invention of the telescope, such instruments were the only ones available to astronomers for measuring the positions of stars and planets.

Quadrantids An annual ➤ *meteor shower* between about January 1 and January 6, peaking on about the 3rd. The radiant of the shower lies in the constellation Boötes, near its border with Hercules and Draco. The name dates from the time when this area of sky was identified as the constellation Quadrans Muralis, now no longer used. The narrow stream of meteors responsible is not associated with any known comet.

quadrature The position of the Moon or of a planet when its angular distance from the Sun, as viewed from the Earth, is 90°.

50000 Quaoar One of the largest known ➤ *Kuiper Belt objects*, discovered in 2002. It follows a near-circular orbit that sometimes brings it nearer than Pluto and sometimes takes it farther away. Its diameter is approximately 1260 km (780 miles).

quasar (quasi-stellar object, QSO) The most powerful type of ➤ *active galactic nucleus*. Quasar is a contraction of "quasi-stellar radio source," the phrase used to describe these objects when they were first discovered in 1963. They

An X-ray image of the quasar 3C273 from the Chandra X-ray Observatory. 3C273 lies in the constellation Virgo and is about 3 billion light years away.

appeared to be star-like with very high ➤ *redshifts*. Although the quasars discovered in the 1960s were all radio sources, most of the many thousands now known are not strong radio sources.

The largest redshifts ever recorded have been measured for quasars and they are the most distant objects observable in the universe. The highest quasar redshift measured by 2007 was 6.43. This quasar is about 13 billion light years away. The fact that we detect such remote objects means that they are giving out far more light and energy than a normal galaxy. The source of their energy is matter falling onto a supermassive ➤ *black hole*.

quasi-stellar object ➤ *quasar*.

quasi-stellar radio source ➤ *quasar*.

quiet Sun The Sun when it is at its minimum level of activity in the ➤ *solar cycle*, with little evidence of ➤ *solar activity*.

R

RA Abbreviation for ➤ *right ascension*.

radar astronomy The use of radar for astronomy. Radar can be used to detect ➤ *meteor showers*, to measure distances to bodies in the solar system and determine their size and shape, and to map the surfaces of planets and moons. Radar signals transmitted by the 305-m (1000-foot) radio telescope at ➤ *Arecibo Observatory* have been used to map Venus and to characterize the size, shape and structure of ➤ *asteroids*. The ➤ *Magellan* spacecraft, placed in orbit around Venus, used ➤ *synthetic aperture radar* to map the planet's surface, which is concealed by opaque cloud, and the ➤ *Cassini* spacecraft used radar to study the surface of ➤ *Titan*. Radar is of fundamental importance for making precise measurements of distance within the solar system.

radial velocity The velocity of an object relative to an observer along the line of sight.

radiant The point on the celestial sphere from which the trails of meteors belonging to a particular ➤ *meteor shower* appear to radiate. Meteors entering Earth's atmosphere from a stream create trails that are almost parallel but perspective makes them seem to diverge from a point in the sky.

radiation belt A ring-shaped region around a planet where electrically charged particles (electrons and protons) are trapped and spiral around the direction of the magnetic field of the planet. The radiation belts surrounding the Earth are known as the ➤ *Van Allen belts*. Similar regions exist around other planets with magnetic fields, such as Jupiter.

radio astronomy The exploration of the universe by detecting radio emission from celestial objects. The principal sources of cosmic radio emission are the ➤ *Sun*, ➤ *Jupiter*, interstellar hydrogen and ionized gas, ➤ *pulsars*, ➤ *quasars*, and the ➤ *cosmic background radiation* of the universe itself. The radio frequencies used for astronomy span a vast range, from 10 MHz to 300 GHz. Several wavebands are protected internationally against interference, such as 1421 MHz (wavelength 21 cm), the natural frequency of atomic hydrogen.

➤ *Radio telescopes* are largely either single steerable dishes, up to 100 m in diameter, or arrays of dishes linked to form ➤ *radio interferometers*. Single dishes have poor angular resolution compared with optical telescopes, so they are mainly used in investigations where positional accuracy is not vital, such as the timing of pulsar signals. For structural details, for example making a map of the emission from a radio galaxy, it is essential to use an interferometer.

Radio astronomy: how the night sky would appear to someone with "radio eyes."

Since its inception in the 1940s, radio astronomy has been directly responsible for the discovery of pulsars, quasars and the microwave background. ➤ *radio galaxy*.

radio galaxy A ➤ *galaxy* that has an ➤ *active galactic nucleus* and is an intense source of radio emission. About one galaxy in a million is a radio galaxy. The radio emission comes from a pair of extended jets emanating from the galaxy and is ➤ *synchrotron radiation*, which is emitted by electrons traveling nearly as fast as light. In ➤ *Cygnus A*, often regarded as the prototype radio galaxy, two huge clouds of radio emission, stretching out symmetrically on each side of a disturbed elliptical galaxy, span more than three million light years.

radioheliograph A radio telescope designed for mapping the distribution of radio emission from the Sun.

radio interferometer A ➤ *radio telescope* in which two or more separate antennas observe the same object simultaneously. The signals received by a pair of antennas are fed into a receiver that multiplies the two voltages. The result depends on the distribution of radio emission from the source being studied. A single measurement of this kind gives little information but, if the spacing and orientation of the interferometer are changed, the varying output can be analyzed by computer to generate maps showing the distribution of radio brightness on the sky. ➤ *aperture synthesis, very-long-baseline interferometry.*

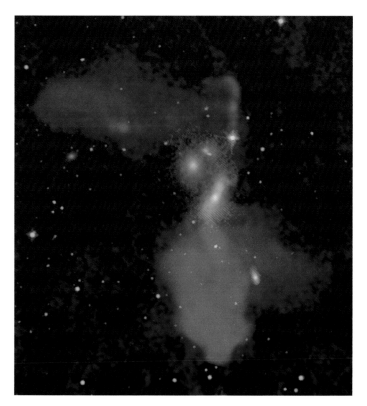

The radio galaxy 3C433. Blue shows the distribution of stars from Hubble Space Telescope observations and red shows the radio radiation as imaged by the Very Large Array.

radiometer Any instrument for measuring the amount of electromagnetic radiation received from an object.

radio source Any natural source of cosmic radio emission. In cosmology, it is often used just to mean ➤ *radio galaxies* and ➤ *quasars*. ➤➤ *radio astronomy*.

radio telescope An instrument for collecting, detecting and analyzing radio waves from any cosmic source. All radio telescopes consist of an antenna that picks up the signals and feeds them first to an amplifier and then to a detector. The large range of frequencies covered by radio astronomy means that radio telescopes vary greatly because different techniques are used for different parts of the spectrum.

A fundamental problem in radio astronomy is getting the best possible resolving power. For resolution of half an arc second, comparable with a good optical telescope, a single dish would need to be 100 km across for observing 21-cm radio waves! However, it is not possible to build fully steerable dishes more than about 100 m across. Single steerable dishes are used mainly for

Ranger 7 took this image, the first picture of the Moon by a US spacecraft, on July, 31 1964, about 17 minutes before crashing onto the lunar surface. The large crater at center right is Alphonsus. Above it is Ptolemaeus and below it Arzachel.

studies of interstellar matter, through the ➤ *twenty-one centimeter line*, and variable sources, such as pulsars. The higher resolution needed to map structure in objects such as ➤ *radio galaxies* and ➤ *quasars* is obtained by linking arrays or networks of telescopes to form a ➤ *radio interferometer*. ➤➤ *aperture synthesis, radio astronomy, very-long-baseline interferometry.*

Ranger A series of nine American lunar spacecraft, launched between 1961 and 1965. Only the last three, Rangers 7, 8 and 9 were successful. Ranger 7, launched in July 1964, returned 4000 images. Thousands more were obtained with Rangers 8 and 9, launched in February and March 1965. They were "hard landers," designed to transmit images during their approach to the Moon until they crash-landed. The three successful Rangers landed successively in the ➤ *Fra Mauro* region, the Mare Tranquillitatis and the crater ➤ *Alphonsus*.

Ras Algethi (Alpha Herculis) A bright variable star in the constellation Hercules. It is a binary, consisting of a red supergiant ➤ *M star* with a sixth-magnitude companion of ➤ *spectral type* F, which appears greenish in contrast. They are 400 light years away. The fainter star is itself a ➤ *spectroscopic binary*. The primary star varies irregularly between magnitudes 3 and 4 in about 128 days. The name Ras Algethi comes from Arabic and means "the kneeler''s head."

2100 Ra-Shalom A small asteroid measuring 3 km (2 miles) across, discovered by Eleanor Helin in 1978. It is the largest known member of the ➤ *Aten* group of asteroids, whose orbits lie wholly within that of the Earth.

RATAN-600 A radio telescope at the ➤ *Special Astrophysical Observatory* of the Russian Academy of Sciences, located at Zelenchukskaya in the Caucasus Mountains. The name is an acronym from "Radio Astronomy Telescope of the Academy of Sciences" in Russian. It consists of 900 parabolic plates forming a circle 600 m (2000 feet) in diameter. It can be used as a whole, or each quarter can operate as a self-contained unit.

ray A light-colored, linear feature extending from a crater on a moon or planet. A number of lunar craters are surrounded by extensive and conspicuous ray systems. They show up particularly well at full Moon, when they can be seen with the naked eye. Ray systems are associated with young craters, such as Tycho and Copernicus. The rays are made of light-colored material blasted out of the crater lying on top of darker terrain around the crater.

R Coronae Borealis The prototype of a class of peculiar variable stars that suddenly and unpredictably dip in brightness by many magnitudes. R Coronae Borealis (R CrB) is normally magnitude 5.8, but every few years dims by up to nine magnitudes. This happens when it throws off a dust cloud. About 40 stars of similar type are known. Typically they are ➤ *supergiant* stars of ➤ *spectral type* F or G.

Reber, Grote (1911–2002) The American Grote Reber was one of the world's first radio astronomers. He trained as a radio engineer but was inspired to take up ➤ *radio astronomy* when he heard of the discoveries made by Karl ➤ *Jansky*. In 1937 he built the first radio telescope to use a parabolic dish. The 9.4-meter (30-foot) dish was mounted on a tilting stand and he was able to detect a number of intense cosmic sources of radio emission. He later constructed other radio telescopes and made the first radio map of the sky in 1941.

red giant An evolved star that has expanded greatly and appears red.

A star becomes a red giant when the hydrogen fuel for nuclear fusion in its central core is exhausted. The core collapses. This releases gravitational energy and a new supply of heat that allows hydrogen fusion to restart, though in a shell around the now inert core. The energy generated by shell burning of hydrogen causes the great expansion of the star's outer layers. As the gas expands, it cools. Regardless of the star's original ➤ *spectral type*, its surface temperature drops until it reaches 4000 K. When the Sun becomes a red giant, it will expand until its diameter is roughly the diameter of the Earth's orbit. Though the light emitted per square meter of a red giant's surface is low compared with hotter stars, they are very luminous overall because of their huge size. All the bright red stars visible to the naked eye are giants or supergiants, such as ➤ *Aldebaran* or ➤ *Betelgeuse*. ➤➤ *stellar evolution*.

Red Planet A popular name for the planet ➤ *Mars*, which looks distinctly red to the naked eye.

redshift (symbol *z*) An increase in the wavelength (λ) of electromagnetic radiation between its emission and its reception. If the increase in wavelength is $\Delta\lambda$, the redshift (*z*) is $\Delta\lambda/\lambda$. Three different mechanisms can cause redshifts to the radiation from astronomical bodies: the ➤ *Doppler effect*, ➤ *gravitational redshift*, and ➤ *cosmological redshift*.

reflection nebula A cool cloud of interstellar gas and dust that shines because the dust scatters the light of stars near to it. A reflection nebula is not itself

A reflection nebula around stars in the constellation Chamaeleon.

luminous. The spectrum of the scattered light is the same as that of the starlight, though blue light is scattered more strongly than red light.

regolith The loose, fine-grained, soil-like material on the surface of the Moon or any other planetary body.

Regulus (Alpha Leonis) The brightest star in the constellation Leo. It is a ➤ *B star* of magnitude 1.4 with two companions of seventh and thirteenth magnitudes. The name is Latin and means ''little king.'' Regulus is 77 light years away.

relativistic Traveling nearly as fast as light.

relativity ➤ *general relativity, special relativity*.

remote sensing The study of Earth or other planetary objects by observation from a distance rather than direct contact and exploration. The term is used particularly for the study of Earth, or other bodies, from orbiting satellites.

The lunar regolith, with a footpad of the Apollo 11 lunar module and astronauts' footprints.

resolution The size of the smallest detail that can be distinguished with an imaging instrument such as a telescope or spectrograph.

resolving power A telescope's ability to distinguish detail. Theoretically, it is limited by the size of the aperture (the diameter of the main light-collecting lens or mirror). In practice, however, the resolving power of a ground-based optical telescope is often limited by the quality of ➤ *seeing* rather than the aperture, unless it is equipped with ➤ *adaptive optics*.

resonance A situation in which one orbiting body is subject to a regular periodic disturbance by the gravity of another. Resonances occur between orbits that have periods in whole-number ratios (e.g. 1:1, 2:1, 3:2). They are responsible for phenomena such as the ➤ *Kirkwood gaps* in the ➤ *asteroid belt* and divisions in Saturn's rings.

Reticulum (The Net) A small southern constellation introduced by Nicolas L. de Lacaille in the mid-eighteenth century. Its two brightest stars are of third magnitude.

retrograde Motion of an object on the celestial sphere in the east–west direction, or, for orbital motion or axial rotation in the solar system, motion that is clockwise as observed from north of the ➤ *ecliptic*. The opposite of retrograde motion is called "direct" motion.

Rhea The second-largest satellite of Saturn, discovered by Giovanni Cassini in 1672. ➤ *Voyager 1* and ➤ *Cassini* images show that Rhea's light-colored, icy surface is saturated with craters and has bright wispy markings. Rhea's low density shows that it consists mainly of ice over a rocky core. Its diameter is 1528 km (949 miles) and its orbit is 527 100 km (427 500 miles) from Saturn.

Rho Ophiuchi cloud A large nebulous region near the star Rho Ophiuchi. It is a mixture of ➤ *reflection nebulae*, ➤ *emission nebulae*, dark ➤ *absorption nebulae* and ➤ *molecular clouds*, and is relatively close, at a distance of about 700 light years. Infrared observations reveal the presence of a cluster of at least 40 stars within the dark cloud. This is a region of very active star formation containing many ➤ *T Tauri stars* and ➤ *Herbig–Haro objects*.

An image of Rhea from the Cassini spacecraft.

Rigel (Beta Orionis) The brightest star in the constellation Orion. At magnitude 0.1, it is slightly brighter than ➤ *Betelgeuse*, which is designated Alpha. Rigel is a supergiant ➤ *B star*, with a seventh-magnitude companion lying 775 light years away. The name Rigel is derived from Arabic and means "leg of the giant."

right ascension (RA) One of the coordinates used to specify positions on the ➤ *celestial sphere* in the equatorial coordinate system. It is the equivalent of longitude on Earth but is measured in hours, minutes and seconds of time eastwards from the zero point, which is the intersection of the celestial equator and the ➤ *ecliptic*, known as the ➤ *First Point of Aries*. One hour of right ascension is equivalent to 15 degrees and is the angle through which the sky appears to turn in one hour of ➤ *sidereal time*. ➤➤ *declination*.

Rigil Kentaurus The star ➤ *Alpha Centauri*. This Arabic name means "foot of the centaur."

rille (rill) A fissure or channel in the lunar surface. Some are vertical faults. Others are collapsed lava tubes, which tend to be more sinuous.

ring galaxy A rare kind of galaxy shaped like a ring. Ring galaxies are thought to result from collisions between galaxies in which one passes right through another. An expanding shock wave of interstellar gas triggers a burst of star formation in a ring around the center of the "target" galaxy.

Ring Nebula (M57; NGC 6720) A bright ➤ *planetary nebula* in the constellation Lyra. In a small telescope it looks like a luminous elliptical ring around a central star, but the observational evidence suggests it has more the shape of a dumbbell, and just happens to be viewed end-on. Its radius is one-third of a light year, and the nebula is 2000 light years away. If it has always expanded at its current rate of 19 km/s, it is about 5500 years old.

The ring galaxy known as Hoag's Object, imaged by the Hubble Space Telescope.

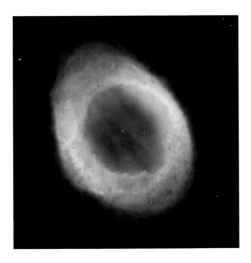

The Ring Nebula.

ring systems Ring structures, composed of numerous individual small bodies and dust, surrounding the four largest outer planets – Jupiter, Saturn, Uranus and Neptune.

The rings of Saturn were discovered by ➤ *Galileo* when he first turned a telescope on the sky in 1610. In 1857, James Clerk Maxwell demonstrated

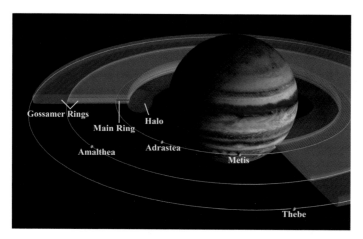

The structure of Jupiter's ring system.

Saturn's ring system. This simulated image depicts the observed ring structure. Purple shows regions where there is a lack of particles under 5 cm (2 inches). Green and blue shades indicate regions where there are particles smaller than 5 cm (2 inches) and 1 cm (0.3 inch). White is a dense region for which accurate data could not be collected. From other evidence, all ring regions appear to be populated by objects up to the size of boulders, from several to many meters across.

theoretically that the rings must be made up of many unconnected particles, and this was later confirmed by spectroscopic observations showing that the inner particles orbit more quickly than the outer ones. In 1977, nine narrow rings around Uranus were detected when the planet occulted a star. In 1979, ➤ *Voyager 1* discovered a faint band around Jupiter and, in the early 1980s, what seemed to be incomplete ring arcs were detected around Neptune, again during an occultation. In 1989, ➤ *Voyager 2* showed that the rings around Neptune are complete but clumpy.

Virtually all the planetary rings lie within their ➤ *Roche limits*. In a disk of debris around a newly formed planet, material beyond the Roche limit could coalesce into moons while nearer to the planet, tidal forces would prevent moons from forming.

The rings around Jupiter are faint and tenuous, and their reflection qualities show that many of the particles can be no bigger than 1 or 2 μm. Dust of this size must be constantly replenished, perhaps by impacts on boulder-sized objects in the ring.

Saturn's rings are far more complex and extensive than any other system. The rings easily visible from Earth were labeled A, B and C, C being the faint inner Crepe ring. The A and B rings are separated by the Cassini Division and there is also a narrow but conspicuous gap towards the outer edge of the A ring, known as the Encke Division or Encke Gap. Voyager 1 detected material inside the C ring, which was called the D ring. Beyond the A ring lie more narrow, tenuous rings, known as the E, F and G rings. The ring particles are thought to be a mixture of water ice and dust, though their composition varies, and they range in size from a few micrometers up to a hundred meters.

Voyager images showed that Saturn's main rings consist of thousands of narrow, closely spaced "ringlets." Many of the observed structures are due to the gravitational action of moons. For example, Pandora and Prometheus act as "shepherds," confining the F ring, and the Cassini Division lies where a satellite with an orbital period half that of Mimas would lie. (This is an example of a ➤ *resonance* phenomenon.) Observations of Saturn's rings from the ➤ *Cassini* spacecraft confirm their variety and complexity. Cassini's discoveries include the fact that the particles in the A ring form into clusters several meters long that assemble and then get torn apart again.

The nine rings of Uranus found in 1977 are labeled, in order of increasing distance from the planet: 6, 5, 4, α, β, η, γ, δ and ε. Two more rings were found by Voyager 2 in 1986 and also a pair of satellites, Ophelia and Cordelia, shepherding the ε ring. In 2003, the Hubble Space Telescope found a further two rings near the moons Portia and Mab. The nine main rings appear to be composed of meter-sized boulders. But in backlighting, *Voyager 2* also saw many slender ringlets composed of dust.

Two main rings orbit Neptune (Leverrier and Adams rings), with a diffuse sheet of material extending outwards from the inner of the two (Plateau). There is also a tenuous third ring nearer the planet (Galle). A faint outward extension to the Leverrier ring has been named Lassell and it is bounded at its outer edge by the Arago ring. The outer ring, Adams, contains three bright arcs, about 8° in extent, which seem to be dominated by dust-sized particles. These have been called Liberty, Equality and Fraternity. It is thought that the gravitational influence of the moon Galatea, orbiting just inside the ring, acts to confine the arcs.

Roche limit The minimum distance from the center of a planet that a hypothetical fluid moon can orbit and not be destroyed by tidal forces. If the planet and satellite have equal densities, the Roche limit is 2.456 times the radius of the planet.

Anywhere inside the Roche limit, a fluid moon would be torn apart. Solid satellites can exist inside the Roche limit because the strength of the rock holds them together. ➤➤ *tides*.

Roche lobes In a ➤ *binary star* system, an hourglass-shaped region between the two stars. Its surface is where the gravitational force exerted by either star is the same. Part way between the stars there is a unique location where the two Roche lobes touch. This is called the "inner ➤ *Lagrangian point*."

When a star in a close binary system expands during the giant phase of its evolution, it may completely fill its Roche lobe. A system like that is said to be "semidetached." Matter then streams through the inner Lagrangian point to the other star. In a ➤ *contact binary*, both stars have completely filled their Roche lobes and gas can be transferred between them.

Rosalind A small moon of Uranus with a diameter of 72 km (45 miles). It was discovered in 1986 by the ➤ *Voyager 2* spacecraft.

ROSAT A German orbiting X-ray astronomy observatory, with participation by NASA and the UK. It was launched in 1990 and operated until December 1998. It carried a large X-ray imaging telescope and a wide-field camera to observe in the extreme ultraviolet.

Rosetta A European Space Agency space mission to a ➤ *comet*, consisting of an orbiter and a lander. It was launched on March 2, 2004 and will reach its target, Comet Churyumov–Gerasimenko, in 2014. Its journey sends it around Earth three times (in 2005, 2007 and 2009), and around Mars once (in 2007) for ➤ *gravity assist*. The orbiter is equipped with 11 instruments and will release a 100-kg lander, which has nine instruments of its own. The orbiter will ultimately be in an orbit 25 km above the comet's nucleus, which is only about 4 km across. The lander will touch down slowly in November 2014 and fire an anchoring harpoon into the comet's surface immediately on impact. The orbiter will stay with the comet as it gets nearer to the Sun and the mission is due to end in December 2015.

Rosette Nebula (NGC 2237, 2238, 2239 and 2246) An ➤ *emission nebula* in the constellation Monoceros, surrounding a young ➤ *open cluster* of stars (NGC 2244). It is roughly circular in shape, with a central hole cleared of dust

Artist's impression of the Rosetta lander anchored to the surface of its target comet.

The Rosette
Nebula.

and gas by the radiation from the stars in the cluster. The nebula's distance is estimated to be 4500 light years.

Rossi, Bruno Benedetti (1905–1993) Rossi is most remembered as a pioneer in the field of ➤ *X-ray astronomy* and he initiated the rocket experiment in 1962 that led to the first discovery of a cosmic source of X-rays, ➤ *Scorpius X-1*. In the 1940s and 1950s, he also advanced the study of ➤ *cosmic rays* at the Massachusetts Institute of Technology, where he was a professor of physics from 1946. He was born and educated in Italy, where he began his career as a physicist, but he left for the United States in 1938. The ➤ *Rossi X-ray Timing Explorer* was named in his honor.

Rossi X-ray Timing Explorer (RXTE) A NASA X-ray astronomy satellite launched in 1995. It was named in honor of the X-ray astronomy pioneer, Bruno B. ➤*Rossi*, in 1996 after launch. The spacecraft was equipped with three instruments. The Proportional Counter Array and the High-Energy X-ray Timing Experiment (HEXTE) working together constituted the largest X-ray telescope flown to date. The third instrument, the All Sky Monitor, was designed to record the long-term behavior of X-ray sources and monitor the sky.

Royal Observatory, Greenwich A UK observatory first established at Greenwich near London in 1675 by King Charles II, to address the problem of finding longitude at sea. ➤ *John Flamsteed*, the first ➤ *Astronomer Royal*, was put in charge. The Observatory subsequently played an important role in positional astronomy and the meridian through Greenwich was adopted as the zero point of longitude in 1884.

In the twentieth century, the emphasis changed to include more astrophysics and researchers moved to Herstmonceux Castle, Sussex, in 1948. The new establishment was known as the Royal Greenwich Observatory (RGO), while the original observatory at Greenwich became part of the UK's National Maritime Museum. The RGO's headquarters moved to Cambridge in 1990 but it closed in 1998.

Royal Observatory, Edinburgh (ROE) A UK astronomy research establishment in Edinburgh. Since 1998 it has comprised the UK's Astronomy Technology Centre and the Institute for Astronomy of the University of Edinburgh.

RR Lyrae star A category of pulsating ➤ *variable star* belonging to ➤ *Population II* (relatively old stars). They are found particularly, but not exclusively, in ➤ *globular clusters*. Though similar to ➤ *Cepheid variables*, they are intrinsically less luminous – by as much as seven magnitudes. All have approximately the same absolute magnitude (+0.5) and this makes them useful distance indicators. Their periods range between a few hours and just over a day, and they vary by anything between 0.2 and 2 magnitudes. Their ➤ *spectral type* is A or F. RR Lyrae, which varies over 0.567 days, is the best-known example.

runaway star A young hot star traveling through space with an unusually high velocity. It is thought that such stars could originally have been in binary or multiple systems where a companion exploded as a ➤ *supernova*.

Three well-known examples are Mu Columbae, AE Aurigae and 53 Arietis. From their speeds and directions of motion it has been calculated that all three were ejected from the same region in the constellation Orion about three million years ago.

Russell, Henry Norris (1877–1957) Russell's greatest achievement was his research on the luminosity of stars, which resulted in the ➤ *Hertzsprung–Russell diagram*. He was a brilliant student at Princeton University where he gained his doctorate, after which he spent three years at King's College, Cambridge. At Cambridge, he determined the distances to stars by measuring their ➤ *parallax* and so discovered the relationships between their ➤ *spectral types* and their luminosity. When describing his discovery, he introduced the terms "giant" and "dwarf" to describe stars for the first time. He spent most of his professional life at Princeton, where he was director of the observatory from 1912 to 1947.

Ryle, Sir Martin (1918–1984) Ryle was a British radio astronomer who developed revolutionary ➤ *radio telescopes* and detectors. Appointed a lecturer in physics at Cambridge University in 1948, he completed the first map of the radio sky in 1950 and discovered some 50 radio sources. His *Third Cambridge Catalogue*, made in 1959, led directly to the discovery of ➤ *quasars* when optical observers tried to match starlike objects with the positions of radio sources Ryle had measured. His third and fourth catalogs comprehensively disproved the ➤ *steady-state theory* advanced by the Cambridge theorist ➤ *Fred Hoyle*. Ryle shared the Nobel Prize for physics in 1974 with a fellow Cambridge radio astronomer, Anthony Hewish.

Henry Norris Russell.

S

Sagan, Carl Edward (1934–1996) The American planetary scientist Carl Sagan was one of the most well-known and influential figures in astronomy in the USA during his lifetime. He was particularly known for his interest in extraterrestial life and he strongly advocated searching for extraterrestrial intelligence. As an advisor to NASA, he contributed to many of NASA's planetary exploration space missions. A gifted popularizer, he wrote several bestselling books. His television series *Cosmos*, first shown in 1980, was an enormous success. In it Sagan examined a wide range of issues in science, particularly the origin of life in the universe. From 1968 he worked at Cornell University and directed the Laboratory for Planetary Studies there.

Sagitta (The Arrow) The third-smallest constellation, but nevertheless a distinctive little group of stars. The two brightest stars are third magnitude. It lies in a rich part of the Milky Way, next to Aquila.

Sagittarius (The Archer) The southernmost constellation of the zodiac. The center of the Galaxy (the Milky Way) lies behind the star clouds in Sagittarius. It is a large constellation, with many bright stars. It also contains a large number of star clusters and nebulae. The ➤ *Messier Catalogue* lists 15 objects in Sagittarius, more than in any other individual constellation. They include the ➤ *Lagoon Nebula*, the ➤ *Trifid Nebula*, the ➤ *Omega Nebula* and the third brightest ➤ *globular cluster* in the sky, M22.

Sagittarius A The strongest radio source in the constellation Sagittarius. It is a complex radio source associated with the galactic center. A bright, compact and variable source embedded within it and known as Sagittarius A* is believed to be the actual center of the Galaxy and to be powered by a supermassive black hole there.

Sagittarius Arm One of the spiral arms of the ➤ *Galaxy*. It lies between the Sun and the center of the Galaxy in the direction of the constellation Sagittarius.

saros The time for a sequence of lunar and solar eclipses before the cycle repeats itself. A saros lasts 6585.32 days (about 18 years). After this period, the Earth, Sun and Moon return to the same relative positions. Succeeding eclipses in a particular ➤ *saros series* occur about 8 hours later and fall nearly 120 degrees of longitude farther west. This was known to the ancient Babylonian and Mayan astronomers and to the builders of ➤ *Stonehenge*.

saros series A series of lunar or solar eclipses occurring at intervals of one ➤*saros*. Since there are up to seven eclipses every year, there are more than 80 saros

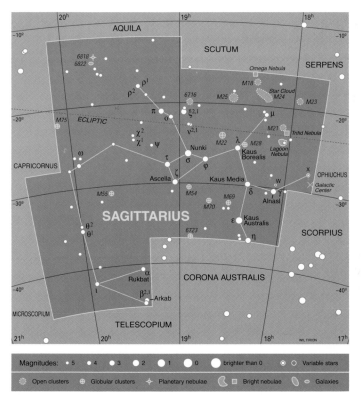

A map of the constellation Sagittarius.

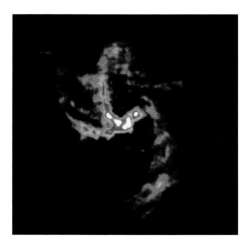

A radiograph of Sagittarius A, which shows emission from gas within the central 10 light years of our Galaxy.

Saturn imaged by the Hubble Space Telescope.

series running concurrently. The eclipses in a series return to the same longitude after an interval of three saroses (54 years). However, each succeeding eclipse in the series moves systematically in latitude (either north or south), from one pole to the other until the series is completed.

satellite Any body in orbit around a larger parent body. Most planets in the solar system have natural satellites, otherwise known as moons.

Saturn The sixth major planet of the solar system in order from the Sun, orbiting at an average distance of 9.54 astronomical units. Saturn is one of the four giant planets, second in size only to Jupiter. Its equatorial diameter is 9.4 times Earth's and its mass is 95 times greater. However, its average density is only 0.7 times that of water. Hydrogen and helium make up the bulk of its mass. There is a central core, 10 or 15 times the mass of Earth, made of rock or a mixture of rock and ice. In the high-pressure region surrounding the core, the hydrogen takes on the form of a metal. The outer half of the planet consists of a deep atmosphere; the visible features of the planet are cloud bands at the top of this atmosphere.

The cloud patterns on Saturn do not normally show much color contrast. However, temporary storms sometimes occur. In late September 1990, a large white spot developed, expanding over a period of weeks to encircle much of the planet's equatorial region. Similar large storms have been seen at intervals of about 27–30 years, in 1876, 1903, 1933 and 1960, close to mid-summer on Saturn in its northern hemisphere.

Images obtained by ➤ *Voyagers 1 and 2* during their encounters in 1980 and 1981 reveal complex circulation currents, similar to those observed on Jupiter. Saturn rotates rapidly, spinning once every 10 hours 32 minutes on average, though the rate varies with latitude. This rapid spin makes Saturn bulge at the equator. Its polar and equatorial diameters differ by 11 percent.

Saturn's most striking feature is its spectacular ➤ *ring system*. The rings lie in the planet's equatorial plane, which is tilted at an angle of 27° to its orbit around the Sun. They are easily visible in a small telescope. As the relative positions of Earth and Saturn change, the rings are presented to us at differing angles, sometimes appearing open, but at other times edge-on so that they disappear from view. There are zones of differing brightness in the ring system, separated by dark divisions. The most marked divisions are Cassini's and Encke's. The Voyager images of the rings showed that they consist of many thousands of narrow ringlets. They are only one kilometer thick and are made up of a huge number of separate pieces of ice and dust, perhaps ranging in size from a hundred meters down to a micrometer.

Before 1980, 10 moons of Saturn were known. By 2006 the number had reached 56 and was increasing every year. ➤➤ *Cassini–Huygens mission.*

Saturn Nebula (NGC 7009) A ➤ *planetary nebula* in the constellation Aquarius. Its unusual shape, with a faint partial outer ring, resembles the planet Saturn. The double ring may be the remains of separate shells thrown off by the central star.

Scheat (Beta Pegasi) The second-brightest star in the constellation Pegasus. It is a ➤ *supergiant* ➤ *M star* 200 light years away and varies in brightness between magnitudes 2.4 and 2.8. The name comes from Arabic and probably means "shoulder."

Schedar (Schedir; Alpha Cassiopeiae) The brightest star in the constellation Cassiopeia. It is a ➤ *supergiant* ➤ *K star* 230 light years away with a magnitude near 2.2, though it is slightly variable. The Arabic name means "breast."

Schiaparelli, Giovanni Virginio (1835–1910) Schiaparelli was director of the Milan Observatory from 1860 to 1900. At the 1877 ➤ *opposition* of Mars, when Earth and Mars were exceptionally close, he produced detailed maps of the martian surface. He described several linear features that he designated as "canali," meaning natural channels or grooves. ➤ *Percival Lowell* in the USA and Camille Flammarion in France subsequently made extravagant claims that they were artificial canals. Schiaparelli also studied Venus and Mercury, and discovered that the tracks in space of ➤ *meteors* follow the orbits of ➤ *comets*.

Schmidt camera A type of wide-field astronomical telescope, designed to take photographs. It was invented by Bernhard Schmidt in 1930 and has a specially shaped glass corrector plate over the end of the telescope tube to ensure the images are sharp. Schmidt cameras produce good images over very wide fields of view – up to tens of degrees across.

Schmidt–Cassegrain telescope A design of optical telescope, incorporating features of both a ➤ *Schmidt camera* and a ➤ *Cassegrain* reflector.

A Schmidt–Cassegrain has a corrector plate like a Schmidt camera but there is a small secondary mirror at the center of the plate, which reflects the

Schröter's Valley photographed by Apollo 15 astronauts.

light back down the tube and through a hole in the primary mirror. The resulting telescope is very compact, so the design is particularly suitable for portable telescopes and is popular with amateurs.

Schröter's Valley A winding valley in the Oceanus Procellarum on the Moon. It starts in a small crater just outside the northern wall of the crater Herodotus, and extends for about 200 km.

Schwarzschild, Karl (1873–1916) Schwarzschild, who was born in Frankfurt, was both an observational astronomer and a noted theorist. In 1909 he became director of the Potsdam Observatory. While serving with the German army in Russia in 1916, he published the first exact solution to complex equations ➤ *Einstein* had published that year in his ➤ *general relativity* theory. He introducing the concept of what is now termed the ➤ *Schwarzschild radius*, which is important in the theory of ➤ *black holes*.

Schwarzschild radius A critical quantity that determines whether a body is a ➤ *black hole*, from which nothing can escape into the outside world. An object that collapses inside its Schwarzschild radius becomes a black hole. The Schwarzschild radius for an object the mass of the Sun is 3 km; for an object the mass of the Earth it is 1 cm.

scintillation Twinkling – rapid variations in the brightness of a star caused by turbulence in Earth's atmosphere. A similar phenomenon affects radio signals from cosmic sources when they travel through Earth's ionosphere.

Scorpio An alternative name for the constellation ➤ *Scorpius*, chiefly used in astrology.

Scorpius (The Scorpion) A large, bright constellation of the southern part of the zodiac. Its brightest star is the first-magnitude ➤ *Antares*.

Scorpius–Centaurus association A loose group of young hot stars about 400 light years away. It includes most of the brightest stars in the constellations ➤ *Scorpius* and ➤ *Centaurus*.

Scorpius X-1 The brightest X-ray source in the sky and the first to be discovered. It is a low-mass ➤ *X-ray binary* star.

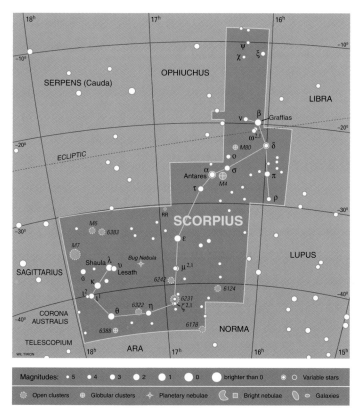

A map of the constellation Scorpius.

Sculptor (The Sculptor's Workshop) An inconspicuous constellation of the southern hemisphere introduced by Nicolas L. de Lacaille in the mid-eighteenth century. Its four brightest stars are fourth magnitude.

Scutum (The Shield) A small constellation near the celestial equator, introduced by Johannes ➤ *Hevelius* in the late seventeenth century with the name Scutum Sobieskii in honor of his patron, King John Sobieski III. It has no stars brighter than fourth magnitude but it contains the ➤ *Wild Duck* star cluster.

seasons The natural cycle of change in the prevailing environmental conditions on a planet, over the course of its orbit around the Sun. A planet experiences seasons if its rotation axis is not at 90° to the plane of its orbit. Both Earth and Mars have marked seasons.

second contact In a total or annular ➤ *eclipse* of the Sun, the point when the edges of the Moon's disk and Sun's ➤ *photosphere* are in contact at the start of totality or the annular phase. In a lunar eclipse, second contact occurs when the Moon just enters the full shadow (umbra) of the Earth completely. The same term can also mean a similar stage in the progress of a ➤ *transit* or ➤ *occultation*.

secular Continuing, or changing in a non-periodic way, over a long period of time.

90377 Sedna One of the larger known ➤ *Kuiper Belt objects*. When discovered in 2003, it was the most distant known object in the solar system. It follows a very elongated elliptical orbit and is estimated to be 1180–1800 km (730–1120 miles) in diameter.

seeing The effect of turbulence in the atmosphere on the quality of the image of an astronomical object. In conditions of good seeing, images are sharp and steady; in poor seeing, they are blurred and appear to be in constant motion.

selenography The study of the surface features and topography of the Moon.

selenology The study of lunar rocks and of the surface and interior of the Moon: the lunar equivalent of geology.

semiregular variable A pulsating variable star with some regularity to its variations but nevertheless unpredictable. Semiregular variables typically vary by only one or two magnitudes over anything from a few days to several years. A number of subgroups have been distinguished.

Serpens (The Serpent) The only constellation split into two parts, which are known as Serpens Caput (the head) and Serpens Cauda (the tail). They lie either side of the constellation Ophiuchus (The Serpent Bearer). The brightest stars in Serpens are third and fourth magnitude.

Setebos A small outer moon of Uranus discovered in 1999. Its diameter is about 30 km (19 miles).

SETI Abbreviation for Search for Extraterrestrial Intelligence. It is used as a general term rather than for any particular experiment. Most SETI projects have involved the search for unnatural narrow-band radio signals. Optical SETI looks for evidence of pulsed laser beams, which could be visible over distances of hundreds of light years.

SETI Institute A private research establishment with headquarters in California dedicated to research in astrobiology and the search for extraterrestrial intelligence.

Seven Sisters A popular name for the ➤ *Pleiades*.

Sextans (The Sextant) A faint constellation of the southern hemisphere introduced in the late seventeenth century by Johannes ➤ *Hevelius*, supposedly to commemorate the instrument he used to make astronomical observations. Its brightest star is magnitude 4.5.

Seyfert galaxy A type of galaxy with a brilliant point-like nucleus and inconspicuous spiral arms, first described by Carl Seyfert in 1943.

Seyfert's Sextet (NGC 6027) A group of galaxies in the constellation Serpens. It consists of five galaxies, together with a large cloud of gas ejected by the principal galaxy in the group. This galaxy, which is a spiral, and three

This Seyfert galaxy is in the southern constellation Circinus and is 13 million light years away. The image was taken by the Hubble Space Telescope.

Seyfert's Sextet imaged by the Hubble Space Telescope.

lenticular galaxies in the group all lie at a distance of 260 million light years and are interacting with each other. The fifth galaxy is a spiral five times farther away, coincidentally lying in the same part of the sky.

shadow bands A phenomenon sometimes observed briefly just before and just after totality at a total solar eclipse. Irregular bands of shadow, a few

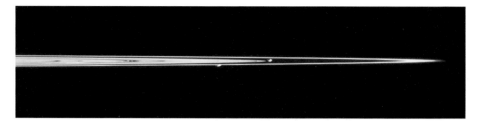

Saturn's moons Pandora (right) and Prometheus (left), which act as shepherd satellites of the F ring (the ring extending farthest right in this image from the Cassini spacecraft). The two moons are separated by about 69 000 km (43 000 miles) in this view.

centimeters wide and up to a meter apart, move over the ground. What causes them is not fully understood. They are seen only if the sky is very clear.

Shaula (Lambda Scorpii) The second-brightest star in the constellation Scorpius, marking the Scorpion's sting. It is a ➤ *B star* of magnitude 1.6 and is 700 light years away.

shepherd satellites Moons, often found in pairs, that hold a planetary ring in place and prevent it from dispersing through the action of their gravity. For example, Prometheus and Pandora are the shepherd satellites for the F-ring in Saturn's ➤ *ring system*.

shield volcano A large volcano with gently sloping sides, built up by successive lava flows from a single vent. Individual layers may amount to only a few meters but they can build up to create a very high mountain. Typically, the slope of the sides is less than 10°. At the top there is a large, shallow, flat-floored crater called a caldera.

Shoemaker–Levy 9, Comet A comet that crashed into the planet Jupiter in July 1994. When it was discovered photographically on March 25, 1993, by

The fragments of Comet Shoemaker–Levy 9 crashing into Jupiter's atmosphere caused the huge dark dust clouds visible in this ultraviolet image from the Hubble Space Telescope.

Carolyn and Eugene Shoemaker and David Levy, it was in an elongated two-year orbit around Jupiter and consisted of a string of about 20 separate fragments. Calculations suggested it had been orbiting Jupiter for several decades, but broke up during a close approach to Jupiter in July 1992. This encounter also set the fragments on their collision course. They struck Jupiter one after another between July 16 and 22, 1994. A number of the impacts produced large dark clouds in Jupiter's atmosphere, as well as powerful fireballs. The dark clouds were discernible for several months before they dispersed.

shooting star A popular name for a ➤ *meteor*.

short-period comet A ➤ *comet* in an elliptical orbit, with a period of several years or decades, comparable with the orbital periods of the planets. Short-period comets have gone into their present orbits after close encounters with planets, especially Jupiter. Two-thirds of all short-period comets are in orbits that extend no more than one astronomical unit beyond the orbit of Jupiter. It is suspected that they were originally in the ➤ *Kuiper Belt*.

Siarnaq A small outer moon of Saturn in a very elliptical orbit, discovered in 2000. It is 32 km (19 miles) across.

Sickle An ➤ *asterism* formed by the stars Alpha, Eta, Gamma, Zeta, Mu and Epsilon in the constellation ➤ *Leo*, named because of its shape.

sidereal Connected with the stars.

sidereal day The Earth's rotation period with respect to the stars. The length of the sidereal day is 23 hours 56 minutes and 4 seconds.

sidereal period The time taken by a planet or satellite to complete an orbit as measured relative to the stars.

sidereal time Time measured by the rotation of the Earth with respect to the stars rather than its rotation relative to the Sun, which is the basis for civil time. The local sidereal time at a particular place is the ➤ *right ascension* on the meridian. Observatories normally have a clock keeping sidereal time.

Siding Spring Observatory ➤ *Mount Stromlo and Siding Spring Observatories*.

Sikhote-Alin shower A major meteorite fall on February 12, 1947, in eastern Siberia. The largest meteorite recovered weighed 1745 kg (3850 lb), but it has been estimated that thousands of pieces fell, weighing up to a total of 100 tonnes. Much of it has never been found.

Sinope A small outer moon of Jupiter discovered in 1914 by Seth B. Nicholson. Its diameter is 38 km (24 miles).

Sirius (Alpha Canis Majoris) The brightest star in the constellation Canis Major and, at magnitude –1.46, the brightest star in the sky. It is a visual binary with an orbital period of 50 years. The primary (A) is an ➤ *A star* and the secondary (B) an eighth-magnitude ➤ *white dwarf*. Sirius B was first seen in 1862, and its nature was deduced from its spectrum in 1925. Sirius lies at a distance of

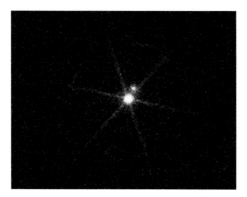

A Chandra Observatory X-ray image of Sirius A and B. The white dwarf, Sirius B, is the brighter in X-rays.

8.7 light years and is the seventh-nearest star to the solar system. The name Sirius is derived from Greek and means "scorching," a reference to its brilliance. Sirius is also known as "the Dog Star," from the constellation in which it lies.

Sirrah An alternative name for the star ➤ *Alpheratz*.

1866 Sisyphus A small asteroid, diameter 7.6 km (4.7 miles), discovered in 1972 by Paul Wild. It is in a comet-like orbit, inclined at 41° to the plane of the solar system, which occasionally brings it relatively close to the Earth. It is the largest known ➤ *Apollo* asteroid.

Skathi A small outer moon of Saturn in a very elliptical orbit. It was discovered in 2000 and its diameter is about 6 km (4 miles).

Skylab An American space station, launched into Earth orbit in May 1973. Three crews, each of three men, were sent to the station for periods of several weeks between 1973 and 1974. The station burnt up on re-entering the atmosphere in 1979.

Skylab in orbit at the end of its mission.

Vesto Slipher.

Slipher, Vesto Melvin (1875–1969) Slipher, an American astronomer, worked for the whole of his career between 1901 and 1952 at the ➤ *Lowell Observatory*, where he was made director in 1926. His skill was in spectroscopy. By 1917, he found that 13 out of 15 spiral "nebulae" he was studying were moving away from the Milky Way at velocities of hundreds of kilometers per second. Many astronomers at the time questioned these findings, and it was not then known that the "nebulae" were in fact galaxies beyond the Milky Way, but in retrospect Slipher's measurements were some of the earliest evidence that the universe is expanding.

Sloan Digital Sky Survey A project to map one quarter of the sky, including measurements of the positions and absolute magnitudes of over 100 million celestial objects. One of its main objectives is to determine the distances of a million of the nearest galaxies, and of 100 000 ➤ *quasars*. The 2.5-m (98-inch) survey telescope, together with a 0.6-m (23.5-inch) support telescope opened in 1997 at the ➤*Apache Point Observatory*.

Small Astronomy Satellites (SAS) Three NASA spacecraft deployed in the 1970s for X-ray and gamma-ray astronomy. SAS-1, launched in 1970, was the first satellite dedicated to ➤ *X-ray astronomy*. After launch, it was given the name Uhuru, which means "freedom" in Swahili, because the launch

SMART-1 took this image of rugged cratered terrain near the Moon's north pole.

date of December 12 coincided with the seventh anniversary of Kenya's independence. SAS-2, launched in 1972, was the first satellite to carry a gamma-ray detector. SAS-3, which followed in 1975, carried further X-ray experiments.

Small Magellanic Cloud ➤ *Magellanic Clouds*.

SMART-1 A European Space Agency mission launched into Earth orbit on September 27, 2003. The main objective was to test technologies for use on future missions. The spacecraft used an ➤ *ion drive* to slowly elongate its orbit and increase its distance from Earth, until it could be transferred into an orbit around the Moon on November 15, 2004. SMART-1 carried three instruments to study the surface composition of the Moon. On September 3, 2006, it was crash-landed on the Moon deliberately.

Smithsonian Astrophysical Observatory (SAO) A research establishment founded by the US Smithsonian Institution in Washington DC in 1890. In 1955, under the directorship of Fred Whipple, its headquarters were moved to the grounds of ➤ *Harvard College Observatory* (HCO). In 1967, an observatory was established at Mount Hopkins in Arizona, now known as the ➤ *Fred Lawrence Whipple Observatory*. In 1973, the Harvard–Smithsonian Center for Astrophysics was formed by combining the SAO and HCO.

SN 1987A ➤ *Supernova 1987A*.

SNC meteorites Unusual ➤ *meteorites*, believed to have come from Mars. SNC are the initials of the three different classes of these meteorites: shergottites, nakhlites and chassignites.

SOFIA An airborne observatory consisting of a 2.5-m (100-inch) telescope mounted in a modified Boeing 747 aircraft. It had been expected to begin operations in about 2006 but the project has suffered delays. The first test flight took place in

April 2007. Its base will be NASA's Ames Research Center in California. SOFIA is an acronym for "Stratospheric Observatory for Infrared Astronomy." It is the successor to the ➤ *Kuiper Airborne Observatory* (KAO).

Flying at between 41 000 and 45 000 feet, SOFIA will operate above 85 percent of Earth's atmosphere and above 99 percent of the water vapor in the atmosphere, which interferes with ➤ *infrared astronomy*. It will be able to observe in the visible, infrared, sub-millimeter and microwave regions of the spectrum. The project is a collaboration between the USA and Germany.

soft gamma repeater A type of rare gamma-ray source that emits infrequent bursts of radiation. Each burst lasts a maximum of a few seconds. Soft gamma repeaters are thought to be a type of ➤ *magnetar*, young, rapidly spinning ➤ *neutron stars*. Two are known to be located within ➤ *supernova remnants*.

SOHO Abbreviation for ➤ *Solar and Heliospheric Observatory*.

Sojourner The name of the small roving vehicle carried by the NASA Mars mission ➤ *Mars Pathfinder*.

sol A martian "day," which is 24 hours 37 minutes and 22.6 seconds in length.

solar Connected with the Sun.

solar activity A variety of phenomena on the Sun that involve disturbances and outbursts of energy. Solar activity varies over time. The most obvious changes in the ➤ *solar cycle* repeat after about 11 years. Phenomena such as ➤ *coronal mass ejections*, ➤ *flares*, ➤ *sunspots*, ➤ *prominences* and ➤ *faculae* are all manifestations of solar activity.

Solar and Heliospheric Observatory (SOHO) A space observatory launched by the ➤ *European Space Agency* on December 2, 1995. It was placed in orbit around the Sun at the point, known as the L_1 ➤ *Lagrangian point*, where the gravitational forces of the Earth and Sun are equal. Its 12 instruments were designed to investigate the solar atmosphere and how it is heated, solar oscillations, how the Sun expels material into space, the structure

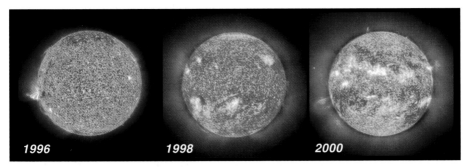

The change in solar activity from minimum to maximum is illustrated by this sequence of three ultraviolet images of the Sun taken by the SOHO spacecraft.

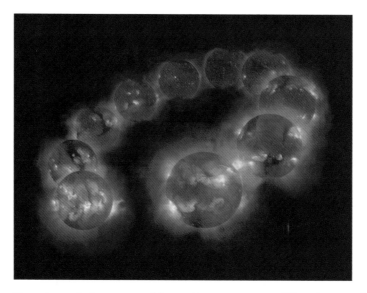

The solar cycle demonstrated by an 11-year sequence of X-ray images of the Sun from the Japanese Yohkoh satellite.

of the Sun and processes operating within it. By 2005 over 1000 comets had been discovered on SOHO images. The mission is due to end in 2009.

solar constant The solar power reaching the top of Earth's atmosphere, at the average Sun–Earth distance. Its value is about 1.35 kW/m^2 but is not in fact constant. It changes slightly over the course of a ➤ *solar cycle* – a large sunspot group decreases it by about 1 percent – and there are also longer-term variations.

solar cycle The periodic variation in the amount of ➤ *solar activity*, particularly the number of ➤ *sunspots*. The cycle is about 11 years long, though it has been closer to 10 years during the twentieth century.

At the start of a new cycle there are few, if any, spots on the Sun. The first ones signalling the new cycle erupt around latitudes 35–45° north and south; over the course of the cycle, subsequent spots appear closer to the equator, finishing at around 7° north and south. On a graph, this pattern of appearance creates a ➤ *butterfly diagram*.

It is generally thought that the solar cycle is caused by an interaction between the "dynamo" responsible for the Sun's magnetism and the Sun's rotation. The Sun does not rotate as a solid body: the equatorial regions rotate fastest and this amplifies the magnetic field, which eventually bursts into the photosphere, causing sunspots. At the end of each cycle the overall magnetic field reverses.

solar nebula The cloud of interstellar gas and dust that condensed to form the Sun and solar system about five billion years ago.

The major planets of the solar system in order from the Sun (not to scale). From left to right: Mercury, Venus, Earth, Mars, Jupiter, Saturn, Uranus and Neptune. The four outer giant planets are shown at about one quarter their real size compared with the inner planets.

solar system The Sun, together with the planets and moons, comets, asteroids, meteoroid streams and interplanetary medium held captive by the Sun's gravity. The solar system is believed to have formed from a rotating disk of gas and dust created around the Sun as it contracted to form a star, about five billion years ago.

solar time Time based on the rotation of Earth. Earth's rotation rate is not precisely constant when checked against an atomic clock so solar time has been superseded by International Atomic Time (TAI). "Leap seconds" are occasionally introduced to keep TAI in step with Earth's rotation.

Since Earth's rotation axis is tilted and its orbit around the Sun is elliptical rather than circular, the Sun's apparent motion through the sky is not uniform over the course of a year. Apparent solar time, as measured by a sundial, differs from mean time by an amount that varies through the year and is known as the ➤ *equation of time.*

solar tower A type of telescope used exclusively for observing the Sun. Close to the ground, the heating effect of the Sun creates a layer of hot, turbulent air, which makes images unsteady. To overcome this problem, the mirror used to send an image of the Sun into the telescope is placed on a tall tower. ➤➤ *vacuum tower telescope.*

solar wind A stream of particles, primarily protons and electrons, flowing outwards from the Sun at up to 900 km/s (2 million mph). The solar wind is essentially the hot solar ➤ *corona* expanding into interplanetary space.

Solis Planum (Solis Lacus) An ancient volcanic plain on Mars, lying to the south of ➤ *Valles Marineris.* To the visual observer, the area has a variable dark spot

The solar tower telescope at Mount Wilson Observatory, California.

(the "lake") whose appearance has earned the feature the nickname of "the eye of Mars."

solstices The two occasions in the year when the Sun reaches its most northerly and most southerly position in the sky, or the Sun's position at these times. The solstices fall approximately midway between the spring and autumn ➤ *equinoxes*.

The solstices occur on about June 21 and December 21. At the summer solstice, the Sun reaches its highest altitude in the sky and the hours of daylight are at their maximum. At the winter solstice the Sun's altitude at noon is at its lowest and the period of daylight is a minimum. The summer solstice in the northern hemisphere (June) is the winter solstice in the southern hemisphere and vice versa.

Sombrero Galaxy (M104; NGC 4594) A spiral galaxy in the constellation Virgo that we see very nearly edge-on. Its marked central bulge and conspicuous lane of dark obscuring dust make it look rather like a wide-brimmed hat.

South African Astronomical Observatory (SAAO) A national optical astronomy facility located at Sutherland, 220 miles from Cape Town, in South Africa. The SAAO was formed in 1972 by merging the old Royal Observatory at Cape Town and the Republic Observatory, Johannesburg. Some of their telescopes were moved to the Sutherland site and, in 1974, a 1.9-m (75-inch) telescope was purchased from the Radcliffe Observatory, Pretoria, and also moved. In 1999, Sutherland was chosen as the site for the ➤ *Southern African Large Telescope*.

An image of the Sombrero Galaxy from the 8.2-m ANTU unit of the European Southern Observatory's Very Large Telescope.

Southern African Large Telescope (SALT) An 11-m telescope opened at Sutherland in South Africa in 2005. The project involves 10 international partners. Its design is closely based on that of the ➤ *Hobby–Eberly Telescope*. The telescope's angle of tilt is fixed and it rotates only at its base. Targets for observing are selected by moving the instruments positioned above the main mirror.

South Atlantic anomaly A region over the South Atlantic Ocean where the lower ➤ *Van Allen belt* of electrically charged particles is particularly close to Earth's surface, creating a hazard for artificial satellites.

Southern Cross Popular name for the cross-shaped ➤ *asterism* in the constellation ➤ *Crux*, formed by the stars Alpha, Beta, Gamma and Delta.

Southern Pleiades Popular name for IC 2602, a large and bright ➤ *open cluster* of stars in the constellation Carina. The star Theta Carinae (magnitude 2.8) is at the center of the cluster and several other member stars are also visible to the naked eye. The whole cluster is about one degree across on the sky.

Soyuz A Soviet spacecraft, used to carry up to three cosmonauts.

space The regions between the planets and stars, excluding their immediate atmospheres.

Space Shuttle A US reusable space vehicle that takes off like a rocket but lands like an aircraft on a runway. A total of five have been built. The first flight was made by *Columbia* on April 12, 1981. The second Shuttle, *Challenger*, was destroyed in an explosion shortly after its tenth launch in 1986. It was replaced by *Endeavour*, which was first launched in 1991. *Discovery* and *Atlantis*, the third

The stars of the Southern Cross (on the right), with the Milky Way and part of the constellation Centaurus.

The Space Shuttle Discovery approaching the International Space Station for docking in July 2006. A module for the Space Station can be seen in the Shuttle's cargo bay.

and fourth Shuttles, first flew in 1984 and 1985, respectively. *Columbia* was destroyed with the loss of its crew on re-entry into the atmosphere on February 1, 2003. Following this disaster, Shuttle launches were suspended for safety improvements. *Discovery* flew again (to the ➤ *International Space Station*) in July

2005 but the next flight was again delayed until July 2006 because of further safety concerns about the remaining craft in the ageing fleet. NASA hopes that Shuttles will be able to continue long enough to see the completion of the International Space Station in 2010 before they are grounded permanently.

space telescope A telescope placed in orbit around the Earth so as to be above the atmosphere. ➤ *Hubble Space Telescope.*

Space Telescope Science Institute A research institute in Baltimore, Maryland, that manages the science program of the ➤ *Hubble Space Telescope*, processes data from the telescope and coordinate its operations with the Space Telescope Operations Control Center.

spacetime A unified framework for specifying the place and time of events and describing the relationships between events in a mathematical way. Spacetime takes account of the fact that the speed of light is always the same even when the emitter and the observer are moving relative to each other. In ➤ *general relativity*, gravity is curvature of spacetime.

Spacewatch A project to search for and study movements of comets and asteroids particularly any that might be a hazard to Earth. It began in the 1980s and carries out full-time surveys with 1.8-meter and 0.9-meter telescopes at ➤ *Kitt Peak.*

space weather The physical conditions in space immediately around Earth and between Earth and the Sun. These conditions vary because of the ➤ *solar wind*, ➤ *coronal mass ejections* and other phenomena related to ➤ *solar activity.*

Special Astrophysical Observatory (SAO) The principal observatory of the Russian Academy of Sciences, for both optical and radio astronomy. It is located in the Caucasus region between the Black Sea and the Caspian Sea.

  The optical observatory near Zelenchukskaya is the location of the Bolshoi Teleskop Azimutalnyi (Large Altazimuth Telescope), which has a 6-m (236-inch) primary mirror. Completed in 1975, it was the first very large telescope to be built on an ➤ *altazimuth mount.* The ➤ *RATAN-600* radio telescope is located on the outskirts of Zelenchukskaya.

special relativity A theory published by Albert ➤ *Einstein* in 1905, which describes how relative motion at a constant speed affects the way physical phenomena are perceived. The special theory of relativity was a direct consequence of the fact that the speed of light in a vacuum is always the same regardless of whether the source of light or the observer is moving. The other underlying principle of the theory is the "principle of relativity." This states that no physical experiment can detect that the "laboratory" is in uniform motion. ➤➤ *general relativity.*

spectral line A sharp peak or dip in a ➤ *spectrum.* Spectral lines result when an atom or ion gives out or absorbs a characteristic amount of energy – a "quantum." Emitting a quantum of energy gives an ➤ *emission line.* Absorbing a quantum of energy makes an ➤ *absorption line* in a continuous spectrum.

spectral type A classification given to a star according to the appearance of its spectrum. A star's spectral type mainly reflects its surface temperature but a luminosity class (whether the star is a dwarf or giant) is often included as well.

The strange order of letters for the classes is a legacy from the first comprehensive attempt at classification, which was undertaken at Harvard College Observatory and published in 1890. The classes were originally designated A–Q but were rationalized and re-ordered as a temperature sequence later, after astrophysicists learned how to interpret spectra correctly. The outcome was the list of basic types still used: O, B, A, F, G, K and M (see the table). The main classes have up to 10 subdivisions, indicated by the numbers 0 to 9 (e.g. A0, K5).

Spectral type	Temperature range	Principal features of visible spectrum
O	> 25 000 K	Relatively few absorption lines. Ionized helium, doubly ionized nitrogen, triply ionized silicon. Hydrogen weak.
B	11 000–25 000 K	Neutral helium, singly ionized oxygen and magnesium. Hydrogen stronger than in O stars.
A	7500–11 000 K	Strong hydrogen. Singly ionized magnesium, silicon, iron, titanium, calcium, etc. and some neutral metals.
F	6000–7500 K	Hydrogen weaker and neutral metals stronger than in A stars. Singly ionized calcium iron, chromium.
G	5000–6000 K	Ionized calcium most conspicuous. Many lines of ionized and neutral metals, and of CH molecule.
K	3500–5000 K	Neutral metals predominate. CH molecule.
M	< 3500 K	Strong lines of neutral metals and of TiO molecule.

The luminosity classes, introduced in 1943, are:

Ia   Luminous supergiants
Ib   Less luminous supergiants
II   Bright giants
III   Normal giants
V   Dwarfs/main sequence

In addition to these basic spectral and luminosity types, which cover the vast majority of stars, additional classifications have been introduced for more

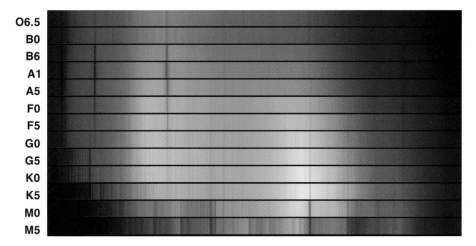

O6.5	
B0	
B6	
A1	
A5	
F0	
F5	
G0	
G5	
K0	
K5	
M0	
M5	

A set of spectra of stars illustrating the range of main spectral types.

unusual stars. These include ➤ *S stars* and ➤ *carbon stars*, ➤ *brown dwarfs* and ➤ *white dwarfs*. ➤➤ *Hertzsprung–Russell diagram.*

spectrogram A permanent record of a ➤ *spectrum*, taken as a photograph or with an electronic detector.

spectrograph An instrument for recording a ➤ *spectrum*.

spectroheliograph An instrument for obtaining images of all or part of the Sun in light of one particular wavelength.

spectrometer An instrument for observing a spectrum and measuring features in it by direct observation.

spectroscope An instrument for observing a ➤ *spectrum* visually.

spectroscopic binary A ➤ *binary star* the double nature of which is revealed by its spectrum even though the two stars cannot be separated visually. In the spectrum of a double-lined spectroscopic binary, features of both stars are detected. Their spectral lines shift relative to each other in a regular way because of the ➤ *Doppler effect*, as the two stars orbit around each other. In a single-lined spectroscopic binary, the two stars differ greatly in luminosity so only the spectrum of the brighter one can be discerned. Its double nature is given away because its lines change wavelength in a periodic way when measured relative to a standard spectrum.

spectroscopy The study and interpretation of ➤ *spectra*.

spectrum (pl. spectra) The result of dispersing a beam of electromagnetic radiation so that components with different wavelengths are spread out in order of increasing or decreasing wavelength. The most familiar example of a spectrum is the natural rainbow, when sunlight is dispersed into its component colors by raindrops. The full spectrum of ➤ *electromagnetic radiation* encompasses, in order of decreasing wavelength, radio waves, microwaves, infrared radiation, visible light, ultraviolet radiation, X-ray radiation and gamma radiation.

There are three principal types of spectra – continuous, emission line and absorption line. If brightness is plotted against wavelength as a graph, a continuous spectrum is a smooth line, with no sharp spikes or dips. Emission lines show as relatively narrow spikes or peaks of intensity. Absorption lines are relatively narrow dips in the intensity of a continuous spectrum. All three kinds can be present in the spectrum of the same object.

Continuous spectra result from ➤ *black body radiation* or ➤ *synchrotron radiation*, for example. Spectral lines are created when discrete packets of energy (quanta), corresponding to precise wavelengths, are emitted or absorbed by atoms or molecules. ➤➤ *absorption line, continuous spectrum, emission line, spectral line.*

Spica (Alpha Virginis) The brightest star in the constellation Virgo, which is about 260 light years away. It is an ➤ *eclipsing binary*, of magnitude 1.0 varying by about 0.1 magnitude with a period of 4.014 days. The primary is a blue-white ➤ *B star* with a mass about 11 times the Sun's. Spica means "an ear of corn."

spicules Spike-like structures in the Sun's ➤ *chromosphere* that may be observed at or near the limb. They change rapidly, having a lifetime of five to ten minutes. Typically, spicules are 1000 km (600 miles) across and 10 000 km (6000 miles) long. They are concentrated along the boundaries of the cells that make up the Sun's ➤ *supergranulation* pattern.

spider The bars supporting the secondary mirror in the tube of a reflecting telescope. Diffraction by the spider causes spikes around images of bright stars.

Spindle Galaxy Popular name for the edge-on galaxy NGC 3115 in the constellation Sextans. Its shape is reminiscent of a spindle wound with yarn. It is a highly evolved galaxy with no obvious evidence of dust.

spiral galaxy A ➤ *galaxy* with spiral arms. Edwin ➤ *Hubble* divided spiral galaxies into two broad groups: those with a central bar (SB galaxies) and those without (S). Each group he additional subdivided into three categories, a, b and c. Sa and SBa galaxies have tightly wound arms and a relatively large central bulge. Sc and SBc galaxies have loose arms and a small central bulge, while types Sb and SBb are intermediate between the two extremes.

Our own ➤ *Galaxy* (the Milky Way) is a barred spiral. Its structure is fairly typical: young stars and interstellar material are concentrated in a disk, particularly in the spiral arms, and around the disk there is a spherical halo containing old stars and globular clusters.

The spiral arms are not permanent rigid structures. As stars and interstellar material orbit around the center of the galaxy, they create temporary regions of enhanced density in a spiral pattern. The arms are

The arms of the nearby spiral galaxy M81 in the constellation Ursa Major
are highlighted in this Spitzer Space Telescope infrared image.

actually made up of different stars and gas clouds at different eras of time.
➤➤ *Hubble classification*.

Spitzer, Lyman (1914–1997) Spitzer was an American astrophysicist who worked
on how matter behaves in interstellar space and inside stars. He was a pioneer in
the study of the ➤ *interstellar medium*. In 1946 he proposed the construction of a
large telescope in space, and advocated the idea for decades. He was the principal
driving force behind the ➤ *Hubble Space Telescope*. Spitzer spent his professional
career at Princeton University, where he succeeded Henry Norris ➤ *Russell* as
director of the Observatory. The ➤ *Spitzer Space Telescope* was named in his honor.

Spitzer Space Telescope A NASA space telescope for ➤ *infrared astronomy*, which
was launched on August 25, 2003, initially on a two-and-a-half-year mission.
With its 0.85-m (33-inch) mirror, it was the largest infrared telescope ever
launched into space. It carried instruments for imaging and spectroscopy in
the wavelength range 3–180 μm. To reduce the amount of coolant it needed to
carry, the telescope was placed in an orbit around the Sun, trailing behind
Earth. In this orbit, compared with being in orbit around Earth, it cooled more
quickly and was less affected by Earth's own infrared emission. The telescope
was named in honor of Lyman ➤ *Spitzer*, Jr (1914–1997), a distinguished
American astrophysicist, who was the first person to propose putting a large
telescope in space. Initially it had been referred to as the Space Infrared
Telescope Facility or SIRTF.

Sponde A small outer moon of Jupiter discovered in 2001. Its diameter is about
2 km (1 mile).

Lyman
Spitzer, Jr

sporadic meteor A ➤ *meteor* that does not belong to an identified
➤ *meteor shower*.

Sputnik Three unmanned spacecraft launched into Earth orbit by the Soviet
Union in 1957 and 1958. Sputnik 1, launched on October 4, 1957, was the
first man-made object to orbit around the Earth.

Square of Pegasus Four bright stars making the shape of a large square. They are
Alpha, Beta and Gamma Pegasi, and Alpha Andromedae. (Alpha Andromedae
was formerly known as Delta Pegasi.)

SS433 A peculiar star, which is number 433 in a catalog compiled by C. Bruce
Stephenson and Nicholas Sanduleak. It is a binary system in which a ➤ *neutron
star* is dragging material from a more massive normal companion.

SS433 is located 18 000 light years away inside the ➤ *supernova remnant*
W50, which believed to be about 40 000 years old. It appears as a fourteenth-
magnitude star in the constellation Aquila. In 1976 it was found to be a source
of X-rays, and radio emission was detected the following year. The optical
spectrum reveals a complex situation with periodic variations and evidence for
a pair of jets traveling at a quarter the speed of light.

The starburst galaxy NGC 908.

Star clouds of the Milky Way in the constellation Sagittarius.

The neutron star has 0.8 times the Sun's mass. Its companion is an ➤ *O star* or ➤ *B star*. It is losing material very quickly and is currently put at 3.2 solar masses. The two orbit each other in 13 days. Material from the larger star streams into an ➤ *accretion disk* around the neutron star. Excessive heating causes some of the material to be blasted out from the central hole of the ring-shaped accretion disk to form the jets. The disk wobbles slightly, like a spinning top, over a period of 164 days. This creates a helical pattern in the radio emission from the jets and a regular 164-day cycle in the apparent velocity of the jets as viewed from Earth.

The open star cluster NGC 625, which lies in the Small Magellanic Cloud.

S star A cool giant star of basic ➤ *spectral type* K or M that, unlike ordinary K and M stars, has absorption in its spectrum caused by the molecule zirconium oxide (ZrO), and often the molecules lanthanum oxide (LaO), yttrium oxide (YO) and vanadium oxide (VO) as well. The zirconium and other heavier elements are the products of nuclear reactions in the star's interior that have risen up to the surface.

star A large, luminous ball of gas generating energy internally by nuclear fusion.
   The minimum amount of material required for a star is about one-twentieth the mass of the Sun. Below this limit, the core cannot get hot enough to sustain fusion reactions. The most massive stars known are about 100 solar masses. A star's mass is the factor that chiefly determines its temperature and luminosity while it is an "ordinary star" on the "main sequence" of the ➤ *Hertzsprung–Russell diagram.*
   Stars are made mostly of hydrogen, with helium as their other major constituent. In the Sun, which is in many ways a typical star, 94 percent of atoms are hydrogen, 5.9 percent helium and less than 0.1 percent other elements. By weight, 73 percent is hydrogen, 25 percent helium, 0.8 percent carbon and 0.3 percent oxygen, the remaining 0.9 percent being all the other elements. ➤➤ *binary star, spectral type, stellar evolution, variable star.*

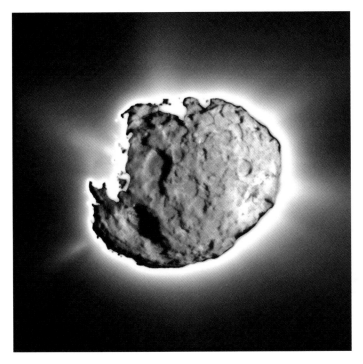

Comet Wild 2 from data collected by the Stardust spacecraft. To create this image, a short-exposure image showing surface detail was overlain on a long-exposure image taken just 10 seconds later showing jets.

starburst galaxy A galaxy with an exceptionally high rate of star formation. Starburst galaxies often emit excessive amounts of infrared radiation, which may account for over 90 percent of the total energy they give out.

star cloud An area of sky, particularly in the ➤ *Milky Way*, where there are large numbers of stars close together.

star cluster A group of stars that formed together. There are two main types, ➤ *open clusters* and ➤ *globular clusters*. Very young stars are often found in loose groupings called stellar ➤ *associations*.

Stardust A NASA mission launched on February 7, 1999, which flew through the ➤ *coma* of comet ➤ *Wild 2*, taking images and collecting a sample of cometary dust. Using an Earth flyby for ➤ *gravity assist*, Stardust flew within 237 km (147 miles) of the comet and collected its samples on January 2, 2004. The sample was successfully returned to Earth in January 2006.

starquake A sudden cracking in the outer crust of a ➤ *neutron star*, similar to an earthquake. A starquake causes an abrupt change in the neutron star's period of rotation, which is called a "glitch."

steady-state theory One of two rival theories of cosmology debated in the mid-twentieth century, the other being the ➤ *Big Bang* theory. Fred ➤ *Hoyle* was one of its main advocates. The steady-state theory assumes that the universe is the same everywhere for all observers at all times. It accounts for the observed expansion of the universe by postulating that new matter is continuously created to fill the voids left as existing galaxies move apart. The discovery of the ➤ *cosmic background radiation* in 1963 was a major setback for the theory and the Big Bang theory is now generally supported by the overwhelming majority of astrophysicists.

stellar evolution The sequence of changes that happen to a star through its lifetime, from its birth out of the interstellar medium to final extinction.

Stars form in clusters in the clouds of gas and dust between the stars. Gravity pulls the material of a newly forming star together. The material falling in releases energy and the center heats up until the temperature is high enough for the nuclear fusion of hydrogen into helium. The time taken for a "protostar" to become a fully fledged star depends on its mass. A star of 10 solar masses takes only 300 thousand years, compared with 30 million years for a star the same mass as the Sun.

The nuclear "burning" of hydrogen in a star's core continues until the hydrogen runs out in the core where it is hot enough. During this phase, the star is on the main sequence of the ➤ *Hertzsprung–Russell diagram*. How long this phase lasts depends on its mass. The Sun will be a main-sequence star for a total of about 10 billion years (of which about half has passed) but a star three times more massive spends only 500 million years in this phase.

When hydrogen burning in the core stops because the fuel is used up, the interior of the star adjusts to the new situation. The now inert helium core

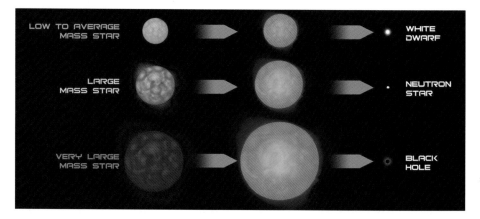

In stellar evolution, the ultimate fate of a star depends on its mass to begin with.

contracts rapidly. In the process, gravitational energy is released, which heats the surrounding layers of hydrogen to the point where hydrogen burning recommences, but in a shell around the core. The result of the new outpouring of energy is to push the outer layers of the star farther and farther outwards. As this gas expands, it cools, and the star becomes a red giant. The combined effect of the increase in size and decrease in temperature is to maintain a more or less constant luminosity overall. Meanwhile, the helium core continues to contract until its temperature reaches a hundred million degrees. This is high enough for the fusion of helium into carbon and oxygen to begin.

Eventually, all the helium in the star's core is consumed. What happens next depends on the mass of the star. In more massive stars, the core contracts again after each fuel has been exhausted. This raises the temperature sufficiently to ignite a new, heavier, nuclear fuel. Ultimately, a situation can be reached in which the central core has been converted to iron, while around the core, in a series of shells, silicon, oxygen, carbon, helium and hydrogen are being burnt simultaneously. Once a star has developed an iron core of about one solar mass no new reactions are possible. At this stage, the core contracts until it implodes catastrophically, setting off a ➤ *supernova* explosion. The naked core that remains becomes a ➤ *neutron star* or ➤ *black hole*.

In lower-mass stars, such as the Sun, the central temperature never gets high enough to progress beyond the burning of hydrogen and helium in shells around the core. The outer layers of the star become unstable and separate from the core to form expanding shells of gas, called a ➤ *planetary nebula*, that gradually disperses into space. In fact, most stars lose a significant amount of material in the form of stellar "winds," which are stronger during the later phases of evolution. The remaining core cools and shrinks, becoming more and more compressed, until it is about the size of Earth. It becomes a ➤ *white dwarf*. A white dwarf has no ongoing internal source of energy. It simply cools down.

This outline of stellar evolution is for single stars. Membership of a binary or multiple system may profoundly influence the course of a star's evolution if it exchanges matter with one of its close companions.

stellar wind Mass loss from a star in the form of a continuous outflow of gas. All stars lose some gas but the rate is highest for the hottest stars. Through winds blowing at hundreds or even thousands of kilometers per second, they can lose a significant fraction of their original mass over their lifetimes. ➤➤ *solar wind*.

Stephano A small outer moon of Uranus discovered in 1999. It is about 20 km (12 miles) across.

Stephan's Quintet (NGC 7317, 7318a and b, 7319 and 7320) A group of five galaxies lying close together in the sky, which were first noted by Edouard Stephan in 1877. Subsequent measurements of their recession velocities show

Stephan's
Quintet.

that NGC 7320 is much closer than the others and just happens to lie along the line of sight by chance. The other four share a common speed and appear to be physically associated.

Steward Observatory The observatory of the University of Arizona. It operates telescopes at a number of sites in Arizona. The largest is the Bok Telescope, a 2.29-m (90-inch) on ➤ *Kitt Peak*, opened in 1969. There is a 1.54-m (60-inch) reflector and a 42-cm (16-inch) ➤ *Schmidt camera* at Mount Bigelow in the Catalina Mountains, 54 km (34 miles) from Tucson. 1.5-m and 1.0-m telescopes are located at Mount Lemmon, and the ➤ *Heinrich Hertz Submillimeter Telescope* together with the 1.8-m Lennon reflector are at the ➤ *Mount Graham International Observatory*.

Stickney The largest crater on Mars's inner moon, ➤ *Phobos*. It is 10 km (6 miles) across, over a third of Phobos's largest diameter of 28 km. Stickney was the maiden name of the wife of Asaph ➤ *Hall* (1829–1907), the American astronomer who discovered Phobos and Mars's other moon, Deimos.

Stonehenge A prehistoric stone monument in the UK, thought to be of astronomical significance. Located 130 km (80 miles) west of London, Stonehenge is one of the finest of all neolithic sites. It was constructed in three phases, commencing with a bank and ditch around 2800 BC. The surviving group of sandstone megaliths in a circle 30 m (100 feet) in diameter was erected in about 2000 BC. Some of the stones form foresight and backsight markers that indicate crucial rising and setting points for the Sun and Moon with considerable accuracy. Astronomers have shown how observations made at Stonehenge would have enabled its users to predict solar and lunar eclipses with certainty. If the astronomical interpretation is a correct one, it implies that the designers of Stonehenge had recorded, or remembered through an oral tradition, observations extending over many centuries.

stony-iron meteorite A major category of ➤ *meteorite*, consisting of a mixture of metallic and stony elements. There are two main types: pallasites and mesosiderites. Pallasites consist of olivine grains enclosed by metal. Typically they contain twice as much olivine as metal by volume. Mesosiderites contain silicates and metal, in roughly equal proportions.

stony meteorite A ➤ *meteorite* consisting entirely of stony material. Stony meteorites are divided into two main classes: ➤ *chondrites* and ➤ *achondrites*. More than 90 percent of the meteorites seen to fall (as opposed to those found by chance) are stony.

stratosphere The region of Earth's atmosphere above the ➤ *troposphere*. It lies between heights of about 15 and 50 km (9 and 30 miles). From the bottom to the top of the stratosphere, the temperature increases from about 240 K to 270 K.

Stratospheric Observatory for Infrared Astronomy ➤ *SOFIA*.

Struve, Otto (1897–1963) Otto belonged to the fourth generation of a notable dynasty of Russian astronomers founded by his great-grandfather Friedrich Struve (1793–1864). He spent his professional career in the USA where he went in 1921. He became director of the ➤ *Yerkes Observatory* (1932–47) and of the ➤ *McDonald Observatory* (1939–50), which he helped to found. As first director of the ➤ *National Radio Astronomy Observatory* (1959–62) he encouraged the search for extraterrestrial intelligence. Struve's astronomical work was mainly concerned with the ➤ *interstellar medium* and he discovered that it contains ➤ *ionized hydrogen*.

Subaru Telescope An 8.3-m telescope at the ➤ *Mauna Kea Observatories* in Hawaii for the National Astronomical Observatory of Japan. Construction began in 1991 and it opened in 1999. It operates in both the visual and infrared spectral regions. Subaru is the Japanese word for the ➤ *Pleiades*.

Submillimeter Array A submillimeter-wave telescope at the ➤ *Mauna Kea Observatories* in Hawaii. The telescope consists of eight movable dishes, each 6-m (20 feet) across.

Submillimeter Telescope Observatory A facility at the ➤ *Mount Graham International Observatory* run jointly by the University of Arizona and the Max-Planck-Institut für Radioastronomie in Bonn, Germany. Its instrument is the 10-m (33-foot) Heinrich Hertz Telescope, which operates in the submillimeter waveband between 0.3 and 1 mm. First observations were made in 1994.

submillimeter-wave astronomy The study of ➤ *electromagnetic radiation* from celestial sources in the wavelength band between 0.3 and 3 millimeters. It uses a combination of techniques from radio astronomy and infrared astronomy.

Sudbury Neutrino Observatory. A view of the structure from the outside.

The telescopes have to be located at particularly dry places and high elevations, because water vapor in Earth's atmosphere absorbs strongly at these wavelengths, and the astronomical signals are mostly weak. However, this region of the spectrum is important for a number of studies in astronomy, including the ➤ *cosmic background radiation*, regions of star formation and molecules in interstellar clouds.

There are only a few submillimeter telescopes in operation. One is the ➤ *James Clerk Maxwell Telescope* situated at the ➤ *Mauna Kea Observatories*, Hawaii, as is the ➤ *Submillimeter Array*. The California Institute of Technology's Submillimeter Observatory (CSO), a 10.4-m (84-foot) telescope with a segmented mirror, is also at the same site. The Swedish–ESO Submillimeter Telescope (SEST) is at the ➤ *European Southern Observatory* (ESO), La Silla, Chile. The reflector has a diameter of 15 m (49 feet) and consists of 176 panels that can be adjusted separately. The ➤ *Heinrich Hertz Submillimeter Telescope* is at Mount Graham in Arizona.

In 1998, the first orbiting observatory for submillimeter wave astronomy was launched from the USA – the ➤ *Submillimeter Wave Astronomy Satellite* (SWAS) – with the prime objective of studying the composition of interstellar clouds.

Submillimeter Wave Astronomy Satellite (SWAS) A small NASA satellite launched in December 1998 to detect and measure the abundance of water, molecular oxygen, carbon monoxide and atomic carbon in a wide variety of astronomical objects. It was put into "hibernation" in July 2004 but reactivated in 2005 to make observations in support of the ➤ *Deep Impact* mission.

subsolar point The point on a body in the solar system from which observers would see the Sun directly overhead.

Sudbury Neutrino Observatory (SNO) A detector for neutrinos from the Sun and other astronomical sources located 2 km (1.25 miles) underground in a nickel

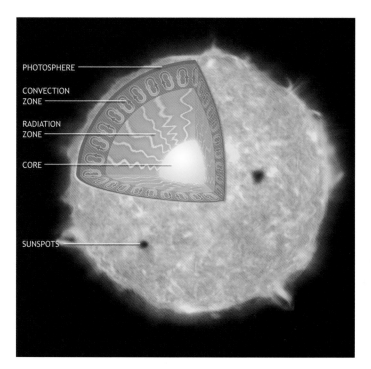

PHOTOSPHERE

CONVECTION
ZONE

RADIATION
ZONE

CORE

SUNSPOTS

The interior structure of the Sun.

mine near Sudbury, Ontario, Canada. It is based on a tank of 1000 tonnes of heavy water and detects faint flashes of light emitted as neutrinos are stopped or scattered. The observatory began operation in 1999. ➤ *neutrino astronomy*.

Summer Triangle The three bright stars ➤ *Vega*, ➤ *Altair* and ➤ *Deneb*, which are particularly conspicuous in the summer evening sky.

Sun The central star of the solar system. Within the range stars cover, the Sun is of medium size and brightness, though the vast majority of stars in the solar neighborhood are smaller and less luminous. It is a dwarf star of ➤ *spectral type* G2 with a surface temperature of about 5700 K. Like all stars, it is a globe of hot gas and its energy source is nuclear fusion taking place in the center, where the temperature is 15 million K. Four million tonnes of solar material are annihilated each second as hydrogen is converted into helium.

Overlying the core is the radiation zone, where high-energy radiation produced in the fusion reactions is converted to light and heat. Over the radiation zone is a convection zone in which currents of gas flow upwards, release energy at the surface, then flown again to be reheated. These circulating currents create the Sun's mottled appearance, or ➤ *granulation*. The surface layers, or ➤ *photosphere*, from which the light we see comes, are some hundreds of kilometers thick.

The layer over the photosphere is the ➤ *chromosphere*, visible as a glowing pinkish ring during a total solar eclipse. ➤ *Spicules* and ➤ *prominences* erupt through the chromosphere. The thinnest, outermost layers, forming the solar ➤ *corona*, merge into the interplanetary medium. ➤➤ *coronal hole, coronal mass ejection, flare, solar activity, solar cycle, solar wind, sunspot.*

Properties of the Sun	
Mass	1.989×10^{30} kg (332 946 Earth masses)
Radius	6.96×10^{5} km (109 Earth radii)
Effective temperature	5785 K
Luminosity	3.9×10^{26} W
Apparent visual magnitude	-26.78
Absolute visual magnitude	4.79
Inclination of equator to ecliptic	$7° 15'$
Synodic rotation period	27.275 days
Sidereal rotation period	25.380 days

sundial A time-keeping instrument consisting of a shadow stick (or gnomon) and a dial on which the shadow cast by the Sun falls. The dial is graduated in hours. A sundial measures apparent ➤ *solar time*. There are many different types of design for sundials of varying degrees of sophistication.

sundog An alternative name for a ➤ *mock Sun* or parhelion.

sungrazer A ➤ *comet* that approaches very close to the Sun and passes through the Sun's outer layers. Around a dozen long-period comets, which have other orbital characteristics in common, form a well-established group of sungrazers. They are also known as the "Kreutz group" after the Dutch astronomer Heinrich Kreutz (1854–1907) who, in 1888, was among the first to note the similarity between the orbits of some of the brightest comets ever observed. Many of the hundreds of comets discovered near the Sun by the orbiting ➤*SOHO* solar observatory are sungrazers.

sunspot A relatively dark region on the Sun where the temperature is lower than in the rest of the ➤ *photosphere*. Sunspots occur where the magnetic field in the photosphere is thousands of times stronger than the average field for the Sun. The strong magnetic field inhibits the upward flow of hot gas, and so has a cooling effect.

Sunspots can occur individually but often form pairs of opposing magnetic polarity or in larger groups. A typical sunspot is around twice the size of Earth and lasts for about a week. Large groups may contain up to 100 spots, extend over hundreds of thousands of kilometers and persist for up to two months.

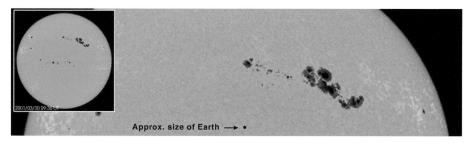

A large sunspot group imaged by the SOHO spacecraft on March, 30 2001.

In the dark central part of the sunspot, called the "umbra," the temperature is about 4200 K compared with the 5700 K of the photosphere. The umbra forms a depression a few hundred kilometers below the general level of the photosphere, giving rise to the ➤ *Wilson effect*. The outer and brighter part of a sunspot, the "penumbra," consists of aligned bright grains on a darker background which look like a border of rays around the spot. The penumbra is the sloping sides of the depression formed by the spot. ➤ *butterfly diagram*, *solar activity*.

super bubble A large, expanding, bubble of luminous interstellar gas hundreds of light years across. Super bubbles contain gas that is tenuous but hot. They are created by ➤ *stellar winds* streaming from massive hot stars, and by ➤ *supernova* explosions.

supercluster A concentration of clusters of galaxies. About 50 are known. On average they contain 12 rich galaxy clusters, though the largest have many more. Superclusters are hundreds of millions of light years across.

supergiant One of the largest, most luminous stars known. Supergiants can be up to 500 times larger than the Sun and many thousands of times more luminous. There are supergiants of all ➤ *spectral types*. They are massive stars (mass greater than about 10 times the Sun's) in an advanced state of ➤ *stellar evolution*. A supergiant is likely to become a ➤ *supernova*.

supergranulation A large-scale pattern of cells of moving gas on the Sun.

Super-Kamiokande A neutrino detector located 1 km (3000 feet) underground at the Kamioka mine in Japan. It is the successor to the earlier Kamiokande (Kamioka nucleon decay experiment). The main element of the detector is a tank containing 50 000 tonnes of water. Sensors record flashes of light, emitted when high-velocity charged particles travel through the water. ➤➤ *neutrino astronomy*.

superior conjunction The position of either Mercury or Venus when it lies on the far side of the Sun as viewed from the Earth.

superior planet Any of the major planets whose orbits lie outside that of Earth's – Mars, Jupiter, Saturn, Uranus and Neptune.

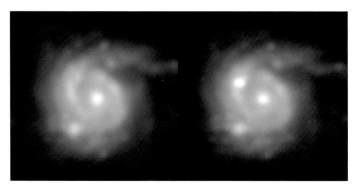

A supernova that exploded near the nucleus of the spiral galaxy NGC3310 in 1991. This picture shows the central region of the galaxy before (left, April 1987) and close to maximum brightness (right, April 5, 1991).

superluminal motion Motion apparently faster than light. The separation on the sky of the components of some radio sources is increasing at a rate that is apparently equivalent to as much as 10 times the speed of light when the distance of the source is taken into account. Speeds in excess of that of light, however, are physically impossible. In reality, the effect is a purely geometrical one caused by one component traveling almost directly towards us at almost the speed of light.

supernova (pl. supernovae) A catastrophic stellar explosion in which so much energy is released that the supernova alone can outshine an entire galaxy of billions of stars. In addition to the energy given off as radiation, 10 times more energy goes into the motion of material blown out by the explosion, and a hundred times as much is carried off by neutrinos.

Supernovae are classified according to their spectra as Type I or Type II. Type II supernovae have hydrogen lines in the spectra while Type I supernovae do not. Type I supernovae are additionally subdivided into Types Ia, Ib and Ic. Type Ia have a strong line due to ionized silicon. Type Ib have strong neutral helium lines. Type Ic have neither.

The progress of all Type I supernovae is very similar: their luminosity increases steadily for about three weeks then declines systematically over six months or longer. However, Types Ib and Ic are fainter than Ia overall. Type II supernovae are more varied. Type Ia supernovae occur in all types of galaxies. Types Ib and Ic are seen only near sites of recent star formation in spiral galaxies. Type II supernovae occur only in the arms of spiral galaxies. This suggests that types Ib and Ic are more closely linked with Type II than with Ia.

The progenitors of Type Ia supernovae are thought to be ➤ *white dwarfs* in binary systems that have dragged material from their companions onto themselves. So much heat is generated it detonates a nuclear explosion that probably destroys the white dwarf. The nuclear reactions create about one solar mass of an unstable isotope of nickel, which decays to iron over a period of months. This radioactive decay accounts for the observed decline in light output.

Type II supernovae appear to be stars of eight solar masses or more that have run the course of ➤ *stellar evolution* and totally exhausted the nuclear fuel available in their cores. At this stage their structure is like that of an onion, consisting of concentric spherical shells in which different nuclear reactions are taking place. Once silicon burning starts in the central core, instability develops within a day because the iron created cannot fuse into heavier elements without an input of energy. With no energy being generated, there is no pressure to balance the weight of the overlying layers.

When the crunch comes, the core collapses in less than a second. The rate of collapse accelerates as iron nuclei break up and neutrons form. However, the implosion cannot continue indefinitely. At the density of nuclear matter, there is sudden resistance. The imploding material bounces back and an outward shock wave is generated. The outer layers of the star are blown outwards at thousands of kilometers per second, leaving the core exposed as a ➤ *neutron star*. In some cases, a stellar ➤ *black hole* is probably created.

The progenitors of Types Ib and Ic are likely to be massive stars that have evolved and already lost their outer envelopes of hydrogen, exposing a core made of heavier elements. As with Type II, they explode when the central iron core collapses.

The material ejected in the explosion forms an expanding ➤ *supernova remnant*. The neutron stars can sometimes be detected as ➤ *pulsars* through their radio emission and, in some cases, by pulsed light and X-ray emission as well.

The explosion of supernovae serves to enrich the chemical composition of the interstellar medium from which subsequent generations of stars are created. Very old stars contain much lower quantities of the elements heavier than hydrogen and helium than are found in the Sun and solar system and many of these heavier elements can be created naturally only in the explosion of a supernova. Supernova shock waves may also trigger star formation as they propagate through interstellar clouds.

Supernovae in our own Galaxy are fairly rare events: only five have been observed visually in the last thousand years. Radio emission from other remnants has been detected, but the light from those outbursts was concealed by dust clouds. The nearest supernova of recent times was ➤ *Supernova 1987A* in the Large Magellanic Cloud.

Supernova 1987A (SN 1987A) A ➤ *supernova* in the Large ➤ *Magellanic Cloud* discovered on February 24, 1987 when it was about sixth magnitude. It was the nearest and brightest supernova observed since 1604. The star that exploded was identified as a twelfth-magnitude blue supergiant, known as Sanduleak – 69° 202. Maximum magnitude, reached in mid-May, was near 2.8.

supernova remnant The expanding shell of material created by the ejection of the outer layers of a star that explodes as a ➤ *supernova*. Some supernova remnants are observable visually; others have been detected through their radio and X-ray emission. A shock wave precedes the material ejected. This collides with the interstellar gas and heats it. The result is a reverse shock, moving inwards, which heats the ejected material and the interstellar material, causing it to emit X-rays. Electrons, accelerated by the shocks, emit radio waves by the ➤ *synchrotron radiation* mechanism. The ejected material breaks up into clumps, so the radiation emitted from the shell often does not make up a uniform ring.

A small proportion of supernova remnants, including the ➤ *Crab Nebula*, have a rather different appearance. In these, the synchrotron radiation coming from within the shell far outshines any from the shell itself. This type of supernova remnant has been termed a plerion. An ongoing supply of electrons traveling at close to the speed of light is needed to account for this emission. In the Crab Nebula, the pulsar produces the electrons. For plerions where no pulsar has been detected, it is assumed that we are observing at the wrong angle to pick up the pulses from the central neutron star. Some other well-known examples of supernova remnants are ➤ *Cassiopeia A*, ➤ *Kepler's Supernova*, ➤ *Tycho's Supernova* and the ➤ *Cygnus Loop*.

surface gravity The local value of the acceleration due to gravity experienced by a free-falling object at the surface of an astronomical body.

Denoted N 63A, this supernova remnant is a member of N 63, a star-forming region in the Large Magellanic Cloud.

The Surveyor III spacecraft is examined on the lunar surface by Charles Conrad Jr., Apollo 12 Commander.

Surveyor A series of seven unmanned American spacecraft launched between 1966 and 1968 to soft-land on the Moon. Five were successful. They conducted experiments to test the Moon's surface for a subsequent manned landing and returned a large number of close-up images of the lunar surface.

Suttungr A small outer moon of Saturn in a very elliptical orbit. It was discovered in 2000 and is about 6 km (3 miles) across.

Swan Nebula An alternative name for the ➤ *Omega Nebula*.

Swift Short for the Swift Gamma-Ray Burst Explorer, a NASA three-telescope space observatory launched on November 20, 2004, for a nominal two-year mission. It was designed to be able to point its gamma-ray, X-ray and ultraviolet telescopes towards a ➤ *gamma-ray burst* within minutes of the detection of a burst. Observations with Swift in 2005 finally resolved the 35-year-old mystery of what causes short gamma-ray bursts.

Sword of Orion The stars Theta and Iota in the constellation ➤ *Orion*, which lie below the three bright stars forming the belt of the mythological figure of Orion.

Sycorax One of two small moons of Uranus discovered in 1997 by Brett Gladman and others, using the ➤ *Hale Telescope*. It is reddish in color and thought to be a captured ➤ *Kuiper Belt* object. Its diameter is estimated to be 190 km (118 miles).

87 Sylvia One of the larger asteroids in the asteroid belt, discovered by Norman R. Pogson in 1866. It is about 280 km (175 miles) across and located in the outer part of the asteroid belt. It has two small moons, discovered in 2001 and 2005, which have been named Romulus and Remus. Romulus is 18 km (11 miles) across and Remus 7 km (4.4 miles). With the discovery of Remus, Sylvia became the first triple asteroid system known. By observing the motion of Romulus and Remus in their orbits, Sylvia's density has been found to be only 20 percent greater than that of water. This means that it is a "rubble-pile" – a loose collection of pieces of rock and ice with empty space between the pieces.

symbiotic stars A ➤ *binary star* system with an unusual type of combination spectrum. The spectrum has features of both a cool star and emission lines characteristic of very-high-temperature gas. The normal interpretation is that the cool star is losing mass to a ➤ *dwarf* or ➤ *white dwarf* companion. Symbiotic stars are variable because mass is transferred irregularly and glowing gas is sometimes eclipsed by the large cool star. They are also known as Z Andromedae stars.

synchronous rotation (captured rotation) The rotation of a moon when its orbital period and the time it takes to rotate on its axis are the same. In synchronous rotation, the moon always has the same face towards its parent planet, as is the case with our own Moon. Synchronous rotation is brought about by tidal interactions over long periods of time.

synchrotron radiation Electromagnetic radiation emitted by an electrically charged particle traveling almost as fast as light through a magnetic field. It was first observed in synchrotron accelerators used by nuclear physicists. It is the major source of radio emission from ➤ *supernova remnants* and ➤ *radio galaxies*. The spectrum of synchrotron radiation has a characteristic profile very different from that of the thermal radiation emitted by hot gas, making synchrotron sources easy to identify.

synodic period For planets, the synodic period is the average time between successive ➤ *conjunctions* of a pair of planets, as observed from the Sun. When only one planet is mentioned, the other is taken to be Earth. For a moon, its synodic period is the average time between successive conjunctions with the Sun, as observed from its parent planet.

Synchrotron radiation is emitted when an electron spirals in a magnetic field.

A view of Mars from Mars Global Surveyor showing the large dark area of Syrtis Major Planum towards the lower right.

synthetic aperture radar (SAR) A radar technique used, for example, by the
➤ *Magellan* mission to Venus, in which the echoes from radar pulsars emitted at
a rate of thousands per second are processed by computer to generate a
detailed picture of the structure of the reflecting surface.

Syrtis Major Planum A cratered volcanic plain on Mars, identified with a dark,
triangular feature (Syrtis Major) easily visible in telescopic views of the planet.

syzygy The rough alignment of the Sun, Earth and Moon, or the Sun, Earth and
another planet. Syzygy thus describes both ➤ *conjunctions* and ➤ *oppositions*.

T

Tarantula Nebula (NGC 2070) A large cloud of glowing ➤ *ionized hydrogen*, 900 light years across, in the Large ➤ *Magellanic Cloud*.

Tarvos A small outer moon of Saturn in a very elliptical orbit. It was discovered in 2000 and its diameter is about 13 km (8 miles).

Tau Ceti A star similar to the Sun 11.7 light years away. It is one of the nearest stars known and, at magnitude 3.5, one of very few stars no more massive than the Sun that is visible to the naked eye. Its ➤ *spectral type* is G8, and it has about 70 percent of the Sun's mass. It was one of the first targets of searches for radio signals from extraterrestrials, but nothing has ever been found and no planets have been found in orbit around it.

Taurids A modest annual ➤ *meteor shower* with twin radiants in the constellation Taurus. It peaks around November 3. The meteoroid stream responsible is associated with Comet ➤ *Encke*.

The Tarantula Nebula imaged by the 8.2-m KUEYEN Telescope of the European Southern Observatory.

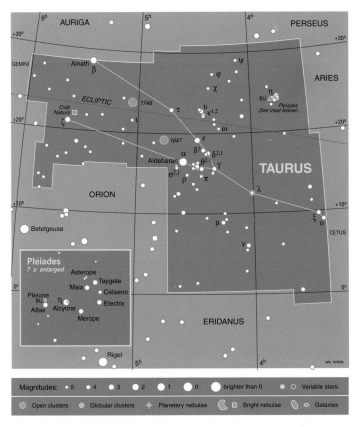

A map of the constellation Taurus.

Taurus (The Bull) A conspicuous zodiacal constellation, traditionally seen as the head and forequarters of a bull. It is possibly one of the most ancient of constellations. The brightest star in Taurus is the first-magnitude ➤ *Aldebaran*, which appears to belong the ➤ *Hyades* cluster, though it is in fact in the foreground. In total, Taurus has 14 stars brighter than fourth magnitude. The ➤ *Pleiades* cluster and the ➤ *Crab Nebula* also lie within the boundaries of Taurus.

Taurus A The radio source associated with the ➤ *Crab Nebula*.

Taurus–Littrow valley The site on the Moon where the Apollo 17 astronauts landed. It is on the south-east border of Mare Serenitatis, in the region of the crater Littrow. The valley is completely surrounded by mountains, some more than 2000 m (6500 feet) high.

Taygeta One of the brighter stars in the ➤ *Pleiades*.

Taygete A small outer moon of Jupiter discovered in 2000. Its diameter is about 5 km (3 miles).

The Taurus–Littrow valley on the Moon. This perspective view was constructed by overlaying a Hubble Space Telescope image with a digital-terrain model acquired by the Apollo program.

Teapot A popular name for an ➤ *asterism* formed by some brighter stars in the constellation Sagittarius.

Teide Observatory An observatory site on the island of Tenerife in the Canary Islands, shared by the Instituto de Astrofisica de Canarias with European partners. The instruments located there include several solar telescopes, a ➤ *spectroheliograph*, a radio telescope for studying the ➤ *cosmic background radiation* and a 1.55-m (61-inch) infrared telescope.

tektite A piece of peculiar natural glass. Tektites are found distributed on Earth's surface in four main areas, called "strewn fields," which are in Australasia, the Ivory Coast, Moravia and Bohemia in the Czech Republic, and Texas and Georgia in the USA. Individual tektites can weigh as much as 15 kg (33 lb). Most, though, are much smaller and their shape and structure suggests that the molten material from which they formed flew through the atmosphere at high speed. The most popular theory for their origin is that they were created from terrestrial material when the impacts of large meteorites melted and ejected rock at the impact sites. Their ages and links to known impact structures support this theory.

telescope An instrument to collect light – or any other kind of ➤ *electromagnetic radiation* – from a distant object, bring it to a focus and produce a magnified image or signal. "Telescope" originally meant an optical instrument but it now has a much broader meaning in astronomy. Telescopes for observing different parts of the spectrum – radio waves or X-rays, for example – employ widely differing designs and techniques.

Optical telescopes fall into two main categories, refractors and reflectors, according to whether a lens or a mirror is used to collect the light. A refracting

Tektites found in Thailand. The largest of these is 7 cm (3 in) long.

telescope has an objective lens at the front of the telescope tube and either an eyepiece or equipment (such as a camera) at the back where the image is formed. In a reflecting telescope a concave mirror at the back of the tube collects the light.

The objective of a refracting telescope is usually a compound lens, made of two or more elements cemented together, with a relatively long focal length. This helps reduce the problems with lenses that affect the quality of the image but it also means that refractors tend to be long and bulky. Small refractors are good for amateurs but very large lenses are difficult to make and mount, and they absorb too much light for astronomical purposes. The world's largest refractor has an objective lens 101 cm (40 inches) in diameter and is at the ➤ *Yerkes Observatory*.

All large astronomical telescopes are reflectors. Reflectors are also popular with amateurs, being less expensive than refractors and easier to make. In a reflector, the light converges towards a focal point in front of the main mirror, called the ➤ *prime focus*. However, it is usually diverted, by means of a secondary mirror, to a more convenient place. Several arrangements are in common use. The ➤ *Newtonian telescope*, ➤ *Cassegrain telescope*, ➤ *coudé focus* and ➤ *Nasmyth focus* all have different applications. Large multi-purpose professional telescopes usually offer observers a choice of foci. The Newtonian focus is used only on amateur visual telescopes.

The primary mirrors in reflecting telescopes are usually made from glass or a ceramic material that does not expand or contract when the temperature changes. The surface must be carefully figured to the required shape, either part of a sphere or part of a paraboloid, to an accuracy of a fraction of the wavelength of light. A thin layer of aluminum is then deposited onto the glass to make the reflecting surface.

In the design of the most modern large telescopes, ➤ *active optics* allow light-weight mirrors to be kept accurately in shape by an array of computer-controlled supports. This also makes it possible to have mirrors composed of a number of separate segments or parts.

Both the light-gathering power and the ➤ *resolving power* of a telescope depend on the diameter of its "aperture" – the mirror or lens that collects the light. Astronomers continually aspire to larger instruments to reach fainter limiting magnitudes and achieve resolution of greater detail, though some of these objectives are also achieved with more sensitive detectors and with ➤ *interferometers.*

Magnifying power is not of great significance, except with small amateur telescopes for visual use. The magnification for visual observing is changed by employing different ➤ *eyepieces.* The maximum magnification is usually governed by ➤ *seeing* conditions rather than the limit of performance of the telescope. The images formed by astronomical telescopes are inverted. Since the introduction of a lens to rectify the image would serve no useful purpose and would absorb valuable light, astronomers prefer to work directly with inverted images.

The mounting of an astronomical telescope is important because it has to be easily pointed at its targets and able to follow them across the sky. Very small amateur telescopes and modern computer-controlled telescopes employ ➤ *altazimuth mountings.* Before the advent of computerized controls, the most practical method was the ➤ *equatorial mounting.* Many older telescopes are on equatorial mounts, and the system remains popular for amateur instruments. ➤➤ *adaptive optics, radio telescope.*

Telescopio Nazionale Galileo A 3.5-m reflecting telescope at the ➤ *Observatorio del Roque de los Muchachos,* in the Canary Islands. It was commissioned by Padua University, Italy, as a national facility for Italian astronomers and completed in 1997. It is modeled on the European Southern Observatory's ➤ *New Technology Telescope.*

Telescopium (The Telescope) An insignificant southern constellation introduced by Nicolas L. de Lacaille in the mid-eighteenth century. It contains only one star as bright as third magnitude.

Telesto A small satellite of Saturn, discovered in 1980 when the planet's rings were edge-on (and thus invisible) as viewed from Earth. Along with another small moon, ➤ *Calypso,* it is in the same orbit as the larger moon ➤ *Tethys,* 294 660 km (183 090 miles) from Saturn. Telesto measures $30 \times 25 \times 15$ km ($19 \times 16 \times 9$ miles).

Tethys imaged by the Cassini spacecraft in natural colour.

terminator The boundary between the illuminated and unilluminated parts of the surface of a planet or moon. Someone at the terminator would be experiencing dusk or dawn.

Terrestrial Planet Finder A proposed NASA space mission for launch after 2010, designed to study the formation, development and characteristics of extrasolar planets.

terrestrial planets The inner rocky planets (Mercury, Venus, Earth and Mars), which are relatively small and made largely of solid rock and metal.

Tethys A moon of Saturn discovered by Giovanni Cassini in 1684. It is almost spherical with a diameter of 1060 km (665 miles). Its low density, only 1.1 times that of water, suggests that at least half of the interior must be ice. Images from the ➤ *Voyager* spacecraft show the surface to be heavily cratered, though there are regions of lower crater density where the surface has been changed in the past. Two notable features are the large crater Odysseus, which is 400 km (250 miles) across, and Ithaca Chasma, a valley more than 2000 km (1250 miles) long that cuts round three-quarters of Tethys. It is 100 km wide and several kilometers deep.

Tethys shares its orbit at a distance of 294 660 km (183 090 miles) from Saturn with two very small moons, Telesto and Calypso.

Thalassa A small moon of Neptune discovered during the flyby of ➤ *Voyager 2* in August 1989. Its diameter is about 80 km (50 miles).

Tharsis Ridge A raised volcanic area on Mars, 10 km (6 miles) above the average level for the planet. Three large volcanoes, with peaks up to 27 km (17 miles) high, lie in a line along the ridge. They are Arsia Mons, Pavonis Mons and Ascraeus Mons.

Thebe A small inner moon of Jupiter, discovered by Stephen P. Synnott in 1979. It measures 110×90 km (68×56 miles).

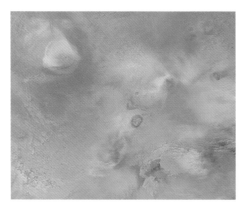

The three volcanoes of the Tharsis Ridge on Mars, and Olympus Mons to the upper left, all veiled with thin cloud.

Thelxinoe A small outer moon of Jupiter discovered in 2003. Its diameter is about 2 km (1 mile).

Themis family One of the ➤ *Hirayama families* of asteroids, located at a distance of 3.13 AU from the Sun. The members of the family are all of the carbonaceous type, suggesting that they all come from the same parent body. Their prototype is 24 Themis, which has a diameter of 228 km (142 miles) and was discovered in 1853 by Annibale de Gasparis.

Themisto A small moon of Jupiter, probably first sighted in 1975 but positively confirmed in 2000. Its diameter is about 8 km (5 miles).

Theophilus A large lunar crater to the north-west of Mare Nectaris, overlapping another large crater, Cyrillus. Theophilus is 100 km (60 miles) in diameter and its terraced walls rise 5 km above the floor. A complex central peak rises to 2.2 km.

third contact In a total or annular ➤ *eclipse* of the Sun, the point when the edges of the Moon's disk and Sun's ➤ *photosphere* are in contact at the end of totality or the annular phase. In a lunar eclipse, third contact occurs when the Moon starts to leave the full shadow (umbra) of the Earth. The term also applies to the similar stage in a ➤ *transit* or ➤ *occultation*.

third quarter The phase of the Moon when half the visible disk of the waning Moon appears illuminated. Third quarter occurs when the celestial ➤ *longitude* of the Moon is 270° greater than the Sun's.

Thrymr A small outer moon of Saturn in a very elliptical orbit. It was discovered in 2000 and is about 6 km (3 miles) across.

Thuban (Alpha Draconis) A third-magnitude star in the constellation Draco, which is 310 light years away. It is a rare example of a white giant star. Despite its designation as Alpha, it is only the seventh-brightest star in Draco. About 5000 years ago Thuban was the nearest bright star to the north celestial pole. Since then the north pole's position among the stars has changed because of ➤ *precession*. Derived from Arabic, Thuban means "dragon."

Thyone A small outer moon of Jupiter discovered in 2003. Its diameter is about 4 km (2.5 miles).

tides The movements of fluids, or stresses induced in solid objects, by a cyclical change in the overall gravitational forces acting upon them. On Earth, fluctuating ocean tides are governed by the daily, monthly and annual variations in the combined gravitational force of the Sun and Moon. These variations arise from Earth's rotation, the Moon's orbital motion around Earth and Earth's orbital motion around the Sun.

time zone A geographical region in which civil time is reckoned to be the same. Time zones are roughly based on longitude bands 15° wide, corresponding to a one-hour difference in local time. There are, however, considerable deviations from regular lines of longitude in the boundaries of time zones in order to take account of where land is and centers of habitation. The difference between most adjacent time zones is one hour, but there are some instances of half-hour differences aimed at minimizing deviations from local time.

Titan The largest moon of Saturn and the second-largest in the solar system (after Ganymede). It was discovered in 1655 by Christiaan Huygens.

Titan is 5151 km (3200 miles) across and orbits Saturn at a distance of 1 221 850 km (759 220 miles). It is surrounded by a thick atmosphere, mainly of nitrogen but also containing methane. The surface pressure is 1.5 times greater than atmospheric pressure at the surface of the Earth and hurricane strength winds blow in the lower atmosphere. The action of sunlight on the methane and other substances in the atmosphere, such as carbon monoxide, produces a layer of opaque orange-colored haze 200 km (125 miles) above the surface. Because of the haze, ordinary telescopes cannot see Titan's surface. However,

Three views of Titan. The first image (left), a natural color composite, shows approximately what Titan would look like to the human eye. The second (center) is a near-infrared image that penetrates through the hazy atmosphere and down to the surface. The third view (right), which is a false color composite, was created by combining two infrared images with a visible light image.

A Voyager 2 image of Titania made in 1986.

infrared radiation can penetrate the haze and be detected, and radar aboard the ➤ *Cassini* spacecraft is being used to create images of Titan's surface. The Huygens probe released by Cassini on January 14, 2005, parachuted down to Titan's surface and returned images and other data.

On Titan, where the temperature is only 95 K (−178 °C), water is frozen solid as rock, but under the conditions there, methane could exist as a liquid. Before the Cassini–Huygens mission, it was thought there might be extensive seas of liquid methane on Titan. Large seas have not been found, but there is much evidence for lakes, shorelines and drainage channels. It is possible that methane rain sometimes falls from the clouds seen in Titan's atmosphere, or that liquid wells up from the ground. Radar images have also shown impact craters and what is thought to be an "ice volcano."

Titania The largest moon of Uranus, discovered by William ➤ *Herschel* in 1787. It measures 1578 km (981 miles) across and orbits Uranus at an average distance of 435 910 km (270 874 miles). The flyby of ➤ *Voyager 2* in 1986 showed Titania to be peppered with numerous craters, though there are regions where the crater density is lower. The surface is also scarred by a large number of valleys and fractures, some of which cut large craters in half. This suggests that some process has dramatically changed Titania's surface in the past.

Titius–Bode law (Bode's law) A mathematical formula that gives approximations to the distances of the planets from the Sun. The formula is: $D = 0.4 + (0.3\,N)$ where D is the distance in ➤ *astronomical units* (AU) and N takes the values 0, 1, 2, 4, 8, ..., doubling for each successive planet. The relationship holds to within a few percent for the seven innermost major planets as long as the value $N = 8$ is taken to represent the largest asteroid, ➤ *Ceres*. However, it breaks down seriously for Neptune and objects beyond.

Clyde Tombaugh, shortly after his discovery of Pluto.

The formula was devised in 1766 by Johann Titius and copied a few years later by Johann E. Bode, who published it. At that time, none of the ➤ *asteroids* had been discovered and the "gap" at 2.8 AU, where the formula predicted that there should be a planet, convinced astronomers that a small planet would be found there, which indeed proved to be the case.

Tokyo Astronomical Observatory The former name of a research institute of the University of Tokyo. In a reorganization in 1988, it was largely incorporated into the new National Astronomical Observatory of Japan, which has its headquarters at the former Tokyo Astronomical Observatory. Its 8-m (300-inch) optical/infrared instrument, the ➤ *Subaru telescope* at the ➤ *Mauna Kea Observatories* in Hawaii, was opened in 1999.

Tombaugh, Clyde (1906–1997) The American astronomer Tombaugh discovered ➤ *Pluto*. Though self-taught and with no formal qualifications, he was taken onto the staff of the ➤ *Lowell Observatory* in 1929 to work on a systematic search for a planet beyond Neptune. On 18 February 1930 he discovered Pluto on photographic plates taken a month earlier.

This computer generated image depicts a view of Earth as seen from the surface of the asteroid Toutatis on 29 November 1996, when it was 5.2 million km (3.3 million miles) away.

Torino scale A numerical scale, from 0 to 10, indicating the seriousness of the risk to Earth presented by a ➤ *near-Earth object*. It was initiated at a meeting in Torino, Italy, in 1999, and revised in 2005.

totality The phase of a solar or lunar ➤ *eclipse* during which the Sun is totally obscured or the Moon totally in the Earth's shadow.

4179 Toutatis An Earth-crossing asteroid discovered in 1989 by Christian Pollas. Radar studies have shown that its shape is very irregular and have revealed the presence of craters and ridges on the surface. It measures 4.7×2.4×1.9 km (2.9×1.5×1.2 miles) and has two distinct lobes. It might even be two bodies in close proximity. It rotates in a very complex manner. Both its shape and rotation are thought to be the result of collisions with other bodies. The plane of Toutatis's orbit is closer to the plane of Earth's orbit than that of any other known Earth-crossing asteroid. It makes frequent close approaches to Earth and in 2004, passed at about four times the Moon's distance.

TRACE Abbreviation for Transition Region and Coronal Explorer, a small NASA satellite launched in April 1998 to study the Sun. Its purpose was to study the connection between fine-scale magnetic fields and physical structures on the Sun.

Tracking and Data Relay Satellite System (TDRSS) A network of seven satellites and a ground station in New Mexico used to track NASA spacecraft and relay their data and commands.

transient lunar phenomenon (TLP) ➤ *lunar transient phenomenon.*

transit (1) The passage of a star or other celestial object across the observer's ➤ *meridian.*

transit (2) The passage of either of the planets Mercury or Venus across the visible disk of the Sun. Transits of Venus and Mercury do not occur every time these

This image of the Sun taken by the TRACE mission shows loops of hot gas in the solar corona which span 30 or more times the diameter of planet Earth.

planets are between the Sun and Earth because their orbits are slightly inclined to the ➤ *ecliptic*.

Transits of Venus occur at a series of intervals that repeat regularly over a 243-year period, though the pattern of intervals within the 243 years changes over timescales of hundreds of years. Since 1631, transits of Venus have been at intervals of 8, 121.5, 8 and 105.5 years, a situation that will continue until 2984.

Transits of Mercury are more frequent, but their pattern of occurrence is more complex. Thirteen or fourteen take place each century, always in either May or November. The dates of transits in the early twenty-first century are May 7, 2003, November 8/9, 2006, May 9, 2016, and November 11, 2019.

Transits of Venus 1631–2125

Date	Start time (UT)	Duration
1631 December 7	03:49	3 h 00 m
1639 December 4	14:56	6 h 58 m
1761 June 6	02:01	6 h 35 m
1769 June 3/4	19:15	6 h 20 m
1874 December 9	01:50	4 h 36 m
1882 December 6	13:57	6 h 13 m
2004 June 8	05:15	6 h 13 m
2012 June 5/6	22:2	6 h 31 m
2117 December 11	00:03	5 h 41 m
2125 December 8	13:19	5 h 35 m

transit (3) The passage of a natural satellite across the disk of its parent planet.

transit circle A telescope mounted so that it is aligned exactly north–south and can rotate up and down about a fixed horizontal axis. Transit circles are used

An image of the Sun projected during the transit of Venus in 2004.

for accurate measurements of the altitudes of stars and for timing their passage across the ➤ *meridian*.

transition region The layer in the Sun's atmosphere above the ➤ *chromosphere* and below the ➤ *corona*. In the transition region, the temperature rises dramatically from 10 000 K in the chromosphere to over 1 million K in the lower corona.

transneptunian object A small planetary body in the outer solar system farther away from the Sun than Neptune (beyond about 30 AU). ➤ *Kuiper Belt*.

Trapezium A popular name for the multiple star system Theta^1 Orionis, which lies at the heart of the ➤ *Orion Nebula* and illuminates it. Four stars, of magnitudes 5.1, 6.7, 6.7 and 8.0, form the trapezium shape and are visible in a small telescope. A larger telescope reveals the presence of two other stars of eleventh magnitude.

Triangulum (The Triangle) A small but distinctive northern constellation between Andromeda and Aries. Its three brightest stars, of magnitudes 3.0, 3.4 and 4.0, form a small, elongated triangle. Triangulum includes the large spiral galaxy, M33, which is a member of the ➤ *Local Group*.

Triangulum Australe (The Southern Triangle) A small but distinctive southern constellation introduced in the 1603 star atlas of Johann Bayer. Its three

An infrared image of the Trapezium star cluster in the Orion Nebula taken by the Hubble Space Telescope.

The Triangulum Galaxy.

brightest stars, magnitudes 1.9, 2.9 and 2.9, form an almost equilateral triangle.

Triangulum Galaxy (M33; NGC 598) A large, nearby, spiral galaxy in the constellation Triangulum. It lies at a distance of 2.7 million light years, and is a member of the ➤ *Local Group*.

Trifid Nebula (M20; NGC 6514) A large luminous cloud of ➤ *ionized hydrogen* around a region of star formation about 8000 light years away. Dust lanes radiating from the center appear to divide the nebula into three parts. It lies in the constellation Sagittarius.

Trinculo A small moon of Uranus discovered in 2001. Its diameter is about 10 km (6 miles).

Triton The largest natural satellite of Neptune. It was discovered in October 1846 by William Lassell, only 17 days after the discovery of Neptune itself. It circles Neptune every 5.9 days at a distance of 354 800 km (220 473 miles) in a retrograde orbit, which is tilted at 23° to the planet's equatorial plane. This unusual orbit has led to speculation that Triton was captured and did not form close to Neptune.

    ➤ *Voyager 2*'s passage within 4000 km (2500 miles) of Triton in August 1989 revealed a wealth of detail. Triton's diameter was found to be 2700 km (1680 miles), slightly less than thought previously. Its gravitational effect on the

The central part of the Trifid Nebula.

spacecraft's trajectory suggests that the bright, icy outer crust and mantle must overlie a substantial core of rock (perhaps even metal) containing two thirds of the satellite's mass. The surface temperature is 38 K, making it the coldest known object in the solar system.

Triton is surrounded by a tenuous atmosphere (surface pressure 15 microbars) of nitrogen, with a trace of methane. The south polar cap is coated with a bright frost, possibly of nitrogen ice, which is gradually evaporating. This region had been in sunlight continuously for nearly 100 years when it was observed. No impact craters were detected there. In the equatorial region, Voyager saw a variety of terrains where it looks as if there is complex volcanic activity, including ➤ *plume eruptions*. Triton's surface is certainly relatively young in astronomical terms.

A Voyager 2 mosaic of Triton made in 1989.

Trojan asteroids Families of ➤ *asteroids* that share the orbit of a planet – especially Jupiter – clustered around the two ➤ *Lagrangian points* 60° away from the planet.

Nearly 2000 Jupiter Trojans are known. They do not stay exactly at the Lagrangian points but oscillate around them over an arc between about 45° and 80° from Jupiter, taking 150–200 years to complete a cycle. The first to be discovered was called ➤ *Achilles*, and it was decided to use names of warriors in the Trojan wars for members of the group identified subsequently.

tropical year The time taken by Earth to travel once round the Sun, measured from equinox to equinox. It is the time it takes for the cycle of seasons to repeat exactly and is 365.242 19 days long.

troposphere The lowest layer of Earth's atmosphere, up to a height of approximately 20 km (12 miles). It is bounded by the tropopause, which marks the transition to the more stable conditions of the stratosphere above.

Tsiolkovskii A crater on the lunar farside, 180 km (110 miles) in diameter. The crater floor is partially flooded by dark lava, through which a central peak protrudes. Tsiolkovskii is one of the most prominent features on the lunar farside, where there are no dark maria areas like on the nearside.

T Tauri star A type of very young star in an early phase of evolution and still contracting. The prototype, T Tauri, lies within a dark dust cloud in the constellation Taurus.

All T Tauri stars vary irregularly and their surface temperatures are in the range 3500–7000 K. They are found in dense interstellar clouds, usually alongside young, hot stars. Large numbers of T Tauri stars have been discovered through the strong infrared radiation they emit. Groups of T Tauri stars are called T associations. Strong ➤ *bipolar outflows* (twin-lobed jets) stream out from T Tauri stars at speeds of several hundred kilometers per second. Where these outflows compress and heat the interstellar gas, they create luminous nebulae called ➤ *Herbig–Haro objects*.

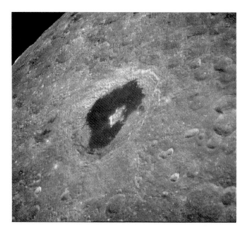

The lunar farside crater Tsiolkovskii photographed by the crew of Apollo 13.

Tucana (The Toucan) A southern constellation introduced in Johann Bayer's 1603 star atlas. Its two brightest stars are third magnitude. The Small ➤ *Magellanic Cloud* lies within its boundaries, and it also includes the large and bright ➤ *globular cluster*, known as ➤ *47 Tucanae*, which is just visibile to the naked eye.

47 Tucanae (NGC 104) The second-brightest ➤ *globular cluster* in the sky (after Omega Centauri). To the naked eye it looks like a fuzzy fifth-magnitude star. It lies at a distance of 13 000 light years and is a relatively young globular cluster.

Tunguska event The violent explosion in Earth's atmosphere of a comet or meteorite in the Tunguska region of Siberia on June 30, 1908. Though the event caused devastation over a large area, no remains of an impacting body or any crater have been discovered. The height of the explosion has been estimated at 8.5 km (5.3 miles). Observers reported seeing a fireball as bright as the Sun. It exploded with a deafening sound and caused a shock wave that shook buildings and caused damage, though there was no loss of human life.

The first expedition to the remote area of the explosion did not take place until 1927. It found that a forest of trees had been snapped in half over a region 30–40 km (20–25 miles) in radius. Over a region of radius 15–18 km (9–11 miles) from the apparent "impact" site, trees had been flattened in a radial pattern and had been stripped of their branches.

twenty-one centimeter line Characteristic radio emission or absorption at a wavelength of 21 centimeters by neutral hydrogen atoms in interstellar space.

Neutral hydrogen is a major component of the ➤ *interstellar medium* and observations of the 21-centimeter line are an important means of finding its distribution, density and velocity in our own Galaxy and thousands of others.

The probability of the small energy change in the hydrogen atom that is responsible for the 21-centimeter emission actually taking place is very low. An individual hydrogen atom with the higher energy level typically waits

The globular star cluster 47 Tucanae.

12 million years to make the change spontaneously. However, the radiation is observed from interstellar hydrogen because of the vast numbers of atoms and because collisions between atoms trigger the emission.

twilight The period before sunrise and after sunset when the sky is illuminated by scattered sunlight. Civil twilight is defined as the period when the center of the Sun's disk is between 90° 50' and 96° below the zenith. Nautical twilight is when it is between 96° and 102° below the zenith and astronomical twilight when it is between 102° and 108° below.

Two Micron All Sky Survey (2MASS) A US collaboration to image the entire sky in near-infrared radiation. The survey is carried out by two automated 1.3-m telescopes. One is located on Mount Hopkins in Arizona and began work on the survey in June 1997. The other is at the ➤ *Cerro Tololo Inter-American Observatory* in Chile, and commenced survey operations in March 1998. Each telescope scans continuously, taking observations in three wavebands simultaneously at 1.25, 1.65 and 2.17 μm.

Tycho A prominent lunar crater in the Moon's southern uplands. It is surrounded by the brightest and most extensive ray system on the Moon, which means it is probably one of the youngest of major lunar features. The terraced walls rise to a height of 4.5 km (15 000 feet) and the central peak to 2.3 km (7500 feet) above the floor of the crater, which is 85 km (53 miles) in diameter.

Tycho Brahe ➤ *Brahe, Tycho*.

Tycho's Supernova (Tycho's Star) A ➤ *supernova* in the constellation Cassiopeia observed by Tycho ➤ *Brahe* in 1572. At its maximum brightness it rivaled Venus and was visible in daylight. The ➤ *supernova remnant* is both an X-ray source and an intense source of radio emission. The expanding shell of gas is faintly visible with powerful optical telescopes.

UBV photometry A method of measuring the colors of stars introduced in the 1950s. It is based on measuring the ➤ *magnitudes* of stars in three regions of the spectrum called *U* (ultraviolet), *B* (blue) and *V* (visual),which are centered on wavelengths 350, 430 and 550 nm, respectively. More bands were added later to extend the scheme into the infrared. These are called *R, I, J, H, K, L, M* and *N*, ranging from 0.7 to 10.2 μm.

UFO Abbreviation for ➤ *unidentified flying object*.

UKIRT Abbreviation for ➤ *United Kingdom Infrared Telescope*.

UKST Abbreviation for ➤ *United Kingdom Schmidt Telescope*.

ultra-luminous infrared galaxy (ULIRG) A galaxy emitting exceptionally strongly in the infrared. ULIRGs were first detected by the ➤ *Infrared Astronomical Satellite* in 1983. They appear to be the results of collisions between two or more galaxies, which trigger immense bursts of star formation. The infrared radiation is emitted by dust, which absorbs energy from hot, newborn stars. Observations suggest that many ULIRGs are the mergers of three or more galaxies.

ultraviolet astronomy The study of electromagnetic radiation from astronomical sources in the wavelength band 10–320 nm. Ultraviolet (UV) radiation is strongly absorbed by Earth's atmosphere, so all UV observations have to be carried out from satellites. The earliest observations were made during brief rocket flights in the 1940s and 1950s. The first satellite to make systematic ultraviolet observations was the first Orbiting Solar Observatory (OSO-1) in 1962. The highly successful International Ultraviolet Explorer (IUE) was launched in 1978 and continued to operate until 1996.

The ultraviolet is often subdivided into the extreme UV (EUV, 10–100 nm), the far UV (FUV, 100–200 nm) and the near UV (NUV, 200–320 nm). The most extreme UV, at the transition to X-radiation around 6–60 nm, is also known as the XUV. At these wavelengths, the techniques of ➤ *X-ray astronomy* are required, but the rest of the UV band can be observed and analyzed by methods similar to those used for visible light. The main difficulty is the limited range of suitable transparent materials and reflective coatings. Glass, for example, absorbs UV strongly, so quartz or fluorite have to be used.

Hotter stars, with surface temperatures in excess of 10 000 K, emit most of their energy in the UV. Even for cooler stars such as the Sun, UV studies are important. The ➤ *interstellar medium* is another important subject for ultraviolet astronomy though, at wavelengths below 91.2 nm, almost all the UV radiation is absorbed by hydrogen, the most widely distributed element in the universe. This makes the detection of distant sources difficult at such short wavelengths.

Ulugh Beg (1393 or 1394–1149) Ulugh Beg was the most important medieval Islamic astronomer. He was the grandson of the conquering warlord Timur (Tamerlane), and was himself a ruler as well as an astronomer. He established Samarkand as an intellectual center with an enormous observatory. He measured the positions of almost 1000 stars, making numerous corrections to the tables made by ➤ *Ptolemy*, which were still in use at that time.

Ulysses A European Space Agency mission, launched on October 6, 1990, to study the interplanetary medium and the solar wind at different solar latitudes. It was the first mission to observe the poles of the Sun. ➤ *Gravity assist* was used to take the orbit of Ulysses out of the plane of the solar system. After an encounter with Jupiter in February 1992, the spacecraft swung back towards the Sun to pass over the solar south pole in 1994 and the north pole in 1995. A second encounter with the Sun took place in September 2000.

umbra (1) An area of total shadow, such as the zone on the surface of the Earth from which totality is observed during a solar ➤ *eclipse.*

umbra (2) The dark central region of a ➤ *sunspot.*

Umbriel A moon of Uranus, discovered by William Lassell in 1851. Its diameter is 1169 km (726 miles). Images from the ➤ *Voyager 2* encounter in 1986 show that Umbriel is much darker than the other four major satellites of Uranus. Its surface seems to have been covered by dark material relatively recently in astronomical terms. It is also pitted with many craters. One of them, 110 km (68 miles) across, is very bright, in marked contrast to the rest of the surface.

unidentified flying object (UFO) Any phenomenon in the sky for which the observer does not have a ready rational explanation. UFO is often used to mean a hypothetical unnatural object from space.

United Kingdom Infrared Telescope (UKIRT) A 3.8-m (150-inch) infrared telescope, located at the ➤ *Mauna Kea Observatories* in Hawaii. It is the largest telescope dedicated solely to infrared astronomy and operates in the wavelength band between 1 and 30 μm.

United Kingdom Schmidt Telescope (UKST) A 1.2-m (48-inch) ➤ *Schmidt camera* located at the ➤ *Anglo-Australian Observatory* and currently administered by the Anglo-Australian Telescope Board. It was opened in 1973.

United States Naval Observatory A US government observatory in Washington, DC, the main purpose of which is to provide astronomical data to the Department of Defense. Its work includes astrometry, the preparation of almanacs, time measurement and the maintenance of the Master Clock for the USA. It has telescopes at Anderson Mesa, near Flagstaff, Arizona, and Black Birch, New Zealand, as well as in Washington.

The observatory was founded in 1830 and given the title US Naval Observatory in 1844. For 50 years it was located at the site now occupied by the Lincoln Memorial. It was moved to its present site, next to the official residence of the Vice President, in 1893. The largest telescope at the site is the 66- cm (26-inch) refractor, dating from 1873, with which Asaph ➤ *Hall* discovered the moons of Mars, Phobos and Deimos, in 1877. The largest telescope belonging to the observatory is a 1.5- m (61-inch) reflector at Flagstaff. Using this instrument, James Christy discovered the moon of Pluto, Charon, in 1978. At the Arizona site, the observatory has also constructed an optical ➤ *interferometer*, the Navy Prototype Optical Interferometer.

Universal Time (UT) A way of keeping time that relates closely to the Sun's daily apparent motion and serves as the basis for civil timekeeping. It is formally defined by a mathematical formula that links it to ➤ *sidereal time*, and is thus ultimately determined from observations of the stars. Time kept precisely by Earth's rotation, which is gradually slowing down and subject to small irregularities, is called UT1. The basis of civil time and broadcast time signals is called coordinated universal time, or UTC. UTC is time as measured by smooth-running atomic clocks but kept to within 0.9 seconds of UT1 by introducing occasional leap seconds.

universe The entirety of all that exists. The size of the observable universe is limited to the distance light has had time to travel since the ➤ *Big Bang*.

Uraniborg The observatory of Tycho ➤ *Brahe* (1546–1601) on the island of Hven, north of Copenhagen, Denmark. It was completed in 1580 and used by Brahe to make accurate astronomical observations for 20 years. Only ruins now remain.

Uranometria A star atlas compiled by Johann Bayer (1572–1625) and published in 1603. In this atlas Bayer introduced the system of labeling stars with Greek letters, which is still in use.

uranometry A largely obsolete term for positional astronomy or ➤ *astrometry*.

Uranus The seventh major planet of the solar system in order from the Sun, discovered by William ➤ *Herschel* in 1781. Its average distance from the Sun is 19.2 AU and it is just bright enough to be seen by the naked eye under good observing conditions. From Earth, it appears as an almost featureless greenish disk. In 1986, the spacecraft ➤ *Voyager 2* passed close to Uranus and its satellites and returned close-up images of them. Ten small satellites were discovered by Voyager 2; five larger satellites were already known: Miranda, Ariel, Umbriel, Titania and Oberon. More recent discoveries have brought the number of moons up to at least 27.

Uranus is one of the four giant planets of the solar system, with a diameter four times Earth's and a mass 15 times greater. Its internal rotation period is 17 hours 14 minutes. It is composed almost entirely of hydrogen and helium.

This infrared view of Uranus, shown in false color, was taken by the Hubble Space Telescope in August 1998. It reveals the main rings, 10 of Uranus's moons and about 20 clouds in the planet's atmosphere.

There is thought to be a small rocky core at the center of the planet, surrounded by a thick icy mantle of frozen water, methane and ammonia. The outermost layer is an atmosphere mostly of hydrogen and helium.

A curious feature of Uranus is that its rotation axis lies almost in the plane of the solar system, rather than being nearly perpendicular to it, as is the case for the other planets. This means that Uranus has greatly exaggerated seasons over its "year" of 84 Earth years. One pole faces the Sun for 20 years while the other side of the planet is in continuous darkness. As spring began in the northern hemisphere of Uranus in the late 1990s, the Hubble Space Telescope observed the atmosphere getting more dynamic as it was warmed by the Sun and many bright clouds appeared.

In 1977, a series of narrow rings was discovered around Uranus. The rings are each only a few kilometers wide and not visible from Earth. They were discovered when Uranus occulted an eighth-magnitude star. The rings caused small dips in the observed brightness of the star just before and just after it was occultated by the disk of the planet. The ring system was subsequently imaged by Voyager 2 in 1986, when two additional rings were discovered, and two

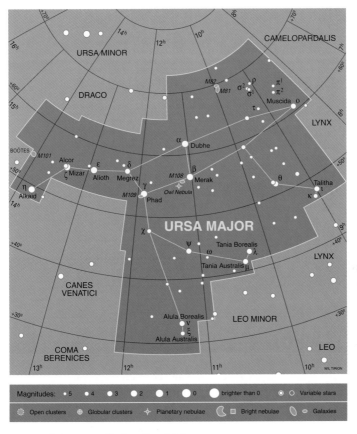

A map of the constellation Ursa Major.

more were found by the Hubble Space Telescope in 2003, bringing the total to 13.

Ursa Major (The Great Bear) One of the most familiar constellations of the northern sky and the third largest in area. It contains 19 stars brighter than fourth magnitude. The shape formed by the seven main stars of the constellation is known as the Big Dipper or the Plough. The two stars Merak and Dubhe in the Big Dipper are known as the Pointers since the line between them points to Polaris. Ursa Major contains a group of galaxies belonging to the ➤ *Local Supercluster*, including the relatively bright spiral galaxy, M81.

Ursa Minor (The Little Bear) The northern constellation that contains the north celestial pole. The brightest star in Ursa Minor, second-magnitude Polaris, is within 1° of the pole. Its main pattern of seven stars is known as the Little Dipper.

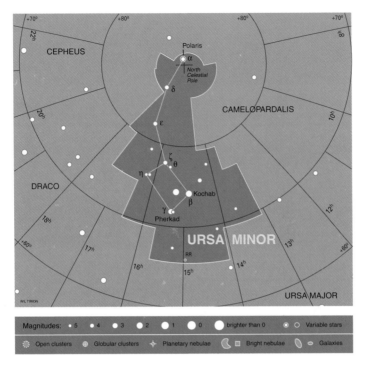

A map of the constellation Ursa Minor.

The Viking 2 lander where it landed on Mars's Utopia Planitia.

UT Abbreviation for ➤ *Universal Time*.

Utopia Planitia An extensive, sparsely cratered plain in the northern hemisphere of Mars. It was the landing site for the ➤ *Viking* 2 spacecraft. Panoramic images returned by the Viking lander showed terrain littered with large numbers of boulders made of open-textured rock.

UV Abbreviation for ultraviolet. ➤ *electromagnetic radiation, ultraviolet astronomy*.

V

V The symbol for visual ➤ *magnitude*. ➤➤ *UBV photometry*.

vacuum tower telescope A design of telescope for observations of the Sun. One at the Sacramento Peak Observatory is typical. In that instrument, sunlight enters the tower 41 m (135 feet) above the ground. A further 67 m (220 feet) of the telescope lie below ground. The entire optical path is virtually air-free, to avoid distortion of the solar image that hot air would cause. In the observing room, an image of the Sun 51 cm (20 inches) in diameter is produced with a resolution better than a quarter of an arc second. Sunlight can be directed into spectrographs or other instruments by tilting the main mirror at the bottom of the central tube.

Valhalla A large circular feature on ➤ *Callisto*, surrounded by 15 concentric rings. The radius of the outermost ring is 1500 km (930 miles). It was caused by an impact, but there is no vertical relief. The rings are like ripples.

The vacuum tower telescope at the Teide Observatory on Tenerife in the Canary Islands.

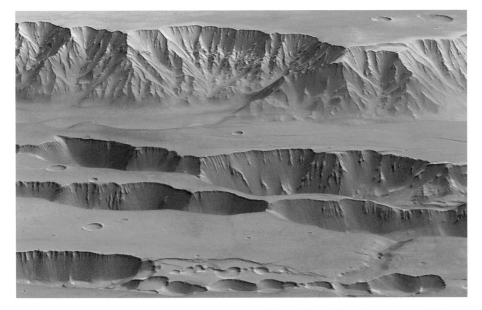

A perspective view across Coprates Chasma, part of the Valles Marineris, constructed from images taken by Mars Express.

Valles Marineris A network of canyons on Mars, extending more than 5000 km (3100 miles) in the east–west direction in the equatorial region. Its western extremity is Noctis Labyrinthus, a complex area of fault valleys that cut the surface into polygonal shapes. The central section consists of several parallel canyons with an average depth of 6 km (3.7 miles). At the center they join up with Melas Chasma, a depression 160 km (100 miles) across. The eastern end of the system is Capri Chasma. Erosion of the canyon sides has deposited debris in the flat valley bottom and revealed layered structure in the surrounding plateau. Flow channels leading into the canyon suggest that erosion by water took place in the remote past. Valles Marineris is thought to have been created by the uplift of the volcanic ➤ *Tharsis Ridge* to the west.

Van Allen belts Two ring-shaped regions around Earth where there are concentrations of high-energy electrons and protons trapped by the Earth's magnetic field. They were discovered by the USA's first successful artificial Earth satellite, Explorer 1, which was launched on January 31, 1958, and named after James Van Allen, the physicist who led the experiment. The inner Van Allen belt lies about 0.8 Earth radii above the equator. The main concentration of the outer belt lies between about two and three Earth radii above the equator, but a broader region, extending from the inner belt out to as far as 10 Earth radii, contains protons and electrons of lower energy, believed to come primarily from the ➤ *solar wind*. Because the magnetic field is

369

Simulated Van Allen belts generated in a NASA laboratory.

offset from Earth's rotation axis, the inner belt dips down towards the surface in the region of the South Atlantic Ocean, off the coast of Brazil. This "South Atlantic Anomaly" presents a potential hazard to the operation of artificial satellites.

variable star A star whose light output varies, whether regularly or irregularly. A graph showing brightness in relation to time for a variable is known as its "light curve."

Eruptive and cataclysmic variables are unpredictabe. Eruptive variables include ➤ *T Tauri stars*, ➤ *luminous blue variables*, and ➤ *flare stars*. Cataclysmic variables undergo explosive processes and include ➤ *novae*, ➤ *dwarf novae* and ➤ *supernovae*. Pulsating variables oscillate in and out because they are unstable internally and vary in a regular way. These include ➤ *Cepheid variables*, ➤ *RR Lyrae stars* and ➤ *Mira stars*. Eclipsing binaries vary because one star periodically passes in front of another. The most well-known example is ➤ *Algol*. Some stars vary as they rotate because their surface is not uniformly bright.

The curious method of naming variable stars is mainly a legacy from Friedrich W. A. Argelander (1799–1875), who used the letters R to Z in conjunction with the constellation name for the nine brightest variables in each constellation. After that, pairs of letters, RR to RZ, SS to SZ and so on to ZZ were used (J being omitted). For further variables, the pairs of letters AA to AZ, BB to BZ and so on were introduced, bringing the number of available designations to 334. Since far more than 334 variables are now known in many

Part of the Veil Nebula.

constellations, the ones in excess of 334 are designated as V335, V336, and so on.

Vega (Alpha Lyrae) The brightest star in the constellation Lyra and the fifth-brightest in the sky. It is an ➤ *A star* of magnitude 0.03 lying 25 light years away.

Veil Nebula (NGC 6960) Part of the ➤ *Cygnus Loop*, which is an old ➤ *supernova remnant*.

Vela (The Sail) A large southern constellation, which is one of the four parts into which Nicolas L. de Lacaille divided the ancient constellation Argo Navis. It lies in part of the Milky Way rich with faint nebulosity and contains 10 stars brighter than fourth magnitude. The stars Delta and Kappa, together with Iota and Epsilon Carinae, make an asterism known as the "false cross" because it is sometimes confused with the constellation Crux, the true Southern Cross.

Vela pulsar A ➤ *pulsar* in the constellation Vela, associated with a 10 000-year-old supernova remnant. It is one of the strongest radio pulsars, and the strongest gamma-ray source in the sky. It was discovered in 1968 and has a short period of 89 milliseconds, which is steadily increasing at a rate of 10.7 nanoseconds a day as the pulsar loses energy. Since observations of it started, the pulsar has also undergone several major ➤ *glitches*.

Venera A series of Soviet spacecraft sent to explore the planet Venus. The first to land successfully was Venera 7 in 1970. There were a further nine Venera probes that returned images of the surface of Venus, and data about the atmosphere and the composition of the planet's crust. In 1982, Venera 13 returned what remain the best ever color images from the surface of Venus, which show a panorama around the landing site.

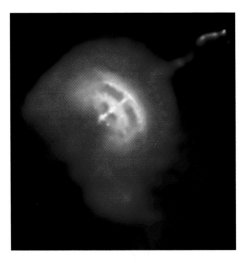

An X-ray image of the Vela pulsar showing jets firing from the pulsar, which is the white dot at the center.

Part of a panorama taken by Venera 13 at its landing site on Venus.

Venus The second major planet of the solar system, in order from the Sun. It is one of the "terrestrial" planets, similar in nature to Earth and only slightly smaller. Like Earth, it is surrounded by a substantial atmosphere.

Venus comes closer to Earth than any other planet and can be the brightest object in the sky (apart from the Sun and Moon). Because its orbit lies inside Earth's, it can never be more than 47° away from the Sun in the sky. This means that Venus can be viewed either in the western sky in the evening, or in the eastern sky in the early morning, but not in the middle of the night. It is sometimes called the "morning star" or the "evening star."

A 3D perspective view of Lavinia Planitia on Venus created from Magellan radar data.

Because it is nearer to the Sun than Earth, Venus goes through a cycle of phases similar to the Moon's. At its brightest and nearest, even a small telescope will show that Venus is actually a crescent.

The surface of Venus is always hidden from us by dense, highly reflecting clouds, which are practically featureless in visible light, though ultraviolet images reveal bands, including a characteristic Y-shape. These clouds consist of droplets of dilute sulfuric acid, created by the action of sunlight on the carbon dioxide, sulfur compounds and water vapor present in the atmosphere.

The atmosphere is almost entirely carbon dioxide, and the surface pressure is more than 90 times the pressure at Earth's surface. The exceptionally high surface temperature of 730 K (450 °C) is a result of the
➤ *greenhouse effect*.

Venus was the target of a large number of Soviet and American probes in the 1970s and 1980s, notably the Soviet ➤ *Venera* series and the American
➤ *Pioneer Venus*. In the extremely high temperature and pressure, many of the probes were destroyed either before returning data or after a relatively short period of operation. Nevertheless, some managed to analyze the chemical composition of some surface rocks and to return limited panoramic views of the surface terrain, showing a rocky desert landscapes.

The first radar maps produced by spacecraft orbiting the planet showed that most of the surface consists of vast plains, above which several large plateaux rise to heights of several kilometers. The two main highland areas are Ishtar Terra in the northern hemisphere and Aphrodite Terra in the equatorial region. The Maxwell Montes are the highest feature, rising to 11 km above the average surface level.

In 1990, the US ➤ *Magellan* spacecraft arrived in orbit around Venus and began mapping the surface by radar in much greater detail than had been achieved previously. The radar maps show much evidence of both impact craters and volcanic activity in the relatively recent past. By solar system standards, the surface of Venus is young: the oldest craters appear to date from 800 million years ago. However, no definite evidence has yet been found of current volcanism.

The thick atmosphere and high surface temperature mean that impact craters are rather different from those on other planets and satellites. Smaller meteorites easily burn up in the atmosphere so there is an absence of smaller craters. The material thrown out in the powerful impacts of larger meteorites did not travel far and tended to melt and spread round the craters.

Large numbers of volcanic features have been identified: lava flows, small domes 2–3 km across, larger volcanic cones hundreds of kilometers across, "coronae" and so-called "arachnoids." The coronae of Venus are circular or oval volcanic structures surrounded by ridges, grooves and radial lines. They appear to be collapsed volcanic domes and are different from any features seen on other planets or satellites. The "arachnoids," which get their informal name from their spider-like appearance, are similar in form to coronae, but generally smaller. The bright lines extending outwards for many kilometers indicate formations that may have been created when magma welled up from the planet's interior, causing the surface to crack.

Venus Express A European Space Agency spacecraft to study the atmosphere of Venus, launched on November 9, 2005. After a journey lasting 153 days it was captured into orbit around Venus and maneuvered into an operating orbit ranging between 250 km (150 miles) and 66 000 km (41 000 miles) from the

Venus Express. An artist's impression of the spacecraft at orbit insertion.

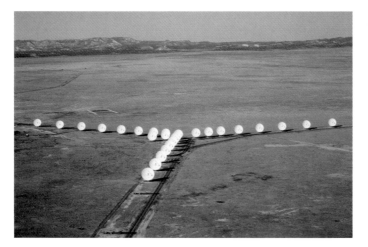

The Very Large Array (VLA) of the National Radio Astronomy Observatory.

planet. It carried seven instruments and was based closely on the ➤ *Mars Express* spacecraft. In 2007, its mission was extended to May 2009.

Very Large Array (VLA) A radio telescope consisting of 27 dishes, each 25 m (82 feet) in diameter. Located near Socorro, New Mexico, it is the world's largest ➤ *aperture synthesis* telescope. The dishes are arranged in a Y-shape with three arms each 21 km (13 miles) long.

Very Large Telescope (VLT) The ➤ *European Southern Observatory*'s set of four, linked 8-m (300-inch) telescopes located at ➤ *Paranal Observatory* in Chile. The light-gathering power of the four telescopes together is equivalent to that of a mirror 16 m (52 feet) in diameter. They were completed in 2001. The four individual telescopes were given names in the local Mapuche language: ANTU (the Sun), KUEYEN (the Moon), MELIPAL (the Southern Cross) and YEPUN (Sirius).

Very Long Baseline Array (VLBA) A network of radio telescopes in North America for ➤ *very-long-baseline interferometry*. It consists of 10 dishes, distributed from Hawaii to St Croix in the US Virgin Islands. The effective diameter is 8000 km (5000 miles) and the resolution that can be achieved is 0.2 milliseconds of arc.

very-long-baseline interferometry (VLBI) A technique in radio astronomy that creates a ➤ *radio interferometer* in which the component antennas are separated by very large distances, typically thousands of kilometers. The antennas are not connected but signals are recorded, together with very accurate timings, at each observing station. The data from each station are brought together later. This technique gives extremely accurate positions for radio sources, resolutions (but not maps) down to a few milliseconds of arc and directly detects continental drift.

The European Southern Observatory's Very Large Telescope (VLT) at Paranal Observatory in Chile. Each enclosure houses an 8.2-m telescope.

The locations of the radio antennas forming the Very Long Baseline Array (VLBA).

Baselines can be made even greater than Earth's diameter by placing radio telescopes in orbit and using them in conjunction with ground-based telescopes. This was done with the Japanese satellite, HALCA launched in February 1997. HALCA was placed in an elliptical orbit, providing a baseline up to three times larger than Earth, and deployed an umbrella-shaped antenna 8 m (26 feet) in diameter. HALCA developed problems in 2003 and its mission was formally terminated in 2005.

4 Vesta An asteroid discovered by Heinrich W. M. Olbers in 1802. It is the third-largest asteroid known, with a diameter 576 km (358 miles), and the brightest of all. It sometimes reaches magnitude 6, making it just detectable by the naked eye under optimum observing conditions. Vesta's brightness is due to its bright surface, which reflects 25 percent of the light falling on it. As it rotates every 5.43 hours, regular changes in the color and spectrum are observed, reflecting the fact that the surface is not uniform.

Vesta appears to be a true mini-planet which has survived largely intact since the solar system formed, rather than being a fragment from a larger body. Hubble Space Telescope images reveal details down to 80 km (50 miles) across, including impact craters. One large crater seems to have torn away part of the crust completely, exposing the mantle below. There is evidence for ancient lava flows dating from four billion years ago when the interior was hot and molten. It is one of the two targets of NASA's ➤ *Dawn* spacecraft, to be launched in 2007. Dawn is scheduled to go into orbit around Vesta in 2011 for eight months before going on to ➤ *Ceres*.

Viking Two identical American spacecraft sent to the planet Mars in 1975. Vikings 1 and 2 both consisted of an orbiter, which remained circling the planet, and a soft lander.

Viking 1 was launched on September 9, 1975 and reached Mars orbit on June 19, 1976. The lander touched down on Chryse Planitia on July 20, 1976. The orbiter's path was adjusted several times to obtain close-up images of Mars's moons, Deimos and Phobos, and to observe different aspects of the martian surface.

Viking 2 was launched on August 20, 1975 and reached Mars orbit on August 7, 1976. The landing was on Utopia Planitia on September 3, 1976.

The orbiters were equipped with two television cameras and instruments to map water vapor and temperature. The landers sampled the upper atmosphere during their descent, made meteorological measurements and carried out experiments on samples of martian soil. One of the mission's prime objectives was to test for the presence of organic material, which might indicate the existence of life, but nothing incontrovertible was found. The orbiters and landers returned thousands of images. The whole of the martian surface was mapped with a resolution of 150–300 m.

The jet streaming from the supermassive black hole at the center of the galaxy M87, which is identified with the radio source Virgo A. The jet consists of electrons and other subatomic particles traveling at close to the speed of light.

The Viking 1 orbiter operated until August 7, 1980, and the Viking 2 orbiter until July 25, 1978. The landers ceased operating in November 1982 and February 1980, respectively. The mission was regarded as very successful, and it had greatly exceeded its expected lifetime.

Virgo (The Virgin) The second-largest constellation in the sky and one of the 12 traditional constellations of the ➤ *zodiac*. The brightest star in Virgo is the first-magnitude ➤ *Spica*, and there are seven others brighter than fourth magnitude. The constellation contains the rich and relatively nearby ➤ *Virgo Cluster* of galaxies.

Virgo A The strongest radio source in the constellation Virgo, identified with the giant elliptical galaxy M87, which dominates the ➤ *Virgo Cluster* of galaxies. The radio emission is associated with a jet 4000 light years long, which is fired out as a consequence of matter falling onto a supermassive black hole in the nucleus of M87.

Virgo Cluster The nearest rich cluster of galaxies at a distance of about 50–60 million light years and the center of the ➤ *Local Supercluster*. It covers 120 square degrees of sky and contains several thousand galaxies. It is an irregular cluster with no central condensation. The giant elliptical galaxy M87 is the most massive member of the cluster.

VISTA Abbreviation for "Visible and Infrared Survey Telescope for Astronomy," a 4-m survey telescope at the European Southern Observatory's ➤ *Paranal Observatory* in Chile. It is due to start operation in 2007.

An infrared mosaic of the central part of the Virgo cluster of galaxies.

visual binary A ➤ *binary star* in which the two components can be resolved as separate images by a telescope.

visual magnitude The ➤ *magnitude* of a celestial object measured over a wavelength band corresponding to the sensitivity of the human eye. ➤ *V*.

Volans (The Flying Fish) A small and faint southern constellation introduced in the 1603 atlas of Johann Bayer with the longer name Piscis Volans, which was later shortened. Its six main stars are third and fourth magnitude.

Voyager 1 One of a pair of almost identical interplanetary spacecraft launched by the USA in 1977. The other was ➤ *Voyager 2*.

The Voyager missions were possible only because of a chance favorable alignment of the outer planets, Jupiter, Saturn, Uranus and Neptune, that happens only once in more than a hundred years. Between them, the two spacecraft were able to explore all these four planets and many of their moons. ➤ *Gravity assist* was used to accelerate the craft from one encounter to the next. The missions were immensely successful, making numerous discoveries and returning huge quantities of data as well as images.

The instruments on the Voyagers consisted of two groups. One set was designed to sample the craft's environment and these remained in operation constantly, even between planetary encounters. The other instruments, to study the target moons and planets, included a wide-angle camera and a

An artist's concept of one of the two identical Voyager spacecraft.

close-up camera. The main communication dish was 3.7 m in diameter and a radioactive power source was used.

Voyager 1 was launched on September 5, 1977. Its closest encounter with Jupiter was on March 5, 1979 at 350 000 km (217 500 miles) and that with Saturn was on November 12, 1980 at 124 000 km (77 000 miles). At Jupiter, it passed close to the moons Io and Callisto and at Saturn got closest to Titan, Rhea and Mimas. After its encounter with Saturn, Voyager 1 left the plane of the solar system and traveled out into interstellar space. ➤ *Voyager Interstellar Mission.*

Voyager 2 One of a pair of interplanetary spacecraft launched by the USA in 1977. It was virtually identical to ➤ *Voyager 1*, except that its power source was designed to last for much longer to survive its longer journey to Uranus and Neptune.

Voyager 2 was launched on August 20, 1977. Its first encounter was with Jupiter on July 9, 1979, and it went within 71 400 km (44 000 miles). It passed close to Europa and Ganymede, complementing the coverage of the ➤ *Galilean moons* obtained by Voyager 1. Voyager 2 arrived at Saturn in August 1981. Closest approach was on August 25, at a distance of 101 000 km (63 000 miles). The spacecraft's trajectory took it near the moons Tethys and Enceladus.

On January 24, 1986 Voyager 2 reached Uranus, which it passed at a distance of 107 000 km (67 000 miles). The mission was completed with the spacecraft's encounter with Neptune and its moon Triton on August 24, 1989, when it passed within 48 000 km (30 000 miles) of the surface of Neptune.

Vulcan A hypothetical planet traveling around the Sun within the orbit of Mercury. Searches were made for such a planet during the late nineteenth century but it is now known that none exists.

Vulpecula (The Fox) A faint constellation, next to Cygnus, introduced by Johann Hevelius in 1690 with the name Vulpecula et Anser – the fox and goose – which was later shortened. It contains no stars brighter than fourth magnitude, but does include the well-known planetary nebula known as the

➤ *Dumbbell Nebula*.

W

walled plain A large, flat-floored lunar ➤ *crater*, particularly one that has been flooded by lava.

waning The part of the cycle of the Moon's phases when the illuminated portion of the visible disk is decreasing. The opposite is waxing.

Water Jar The group of stars Gamma, Eta, Zeta and Pi in the constellation Aquarius, normally shown as the Water Carrier's jar in representations of the mythological figure associated with the constellation.

waxing The part of the cycle of the Moon's phases when the illuminated portion of the visible disk is increasing. The opposite is "waning."

Westerbork Observatory A Dutch national radio astronomy observatory that is part of the Netherlands Foundation for Research in Astronomy. Its administrative headquarters are at ➤ *Dwingeloo Observatory*. The instrument at Westerbork Observatory is called the Westerbork Synthesis Radio Telescope (WSRT). It is a 14-dish ➤ *aperture synthesis* telescope, and came into operation in 1970.

Whipple Observatory ➤ *Fred Lawrence Whipple Observatory*.

Whirlpool Galaxy (M51; NGC 5194) A ➤ *spiral galaxy* in the constellation Canes Venatici, which we see face-on. It is 13 million light years away. This galaxy was the first to be recognized as having spiral structure. The discovery was

Part of the Westerbork Observatory's Synthesis Radio Telescope.

The central part of the Whirlpool galaxy. The pink regions reveal where new bright stars are forming. This picture combines images from the Hubble Space Telescope and the ground-based National Optical Astronomy Observatories.

made by Lord Rosse in 1845. It is accompanied by a much smaller irregular galaxy, NGC 5195, which is in orbit around it.

white dwarf The remains of a star in an advanced state of ➤ *stellar evolution*, composed of ➤ *degenerate matter*, in which atomic nuclei and electrons are packed in close together. A white dwarf is created when a star finally runs out of fuel for nuclear fusion. Its outer layers blow off and form a ➤ *planetary nebula* and its core collapses under its own gravity. The process stops when the electrons in the core cannot be compacted further and instead resist the collapse. Subramanyan ➤ *Chandrasekhar* demonstrated theoretically that the upper mass limit for white dwarfs is 1.4 times the mass of the Sun. If a more massive stellar core collapses, it must become a ➤ *neutron star* or ➤ *black hole*.

The size of the white dwarf Sirius B, which has a mass similar to the Sun's, compared with the size of Earth.

The first white dwarf was recognized in 1910. It was the star 40 Eridani B, which was shown to have a surface temperature of 17 000 K but a total luminosity so low that it must be smaller than Earth. Other well-known white dwarfs include van Maanen's star and Sirius B, a faint companion to the brightest star in the sky. Sirius B, first seen in 1862, has a mass about the same as the Sun's concentrated in a ball slightly smaller than Earth. It is 10 000 times fainter than Sirius A, which is a normal ➤ A star.

Though called "white" dwarfs as a group, these degenerate stars actually cover a range of temperatures and colors from the hottest, which are white and have surface temperatures as high as 100 000 K, to cool red objects at only 4000 K. Since they have no internal source of energy, white dwarfs are in a long process of gradually cooling off, during which their temperature declines. Their ultimate fate is to become a black dwarf – a non-luminous dead star.

The spectra of white dwarfs are bewilderingly complex, reflecting a range of temperature and composition. Typically, their spectra contain very broad absorption lines, though some show no lines at all. The line-forming region is only a few hundred meters thick. Some white dwarfs show evidence only for hydrogen, presumably because the helium and heavier elements have sunk in the strong gravity. In other cases, helium and heavier elements are seen but no hydrogen.

A new classification scheme for white dwarfs was adopted from 1983. Each star's designation consists of three capital letters, the first being D for degenerate. The other two letters depend on features seen in the spectrum.

Wild 2, Comet A periodic comet discovered in 1978 by the Swiss astronomer Paul Wild, observing near Berne. Its orbital period is 6.4 years. On January 2, 2004, the spacecraft ➤ *Stardust* collected a sample of material from the ➤ *coma* of Comet Wild 2 and returned images of its nucleus, which measures about 5 km (3 miles) across.

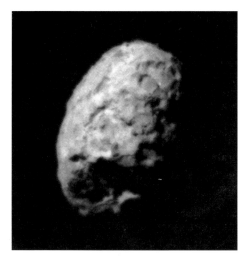

This image of Comet Wild 2 was taken during the Stardust spacecraft's close approach in January 2004. It is a distant side view of the roughly spherical comet nucleus. One hemisphere is in sunlight and the other is in shadow.

Wild Duck Cluster (M11; NGC 6705) An ➤ *open cluster* of about 200 stars in the constellation Scutum. Its shape as seen in a small telescope is similar to a flight of wild ducks.

Wilkinson Microwave Anisotropy Probe (WMAP) A NASA space mission launched in late 2000 to measure the properties of the ➤ *cosmic background radiation* at microwave wavelengths over the whole sky. It was placed in an orbit around the Sun, in a halo orbit around the L_2 ➤ *Lagrangian point*. Its ability to resolve detail was much greater than that of its predecessor, the ➤ *Cosmic Background Explorer* (COBE). The results from WMAP strongly support the ➤ *Big Bang* theory of the universe and give an age for the universe of 13.7 billion

The dome of the William Herschel Telescope.

385

years. They also show that the geometry of the universe is flat (rather than curved), which supports the idea of "inflation" – that the universe expanded very rapidly soon after it began.

William Herschel Telescope A 4.2-m (160-inch) reflecting telescope in the Isaac Newton Group at the ➤ *Observatorio del Roque de los Muchachos*, La Palma, Canary Islands. Observing time is shared between the collaborating countries – the UK, Spain and the Netherlands. It is a general-purpose telescope, equipped with a large range of instruments, and came into operation in 1987.

Wilson effect A change in the appearance of a ➤ *sunspot* as the Sun's rotation carries it close to the edge of the Sun's visible disk (the "limb"). The penumbra of the spot nearest the limb appears wider than that on the other side of the spot. This is because the sunspot is a depression. The phenomenon was first observed by the Scottish astronomer Alexander Wilson (1714–86) in 1769.

WIYN Telescope A 3.5-m telescope at ➤ *Kitt Peak*, opened in 1994. It is operated jointly by the University of Wisconsin, Indiana University, Yale University, and the National Optical Astronomy Observatories.

WMAP Abbreviation for ➤ *Wilkinson Microwave Anisotropy Probe*.

Wolf, Johann Rudolf (1816–1893) The Swiss astronomer Wolf is remembered for his comprehensive pioneering work on ➤ *sunspots* and the ➤ *solar cycle*. While director of the Zurich Observatory, he used historical data to discover that the length of the solar cycle is 11.1 years on average. Under his direction, Zurich became a world center for information on sunspots and he developed a formula, based on the number of sunspots visible and their size, to indicate the level of sunspot activity at any time.

Wolf–Rayet star A rare type of exceptionally hot star with surface temperatures between 20 000 K and 50 000 K. The spectra of Wolf–Rayet stars contain strong, broad emission lines. The emission lines are thought to come from an expanding envelope of gas flowing off the star. Some are the central stars of ➤ *planetary nebulae*. Their name comes from two nineteenth-century French astronomers, Charles Wolf and Georges Rayet.

X

Xena A temporary nickname given by its discoverers to the ➤ *dwarf planet*, now formally named ➤ *Eris*.

XMM-Newton Observatory An X-ray astronomy observatory launched by the ➤ *European Space Agency* in January 2000 into a 48-hour elliptical orbit around Earth. The nominal mission was two years, though it was designed to operate for up to 10 years. The satellite carries three identical X-ray telescopes, each consisting of 58 nested precision reflectors, together with a 30-cm optical/ultraviolet telescope. There are a total of nine instruments for imaging and spectroscopy. Originally known only as XMM, it was renamed after launch in honor of Isaac ➤ *Newton*.

X-ray astronomy The study of X-radiation from astronomical sources. The X-ray waveband is usually considered to be from about 10 to 0.01 nm, between the extreme ultraviolet (XUV) and gamma rays.

No X-rays from space can penetrate the atmosphere to the ground, so all X-ray astronomy is carried out with instruments on rockets or satellites. X-rays

An artist's impression of the XMM-Newton spacecraft.

from the Sun were detected during rocket flights in the 1950s. The first X-ray source beyond the solar system to be discovered was ➤ *Scorpius X-1*, found in 1962 by a group led by Ricardo Giacconi. By 1970, there were more than 40 known X-ray sources detected during rocket-borne experiments. However, satellites were needed to conduct more extensive surveys.

The first satellite dedicated to X-ray astronomy was Uhuru (1970), the first of NASA's ➤ *Small Astronomy Satellite* series. In 1973, a telescope capable of producing X-ray images was used successfully to image the Sun during the Skylab mission. This X-ray telescope used an array of concentric, cylindrical mirrors to reflect the X-rays at grazing incidence and bring them to a focus, and detectors capable of recording the positions of arrival of the photons over a field of view. Such an imaging X-ray telescope was used for objects other than the Sun for the first time by the ➤ *Einstein Observatory*. In 1985, a different type of X-ray telescope, using the "coded mask" technique, was deployed in orbit on Spacelab 2. This incorporates a diaphragm with a complex pattern of holes. Other important X-ray astronomy satellites include ➤ *ROSAT* (1990), ➤ *BeppoSAX* (1996), the ➤ *Chandra X-ray Observatory* (1999), and *XMM-Newton* (2000).

➤ *Black body radiation* in the X-ray band comes from sources at temperatures in excess of one million degrees. However, much of the X-ray emission detected from astronomical sources is generated in other ways, such as nuclear reactions in interacting binary star systems.

Most bright X-ray sources are ➤ *X-ray binaries*, which are interacting binary stars. The other main sources of astronomical X-rays are the hot diffuse gas surrounding galaxies and between the galaxies in clusters, ➤ *supernova remnants*, and ➤ *active galactic nuclei*. In 1996, X-rays were for the first time detected from several ➤ *comets*. ➤➤ *Hinode, Yohkoh*.

X-ray binary An interacting binary star system in which one component is a degenerate star – a ➤ *white dwarf*, a ➤ *neutron star* or a ➤ *black hole*. There are two kinds. In high-mass X-ray binaries (HMXB), the degenerate star's companion is a star of 10 or 20 solar masses and matter from its extended envelope flows directly onto the degenerate star. In low-mass binaries (LMXB) the two components are of similar mass and material is transferred to the degenerate star via an ➤ *accretion disk*. As it gains gravitational energy, the material flowing between the stars reaches temperatures high enough for it to emit X-rays. X-ray binaries often vary. The timescales for the variations may reflect the orbital period of the stars around each other, the rotation period of the degenerate star or a "wobble" of the accretion disk. Their X-ray luminosity ranges from 100 to 100 000 times the total luminosity of the Sun. Some systems, called ➤ *X-ray bursters*, show much more dramatic and random variations.

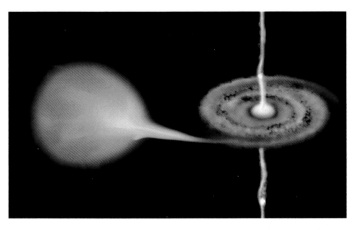

This artist's concept of an X-ray binary shows a double star system with a normal Sun-like star in orbit around a black hole. As gas is pulled from the normal star, it forms a disk around the black hole and is heated to temperatures of millions of degrees. Intense electromagnetic forces in the disk can expel jets of high-energy particles.

X-ray burster A stellar X-ray source that has violent and random changes to its emission.

X-ray bursters were discovered by a Dutch satellite in 1976. The bursts may last for several days and may recur, but are not regular. A rapid burster repeats at intervals no longer than 10 seconds. The generally accepted explanation is that X-ray bursters are interacting binary systems, similar to a ➤ *nova*, except that material falls onto a ➤ *neutron star* rather than a ➤ *white dwarf*, and the gas transferred is predominantly helium rather than hydrogen. The X-ray burst occurs when the accumulation of transferred material reaches the critical temperature and density to detonate a nuclear explosion. ➤➤ *X-ray astronomy*.

X-ray nova An ➤ *X-ray binary* system that suddenly becomes a temporary very intense source of X-rays.

X-ray pulsar A ➤ *pulsar* that emits X-rays.

XUV A term sometimes applied to the short-wavelength end of the ultraviolet region of the ➤ *electromagnetic spectrum* in the range 6–60 nm, where it merges with the X-ray band. It overlaps the region also known as the extreme ultraviolet (EUV). ➤➤ *ultraviolet astronomy*.

Y

year The period of time taken for the Earth to orbit the Sun. The exact length of the year depends on the reference point taken.

Types of years

Year type	How defined	Length in days
Tropical	From equinox to equinox	365.242 19
Sidereal	Relative to the stars	365.256 36
Anomalistic	Between successive perihelion passages of Earth	365.259 64
Eclipse	Time for intersection of Moon and Earth's orbits to return to same alignment relative to Sun	346.620 05
Gaussian	Applying the third of ➤ *Kepler's laws* to Earth's orbit	365.256 90

Yerkes Observatory An observatory in Williams Bay, Wisconsin. The observatory has the largest refracting telescope ever built, with an objective lens 1 m (40 inches) in diameter. It was constructed between 1895 and 1897. The project was largely the brainchild of ➤ *George Ellery Hale*, who persuaded the Chicago millionaire Charles Yerkes to finance it.

Ymir A small outer moon of Saturn in a very elliptical orbit. It was discovered in 2000 and is 16 km (10 miles) across.

Yohkoh A Japanese astronomy satellite launched in August 1991 to study X-rays and gamma rays from the Sun. It operated until December 2001. Yohkoh means "sunbeam" in Japanese.

Z

z The symbol normally used for ➤ *redshift*.

zenith The point directly overhead.

zenithal hourly rate (ZHR) The hypothetical rate at which meteors belonging to a particular ➤ *meteor shower* would be observed by an experienced observer, watching a clear sky with limiting magnitude 6.5, if the radiant were located in the zenith. In practice, observed rates are always lower, because fewer meteors are detected when the radiant is lower and skies are rarely so ideally clear.

zodiac A belt of 12 constellations through which the Sun's path in the sky – the ➤ *ecliptic* – passes. They are Aries, Taurus, Gemini, Cancer, Leo, Virgo, Libra, Scorpius, Sagittarius, Capricornus, Aquarius and Pisces. Formerly, the ecliptic went through only these 12 constellations, but the effects of ➤ *precession* and the precise definitions of constellation boundaries mean that it now also goes through a thirteenth, Ophiuchus. Since the orbits of the planets lie very nearly in a plane, their apparent paths remain in or close to the zodiacal constellations.

In traditional astrology, the zodiac is divided into 12 equal 30° portions, each of which is allocated to a "sign," but these do not correspond exactly to the astronomical constellations, which are of varying sizes. The effect of precession has also contributed to increasing disparity between the astrological signs and the astronomical constellations.

zodiacal light A faint cone of light in the sky extending along the ecliptic. It is visible on clear moonless nights in the west following sunset, and in the east just before sunrise. It is caused by sunlight scattered from micrometer-sized dust particles between the planets. The zodiacal light is dimly present all round the ecliptic. There is a brighter patch directly opposite the Sun. This is known as the "gegenschein," or "counterglow."

Picture credits

Constellation maps by Wil Tirion

absorption nebula: Atlas Image mosaic courtesy of 2MASS/UMass/ IPAC-Caltech/NASA/NSF, image mosaic by S. Van Dyk (IPAC)

accretion disk: NASA/CXC/A. Hobart

active galaxy: NASA/CXC/SAO

active optics: J. Mitton

Allen Telescope Array: UC Berkeley/Isaac Gary

ALSEP: NASA

ALMA: NRAO/AUI and ESO

Amalthea: NASA/JPL-Caltech

Andromeda Galaxy: T. A. Rector and B. A. Wolpa/NOAO/AURA/NSF

Antennae Galaxies: NASA/JPL-Caltech/ Harvard-Smithsonian CfA/NOAO/ AURA

Arecibo Observatory: NAIC/Arecibo Observatory, a facility of the NSF

Apollo program: NASA

Ariel: NASA/JPL-Caltech

armillary sphere: Tycho Brahe, *Mechanica*

astrolabe: J. Mitton

atmospheric window: CXC

aurora: SOHO EIT and LASCO consortia/ ESA/NASA/Jan Curtis

Australia Telescope National Facility: CSIRO

Barnard's Galaxy: 2MASS/Umass/IPAC-Caltech/NASA/NSF

barred spiral galaxy: ESO

Beta Pictoris: NASA, ESA, D. Golimowski (Johns Hopkins University), D. Ardila (IPAC), J. Krist (JPL), M. Clampin (GSFC), H. Ford (JHU), and G. Illingworth (UCO/Lick) and the ACS Science Team

bipolar outflow: NASA, ESA and The Hubble Heritage Team (STScI/AURA)

Black-eye galaxy: NOAO/AURA/NSF

Brahe, Tycho: Tycho Brahe, *Mechanica*

brown dwarf: NASA/IPAC/R. Hurt

Bubble Nebula: NASA, Donald Walter (South Carolina State University), Paul Scowen and Brian Moore (Arizona State University)

Bug Nebula: NASA, ESA and A.Zijlstra (UMIST, Manchester, UK)

butterfly diagram: David Hathaway/ NASA

Callisto: NASA/JPL-Caltech

Calypso: NASA/JPL/Space Science Institute

Carina Nebula: Nathan Smith, University of Minnesota/NOAO/AURA/NSF

Cartwheel Galaxy: NASA/JPL/Caltech/ P. Appleton *et al.*/CXC/A. Wolter and G. Trinchieri *et al.*

Cassini-Huygens: NASA/JPL

Cassiopeia A: X-ray: NASA/CXC/SAO; Optical: NASA/STScI; Infrared: NASA/ JPL-Caltech

Cat's Eye nebula: NASA, ESA, HEIC, and The Hubble Heritage Team (STScI/AURA)

Centauarus A: European Southern Observatory

Chandrasekhar, S.: AIP/CXC

Chandra X-ray Observatory: CXC/NGST

circumstellar disk: NASA, ESA, and P. Kalas (University of California, Berkeley)

cluster of galaxies: NASA/CXC/Columbia U./C. Scharf *et al.*

Coma cluster: Isaac Newton Group of Telescopes, La Palma/Duncan A. Forbes (Swinburne University, Australia)

comet: NASA, ESA, H. Weaver (JHU/APL), M. Mutchler and Z. Levay (STScI)

Cone Nebula: NASA, H. Ford (JHU), G. Illingworth (UCSC/LO), M. Clampin

(STScI), G. Hartig (STScI), the ACS Science Team, and ESA

Copernicus: John Caldwell (York University, Ontario), Alex Storrs (STScI) and NASA

corona: NSO/AURA/NSF

coronal mass ejection: SOHO LASCO consortium/ESA/NASA

Crab Nebula: NASA, ESA and Allison Loll/ Jeff Hester (Arizona State University).

crater: NASA

Crescent Nebula: T.A. Rector (NRAO/AUI/ NSF) and NOAO/AURA/NSF

Cygnus A: NRAO/AUI

Cygnus X-1: ESA/Martin Kornmesser

Deep Impact: NASA/JPL-Caltech/UMD

diamond ring effect: Bill Livingston/NSO/ AURA/NSF

Dione: NASA/JPL/Space Science Institute

Draper, Henry: Yerkes Observatory

Dumbbell Nebula: Michael Pierce, Robert Berrington (Indiana University), Nigel Sharp, Mark Hanna (NOAO)/ WIYN/NSF

Eagle Nebula: Bill Schoening/NOAO/AURA/ NSF

Earth: NASA

Eclipse: NASA-KSC

EGG: Jeff Hester and Paul Scowen (Arizona State University), and NASA

Egg Nebula: Raghvendra Sahai and John Trauger (JPL), the HST WFPC2 science team, and NASA

Einstein, A.: AIP Emilio Segrè Visual Archives

Einstein Cross: NASA/ESA

Einstein Rings: NASA, ESA, and the SLACS Survey team: A. Bolton (Harvard/ Smithsonian), S. Burles (MIT), L. Koopmans (Kapteyn), T. Treu (UCSB), and L. Moustakas (JPL/Caltech)

Elliptical galaxy: NOAO/AURA/NSF

Electromagnetic radiation: CXC/S. Lee

Elysium Planitia: NASA/JPL/Malin Space Science Systems/NSSDC

Emission nebula: NASA, ESA, and A. Nota (STScI/ESA)

Enceladus: NASA

Epimetheus: NASA/JPL/Space Science Institute

Eros: NASA and APL (Johns Hopkins University)

Eskimo Nebula: NASA, A. Fruchter and the ERO Team (STScI)

Eta Carinae: Jon Morse (University of Colorado), and NASA

Etched Hourglass: Raghvendra Sahai and John Trauger (JPL), the HST WFPC2 science team, and NASA

Europa: NASA/JPL-Caltech

flare: SOHO EIT consortium/ESA/NASA

Fomalhaut: NASA/JPL-Caltech/K. Stapelfeldt (JPL)

galactic center: NASA/JPL-Caltech/S. Stolovy (Spitzer Science Center/ Caltech)

Galaxy: NASA/CXC/M.Weiss

Galilei, Galileo: Yerkes Observatory

Galilean moons: NASA/JPL-Caltech

Ganymede: NASA/JPL-Caltech

Gaspra: NASA/JPL-Caltech

Gemini Telescopes: Gemini Observatory

globular cluster: NASA and the Hubble Heritage Team (STScI/AURA)

globules: NASA and the Hubble Heritage Team (STScI/AURA)

gravitational lens: NASA, Andrew Fruchter (STScI), and the ERO team (STScI, ST-ECF)

granulation: Royal Swedish Academy of Sciences

Great Dark Spot: NASA/JPL

Great Red Spot: Hubble Heritage Team (STScI/AURA/NASA) and Amy Simon (Cornell University)

Greenbank Telescope: NRAO/AUI

Hadley Rille: NASA

h and chi Persei: N. A. Sharp/NOAO/AURA/ NSF

Hale–Bopp, Comet: European Southern Observatory

Halley, Comet: ESA

Halley, Edmond: J. Mitton

Hellas Planitia: ESA/DLR/FU Berlin (G.Neukum)

Helix Nebula: NASA/JPL-Caltech/SSC

Herbig–Haro object: NASA/JPL-Caltech/A. Noriega-Crespo (SSC/Caltech), Digital Sky Survey

Hertzsprung–Russell diagram: Michael Perryman *et al.* data as published in *Astronomy & Astrophysics*, 1997, vol. 323, L49

Horsehead Nebula: European Southern Observatory

Hoyle, Fred: S. Mitton

Hubble classification: STScI

Hubble Space Telescope: STScI and NASA

Hyperion: NASA/JPL/Space Science Institute

Iapetus: NASA/JPL/Space Science Institute

Ida: NASA/JPL-Caltech

infrared astronomy: 2MASS/J. Carpenter, M. Skrutskie, R. Hurt

interacting galaxies: ESO

INTEGRAL: ESA

International Space Station: STS-115 Shuttle Crew/NASA

Io: NASA/JPL-Caltech

ion drive: ESA

irregular galaxy: ESO

Itokawa: JAXA

James Clerk Maxwell Telescope: Joint Astronomy Center, Hawaii

James Webb Space Telescope: ESA

Jansky, Karl: Bell Telephone Laboratories.

Janus: NASA/JPL/Space Science Institute

Jewel Box: NOAO/AURA/NSF

Jodrell Bank Observatory: Jodrell Bank, University of Manchester/Anthony Holloway

Jupiter: NASA/JPL/University of Arizona

Keck Observatory: NASA/JPL

Kennedy Space Center: NASA

Kepler, Johannes: Courtesy of the Archives, California Institute of Technology

Kepler's Supernova: NASA/ESA/JHU/R. Sankrit and W. Blair

Keyhole Nebula: 2MASS/G. Kopan

Kitt Peak: J. Mitton

Kuiper Belt Object: NASA, ESA, and A. Feild (STScI)

Lagrangian points: NASA

Laplace, S.-P.: J. Mitton

Large Binocular Telescope: Max Planck Institute for Astronomy

La Silla Observatory: ESO

lenticular galaxy: NASA, ESA, and The Hubble Heritage Team (STScI/AURA)

Leverrier, U.-J.-J.: Yerkes Observatory

light echo: NASA, ESA, and The Hubble Heritage Team (STScI/AURA)

Little Dumbbell: N.A.Sharp, NOAO/AURA/ NSF

Loki: NASA/JPL/USGS

Lowell, Percival: Lowell Observatory

Lunar Roving Vehicle: NASA

Lunik: NASA

Magellanic Clouds: NASA/JPL-Caltech/ M. Meixner (STScI) and the SAGE Legacy Team

magnetosphere: NASA/CXC/M. Weiss

mare: NASA

Mariner: NASA/JPL

Mars: NASA and the Hubble Heritage Team (STScI/AURA)

Mars Exploration Rovers: JPL

Mars Reconnaissance Orbiter: NASA/JPL

Mathilde: Mark Robinson, Northwestern University, and Scott Murchie, Johns Hopkins University Applied Physics Laboratory.

Mauna Kea Observatory: Copyright 1999, Neelon Crawford – Polar Fine Arts, courtesy of Gemini Observatory and National Science Foundation

McMath–Pierce Solar Telescope Facility: J. Mitton

Mercury: NASA/JPL-Caltech

meteor: Isaac Newton Group of Telescopes, La Palma/Alan Fitzsimmons

Meteor crater: NASA

meteorite: J. Mitton

Mice: NASA, H. Ford (JHU), G. Illingworth (UCSC/LO), M.Clampin (STScI), G. Hartig (STScI), the ACS Science Team, and ESA

Mimas: NASA/JPL/Space Science Institute

Mir: STS-89 Crew/NASA

Mira: X-ray: NASA/CXC/SAO/M. Karovska *et al.*; Illustration: CXC/M. Weiss

Miranda: NASA/JPL-Caltech

Mitchell, Maria: J. Mitton

Moon: NASA

Neptune: NASA, L. Sromovsky, and P. Fry (University of Wisconsin-Madison)

New Horizons: Johns Hopkins University Applied Physics Laboratory/ Southwest Research Institute (JHUAPL/SwRI)

Newton, Isaac: Yerkes Observatory

nova: NASA/CXC/M. Weiss

Oberon: NASA/JPL-Caltech

Olympus Mons: NASA/USGS

Omega Centauri: 2MASS/UMass/ IPAC-Caltech/NASA/NSF

Omega Nebula: 2MASS/Umass/ IPAC-Caltech/NASA/NSF

open cluster: Heidi Schweiker/NOAO/AURA/ NSF

Orientale Basin: NRAO/AUI and Bruce Campbell/Smithsonian Institution

Orion Nebula: NASA, ESA, M. Robberto (Space Telescope Science Institute/ ESA) and the Hubble Space Telescope Orion Treasury Project Team

Owl Nebula: Nordic Optical Telescope/ Magnus Galfalk

Pandora: NASA/JPL/Space Science Institute

Parkes Observatory: CSIRO

Pavonis Mons: Credits: ESA/DLR/FU Berlin (G. Neukum)

Payne-Gaposchkin, C.: Yerkes Observatory

Pele: NASA/JPL

Pelican Nebula: John Bally (University of Colorado), NOAO/AURA/NSF

Phobos: ESA/DLR/FU Berlin (G. Neukum)

Phoebe: NASA/JPL/Space Science Institute

Pinwheel galaxy: George Jacoby, Bruce Bohannan, Mark Hanna/NOAO/ AURA/NSF

Pioneer: NASA

Pistol Star: Don F. Figer (UCLA) and NASA

planetary nebula: NASA and the Hubble Heritage Team (STScI/AURA)

Pleiades: NASA, ESA and AURA/Caltech

Pluto: Gemini Observatory

polar cap: NASA

polar-ring galaxy: NASA and the Hubble Heritage Team (STScI/AURA)

Prometheus: NASA/JPL/Space Science Institute

prominence: SOHO EIT consortium/ESA/ NASA

proplyd: C.R. O'Dell/Rice University; NASA

Ptolemaic system: Peter Apian, *Cosmographia* (1539)

quasar: NASA/CXC/SAO

radio astronomy: Image courtesy of NRAO/ AUI

radio galaxy: NRAO/AUI

Ranger: NASA

regolith: NASA

reflection nebula: ESO

Rhea: NASA/JPL/Space Science Institute

ring galaxy: NASA and the Hubble Heritage Team (STScI/AURA)

Ring Nebula: The Hubble Heritage Team (AURA/STScI/NASA)

ring system (Jupiter): NASA/JPL-Caltech

ring system (Saturn): NASA/JPL

Rosetta: ESA

Rosette Nebula: N. A. Sharp/NOAO/AURA/ NSF

Russell, H. N.: Yerkes Observatory

Sagittarius A: NRAO/AUI

Saturn: NASA, ESA and E. Karkoschka (University of Arizona)

Schröter's Valley: NASA

Seyfert galaxy: NASA, Andrew S. Wilson (University of Maryland), Patrick L. Shopbell (Caltech), Chris Simpson (Subaru Telescope), Thaisa Storchi-Bergmann and F. K. B. Barbosa (UFRGS, Brazil) and Martin J. Ward (University of Leicester, UK).

Seyfert's Sextet: NASA, J. English (University of Manitoba), S. Hunsberger, S. Zonak, J. Charlton, S. Gallagher (PSU), and L. Frattare (STScI)

shepherd satellite: NASA/JPL/Space Science Institute

Shoemaker-Levy 9: The HST Comet Team

Sirius: NASA/SAO/CXC

Skylab: NASA

Slipher, Vesto: Lowell Observatory

SMART-1: ESA

solar activity: SOHO EIT consortium/ESA/ NASA

solar cycle: ISAS

solar system: ESA

solar tower telescope: J. Mitton

Southern Cross: ESO

Space Shuttle: NASA

spectral type: NOAO/AURA/NSF

spiral galaxy: NASA/JPL-Caltech/S. Willner (Harvard-Smithsonian Center for Astrophysics)

Spitzer, Lyman: Denise Applewhite/ Princeton University

starburst galaxy: ESO

star cloud: Vanessa Harvey, REU program/ NOAO/AURA/NSF

star cluster: ESA and NASA

Stardust: NASA/JPL-Caltech

Stephan's Quintet: Gemini Observatory/ AURA/Travis Rector, University of Alaska, Anchorage

stellar evolution: NASA/CXC/M. Weiss

Sun: NASA/CXC/M. Weiss

sunspot: SOHO MDI consortium/ESA/ NASA

supernova: N. A. Sharp, G. J. Jacoby/NOAO/ AURA/NSF

supernova remnant: NASA, ESA, HEIC, and The Hubble Heritage Team (STScI/AURA)

Sombrero Galaxy: European Southern Observatory

Sudbury Neutrino Observatory: Ernest Orlando Lawrence Berkeley National Laboratory.

Surveyor: NASA

synchrotron radiation: CXC/S. Lee

Syrtis Major Planum: NASA/MGS/MSSS

Tarantula Nebula: ESO

Taurus-Littrow valley: NASA, ESA, and J. Garvin (NASA/GSFC)

tektites: J. Mitton

Tethys: NASA/JPL/Space Science Institute

Tharsis Ridge: NASA/MGS/MSSS

Titan: NASA/JPL/Space Science Institute

Titania: NASA/JPL-Caltech

Tombaugh, Clyde: Lowell Observatory

Toutatis: NASA

TRACE: M. Aschwanden *et al.* (LMSAL), TRACE, NASA

transit: J. Mitton

Trapezium: NASA; K.L. Luhman (Harvard-Smithsonian Center for Astrophysics, Cambridge, Mass.) and G. Schneider, E. Young, G. Rieke, A. Cotera, H. Chen, M. Rieke, R. Thompson (Steward Observatory, University of Arizona, Tucson, Ariz.)

Triangulum Galaxy: N. Caldwell, B. McLeod, and A. Szentgyorgyi (SAO)

Trifid Nebula: Gemini Observatory/GMOS Image

Triton: NASA/JPL-Caltech

Tsiolkovskii: NASA

47 Tuc: ESO

Uranus: Erich Karkoschka (University of Arizona) and NASA

Utopia Planitia: NASA/JPL

vacuum tower telescope: J. Mitton

Valles Marineris: ESA/DLR/FU Berlin (G. Neukum)

Van Allen belts: NASA

Veil Nebula: T. Rector/University of Alaska Anchorage and WIYN/NOAO/AURA/ NSF

Vela pulsar: NASA/CXC/PSU/G. Pavlov et al.

Venera: Russian Space Agency

Venus: NASA/JPL-Caltech

Venus Express: ESA

Very Large Array: NRAO/AUI

Very Large Telescope: European SouthernObservatory

Very Long Baseline Array: NRAO/AUI

Virgo A: NASA and the Hubble Heritage Team (STScI/AURA)

Virgo cluster: Atlas Image mosaic courtesy of 2MASS/UMass/IPAC-Caltech/ NASA/NSF. Image mosaic by T. Jarrett (IPAC)

Voyager: NASA/JPL

Westerbork Observatory: J. Mitton

Whirlpool galaxy: NASA and the Hubble Heritage Team (STScI/AURA)

white dwarf: ESA/NASA

Wild 2: NASA/JPL-Caltech

William Herschel Telescope: PPARC, Nik Szymanek and Ian King

XMM-Newton: ESA

X-ray binary: NASA/CXC/M. Weiss